*Transatlantic
Industrial Revolution*

Interior of Thomas Robinson's power loom factory at
Stockport, Cheshire, England, in 1835. The looms are of a
Sharp & Roberts design. (Source: Ure, *The Philosophy of
Manufactures*. Courtesy Merrimack Valley Textile
Museum.)

Merrimack Valley Textile Museum
North Andover, Massachusetts

The MIT Press
Cambridge, Massachusetts

Transatlantic
Industrial Revolution:
The Diffusion of
Textile Technologies
Between Britain and
America, 1790–1830s

David J. Jeremy

This book was set in Palatino by Achorn Graphic Services
and printed and bound by Halliday Lithograph in the
United States of America.

Library of Congress Cataloging in Publication Data

Jeremy, David John, 1939–
  Transatlantic industrial revolution.

  Based on the author's thesis (Ph.D.)—University
of London, 1978.
  Bibliography: p.
  Includes index.
  1. Textile industry—United States—Technological
innovations—History. 2. Textile industry—Great
Britain—Technological innovations—History.
3. Diffusion of innovations—United States—History.
4. Diffusion of innovations—Great Britain—History.
5. Technology transfer—History. I. Title.
HD9855.J47   338.4'5677'00941   81-517
ISBN 0-262-10022-3   AACR2

To Theresa, my wife,
who made this study possible

She seeketh wool, and flax,
and worketh willingly with her hands.
Proverbs 31:13

Contents

Contents

List of Illustrations and Figures

Illustrations

Frontispiece: Interior of Thomas Robinson's power loom factory, Stockport, Cheshire, England, 1835.

Title page: British emigrants' transport across the Atlantic: the brig *Frances* in a hurricane, with topgallant masts sent down, 1824. (Source: Capt. Joseph Pattinson's log.)

List of Tables

Acknowledgments

This book is a revised version of a study accepted by the University of London as a thesis for the degree of doctor of philosophy in July 1978. I am indebted to a large number of institutions and individuals on both sides of the Atlantic in writing this study, which was prepared in two stages. When I was in America between 1967 and 1973, I was privileged to embark on the topic when working in a number of learned institutions and reading in some of the major manuscript collections. Over these years I became indebted to the University of Delaware (John A. Munroe), the Eleutherian Mills Historical Library (Richmond D. Williams, Norman B. Wilkinson, Eugene S. Ferguson, and Harold B. Hancock), and the Smithsonian Institution's National Museum of History and Technology (especially Philip W. Bishop, Robert M. Vogel, Edwin A. Battison, Grace Rogers Cooper, and Rita J. Adrosko). In New England my research opportunities were further extended while I was curator of the Merrimack Valley Textile Museum (1970–1973), an institution with one of the finest collections of textile artifacts, books, and manuscripts in America. Here my colleagues—Thomas W. Leavitt, Helena Wright, Joyce Messer, Betty Goddard, Robert Hauser, and Marion Hyde—and the founders of that splendid museum, Horatio Rogers and Caroline Stevens Rogers, generously cooperated with an intruding Englishman. On my weekly visits to Boston, I was able to read in Harvard's Baker Library, where Robert W. Lovett, Kenneth E. Carpenter, and Glenn Porter proved good lunchtime companions as well as helpful scholars. While I was in North Andover at the museum, I had the chance to dig into the patent litigation buried in circuit court case files in the Federal Record Center at Waltham, where James K. Owens facilitated my work. In Rhode Island, Paul E. Rivard (then director of the Old Slater Mill Museum) directed my attention to various primary sources.

The results of my American activities were three articles and plans to write a comprehensive study of the subject of this book. The plans came to fruition after I returned to teaching in England. Providentially a post near London opened, and Charlotte J. Erickson of the London School of Economics and Political Science (LSE) accepted me as a graduate student. Under her guidance I turned to the U.S. Customs passenger lists and developed an economic dimension to my understanding of the technologies. In supervising the thesis, Professor Erickson proved to be an incisive and encouraging critic who spurred me on when the going was rough. Theo C. Barker of the LSE also helped by commenting on the early chapters, directing my attention to the journal *Repertory of Arts*. Since my return to England, Anthony F. C. Wallace of the University of Pennsylvania generously invited me to share in his study of William Pollard, from which I learned a great deal, and Bob Lovett has kindly kept me up to date with American research in the field. Other scholars who have gladly helped me in England include Julia de L. Mann, R. S. Fitton, and Negley Harte.

While registered at the LSE, I received grants-in-aid from the Pasold Research Fund and the Merrimack Valley Textile Museum. The latter supported me during a term's unpaid leave of absence from my teaching post at the Cecil Jones High School, Southend-on-Sea, in summer 1977. To these institutions I wish to record my gratitude.

Thesis revision has brought further indebtedness. A. E. Musson of Manchester University, the external examiner of my thesis, made searching criticisms. Seymour Broadbridge of Liverpool University and John R. Harris of Birmingham University kindly invited me to air parts of my work in their departmental research seminars. Professor Harris's seminar led me to rewrite the section dealing with the evidence of textile patents.

My obligations to the trustees and staff of the Merrimack Valley Textile Museum have continued right up to publication. To the privilege of working and learning at the museum and the benefit of research support, they have now added the honor of publication sponsorship. For the confidence he has reposed in me and for the part he has played in all this, I especially want to thank the museum director, Thomas W. Leavitt. I am also very grateful to Helena Wright for her unfailingly prompt responses to my requests for references and information and for assembling prints of the illustrations selected for the book. Proofreading and indexing were completed soon after I joined the LSE's Business History Unit, and I am grateful to the Unit's director, Leslie Hannah, for his interest and support. I also thank the staff of the MIT Press for their full cooperation.

Finally, words cannot express my debt, or gratitude, to my family: to Joanna and Rebecca, who patiently tolerated their father's distraction with studies, and to my wife, Theresa, to whom this book is dedicated. Theresa steeled my nerve, helped out with masses of rough typing, checked script and tables and did numerous other chores, and shared the frustrations (and fewer of the joys) of research and writing with both impatience and love.

Transatlantic
Industrial Revolution

Introduction:
A Note on Theory,
Historiography,
and Methodology

The diffusion of innovations, a process as important to economic development as invention, has received little attention from historians until relatively recently. Anthropologists and sociologists led the way with research examining the spread of practices as diverse as water boiling and birth control and of products ranging from ham radio sets to hybrid wheat strains and new drugs.[1] Economists increasingly moved into the field during the 1950s and 1960s, seeking explanations for differences in economic growth rates and reasons for failures to transplant American technology abroad.

From this work two different, but not necessarily exclusive, methodological approaches have emerged. On the one hand there is the contention that "the diffusion of inventions is an essentially economic phenomenon, the timing of which can be largely explained by expected profits."[2] On the other hand social factors clearly play a part, and often a crucial one as the varied success of American foreign aid programs indicates.[3] The relation between the studies of economists and sociologists on diffusion has been likened to the parable of the blind men and the elephant: "In effect, the economists are studying the demand for information by potential innovators and sociologists the problems in the supply of communication channels."[4]

Although no general theory has been developed to fit all cases of technology diffusion in both its economic and social dimensions, two economic concepts have been broadly accepted: pioneering economies are not necessarily overhauled by imitators, and followers do not necessarily take the same route as have leaders.[5] Such general propositions imply a wide diversity in diffusion experience. Even so, a number of common constraints affecting rates of transfer have been distilled by Rosenberg from earlier work on technology transfer.[6] Following Usher, who viewed inventive activity as continuous and cumulative, Rosenberg observed that diffusion is likely to be a gradual process, depending on the pace at which secondary improvements perfect the original invention. Further, he noted that the level of human skills in the receptor economy is one of several factors influencing the length of the imitator's learning period; that machine-making skills are a crucial element in a receiving economy; that bottlenecks in other productive activities will have to be removed before a new invention can spread (Rosenberg calls this the problem of complementarities); that old techniques may be improved so as to extend their economic viability; and that social and institutional constraints may inhibit diffusion.

Of the growing number of historical case studies of technology diffusion appearing in recent years, most have focused on a single technology or a single country. Some, like Pursell's on the stationary steam engine or Brittain's on electrical power technology, have examined the international movement of specific technologies.[7] More—Harris's on coal-mining skills, Feller's on the Draper loom, and Temin's on steam and water power—have dealt with a given technology in a single economy, though Coleman points out the hazards of confining a diffusion study to the national market economy or to a single industry.[8] Very few studies—Robinson's on Russia is one—have explored the transfer of a range of technologies to a particular country.[9] A few have treated regions or firms: Wilkinson's on the Brandywine valley, Sloat's on the Dover Manufacturing Company, and Harris's on two glass-making firms, one in England and one in France.[10] One of the latest historical accounts of technology diffusion treats a single event: the construction of the Britannia Bridge in Wales.[11]

Other historical studies have concentrated on the diffusion process itself, especially on technology carriers. Scoville, one of the first historians in the field, investigated a minority group and concluded that such groups were unlikely to effect the migration of a modern industrial technology.[12] Lately other carriers have been emphasized: the capital goods firm inside the market economy; the multinational firm across national boundaries; and individual artisans and inanimate media, internally and internationally.[13]

A case study of technology diffusion in early industrial textile manufacturing needs little defense.[14] The classical industrial revolution in both Britain and New England, prior to the coming of railways, derived much of its impetus from the transforming cotton industry. Ship construction, port facilities, inland transportation systems, coal and iron production, chemical engineering, machine making, factory organization, urban building and living, financial and marketing structures, and power sources—in short, many of the techniques and institutions of modern industrial capitalism—were all affected in varying ways and degrees by an expanding cotton industry. Yet no comprehensive analysis of the international diffusion of early industrial textile technology has been written. In a sense, this present study is only half of the story. Diffusion eastward from Britain still awaits thorough treatment.

Although I have not been unmindful of the numerous insights and questions of other workers in the diffusion-of-technology field, just two questions have preoccupied my attention. First, what circumstances hindered or promoted the spread of cotton and woolen technologies across the Atlantic and their establishment in America? Second, what conditions in America prompted the reshaping of the imported technologies? Before these questions could be addressed, however, I had to define and formulate a basic vocabulary, analytical tools, and terms of reference.

By *diffusion* or *transmission* (employed interchangeably) is meant the spread of an innovation from its originating firm or economy to a host firm or economy. Technology is defined as "a spectrum, with ideas at one end and techniques and things at the other, with design as a middle term."[15] From this perspective, technology embraces operating rules, like warp tables and production formulas; implements and machinery; patent specifications, drawings, plans, and verbal descriptions; models and parts of machines; arrangements in the organization and management of production; and human skills. In part, much of the textile technology of this period resided in people; elsewhere, and over time generally, it became increasingly incorporated in inanimate media.

Apart from the broad insights of economists, I have employed a variation of Mira Wilkins's stage analysis to understand the diffusion process.[16] For the purposes of this study, there were limitations on the usefulness of her two-stage model. The criterion of firm ownership is inappropriate for the early industrial period. Immigrants then worked on their own account, not as agents of multinational companies. Further, it is impossible to discover the identity or national origins of all mill owners. Here I assume that immigrant mill owners, managers, operatives, and machine makers entering the American textile trades had economic interests akin to those of native-born capitalists and workers. Yet a stage model seemed to be the most useful tool with which to analyze the realities of technology diffusion.

The establishment and persistence of commercially successful businesses employing the new technology represented a major stage in diffusion. When firms in the host economy modified the new technology to match their immediate economic or social circumstances, the diffusion process was completed. To encompass these stages and the intermediary steps, I amplified Wilkins's model into four stages: (1) creating the potential for diffusion, with the arrival of immigrants, machines, plans, and so forth, from the originating economy; (2) establishing pilot production units in the host economy; (3) spreading or internally diffusing the imported technology within the host economy; and (4) modifying the imported technology to suit the factor endowment and social structure of the host economy.

The textile technologies surveyed in this book primarily cover the mechanical manufacture of yarn and woven cloth in the cotton and woolen industries. Excluded are chemical technology, as in dyeing; fabric-making branches other than cotton goods and woolen cloth weaving, like knitting, lace, carpet making, smallwares, or hatting; the worsted side of the wool industry; and the technology of other textile fibers, like flax, jute, and silk.

I have not supplied thorough descriptions of the basic mechanical technology. Such descriptions would only distract attention from the focus on the diffusion process. Readers will be helped, however, by the large number of illustrations and their captions; by the full glossary of technical terms; and by the table shown here displaying the sequence of major cotton and woolen manufacturing processes, machines, and devices referred to in the text.

Early Industrial Textile Technology: Major Processes, Machines, and Devices Used in Britain and the United States, 1790–1830s

	Cotton			Wool		
Stage	Process	Fiber State	Machine[a]	Process	Fiber State	Machine[a]
I: Preparatory	Sorting	Raw cotton	. . .	Sorting	Wool (raw)	. . .
	Picking	Loose cotton	Picker	Scouring	Wool (raw)	. . .
	Mixing	Loose cotton	. . .	Dyeing	Wool (raw)	. . .
	Dyeing	Loose cotton	. . .	Picking	Wool (raw)	Picker
	Carding–breaker card	Sheet or lap	Card (carding machine)	Oiling	Wool (raw)	. . .
	Carding–finisher card	Sliver (endless tubular length)	Card	Carding–breaker card	Sheet or lap	Card (carding machine)
	Drawing	Doubled and twisted end or sliver	Drawing frame	Carding–finisher card	Short rolags	Card
II: Spinning	Roving	Drawn and twisted end or sliver	Roving can frame, stretchers, speeders	Slubbing	Continuous sliver or roping	Billy, condenser
	Intermittent spinning	Yarn	Jenny, mule, self-acting mule	Intermittent spinning	Yarn	Jenny, jack
	Continuous spinning	Yarn	Water frame, Throstle, cap and ring frames	Reeling	Yarn	Reel
	Reeling	Yarn or thread	Reel	Dyeing	Yarn	. . .

Stage	Cotton			Wool		
	Process	Fiber State	Machine[a]	Process	Fiber State	Machine[a]
	Twisting	Plyed yarn (thread)	Twisting frame			
	Bleaching	Yarn or thread	...	Twisting	Plyed yarn (thread)	Twisting frame
	Dyeing	Yarn or thread	...			
III: Weaving	Spooling/ winding	Yarn or thread	...	Spooling/ winding	Yarn or thread	...
Warp preparation	Warping	Yarn or thread	Warper	Warping	Yarn or thread	Warping mill
	Dressing (slashing)	Yarn or thread	Dresser	Dressing	Yarn or thread	...
	Drawing in	Yarn or thread	...	Drawing-in	Yarn or thread	...
	Reeding	Yarn or thread	...	Reeding	Yarn or thread	...
Filling preparation	Quill winding	Yarn	Filling frame	Quill winding	Yarn	...
Weaving	Weaving	Cotton goods	Power loom:[1] Shedding Picking Beating up Letting off Taking up Stop motions: Warp Weft Warp-protector	Weaving	Cloth	Fly shuttle handloom Satinet power loom Broadcloth power loom
	Bleaching	Cotton goods	...	Scouring	Cloth	...
				Dyeing	Cloth	...
IV: Finishing	Printing	Cotton goods	Roller and surface printing machines	Fulling	Cloth	Fulling mill
	Fustian cutting	Cotton goods	...	Tenter drying	Cloth	...

Stage	Cotton			Wool		
	Process	Fiber State	Machine[a]	Process	Fiber State	Machine[a]
	Dyeing	Cotton goods	Padding machine	Raising	Cloth	Gig mill
				Shearing	Cloth	Shearing frame
	Beetling	Cotton goods	Beetling machine	Brushing	Cloth	Dresser
	Calendering	Cotton goods	Calender	Burling	Cloth	. . .
	Pressing	Cotton goods	Press	Pressing	Cloth	Screw press

[a] . . indicates no machine used; old manual implements still employed.
[b] Motions for plain and twill power looms.

I
Constraints on the Diffusion of the Technologies

1
America's Manufacturing Potential and the State of British and American Textile Manufactures in 1790

The cloudy but dry Independence Day of 1788 saw a spectacular procession wind through Philadelphia. Organized to celebrate the ratification of the Constitution and to display the new spirit of unity, the five-thousand-strong parade of political leaders, civic officials, militia, and tradesmen also demonstrated the varied skills of Pennsylvania's manufactures. Some floats, however, announced more than local achievements.

On the Manufacturing Society's carriage, a third of the way along the parade, America's first fruits of the international movement of men and machines were brought to public view. Two men powered a carding machine processing cotton at the rate of fifty pounds a day; a woman spun yarn on an eighty-spindle jenny; a weaver worked a lace loom and another wove jean on a fly-shuttle hand loom; and while a block cutter demonstrated his skills, the renowned hand calico printer John Hewson, an English immigrant, busily handprinted cotton goods with the assistance of his wife and four daughters. Above the float, a brisk wind spread the calico printers' flag: "In the center 13 stars on a blue field and 13 red stripes on a white field; round the borders of the flag were printed 37 different patterns of various colors. . . . Motto—*'May the union government protect the manufactures of America.'* " Immediately behind the carriage tramped a hundred weavers following their own flag, "a rampant lion on a green field, holding a shuttle in his dexter paw. Motto—*'May government protect us.'* " Cotton card makers marched with the weavers. Other floats in the procession included those organized by ropemakers and chandlers; fringe and ribband weavers; turners, instrument makers, and spinning wheel makers; blacksmiths, whitesmiths, and nailors; tin-plate workers; hatters; and stocking manufacturers.[1] This display of patriotic sentiment signaled the beginning of the transfer of industrial textile technology across the Atlantic.

By 1790 the United States apparently offered conditions parallel to those that had fostered Britain's new textile technologies. With the obvious difference that the United States possessed far greater natural resources, the two countries shared a common language, similar social and legal institutions, and the same attitudes to work, summed up in the Puritan ethic. Furthermore, the will to economize and make profits was well established by the merchants at the top of the economic class structure in the 1780s. Among the northern merchants at least, the class was not a closed one.[2] Moreover, the political settlement of 1788 promised to reconcile state rivalries and create a climate of political confidence necessary for industrial expansion. Visible results of the federal government's efforts by 1790 were the organization of a national revenue and public securities, patent and copyright laws to secure intellectual property, and plans for a national bank and coinage. Besides a commercial infrastructure essential to industrial activity, these measures gave some slight tariff protection to domestic manufactures, which encouraged infant industries to hope for more.[3]

Ranged against these assets were powerful—and serious—constraints on capital investment and manufacturing. Two related to land. The abundance of land made it cheap. Most states during the Revolution and the 1780s disposed of their lands quickly, deflating land prices drastically. Georgia alone offered 25 million acres to four land com-

panies at less than half a cent an acre. Cheap land tended to divert labor and investment away from industry.[4] Second, most American farming was at a subsistence level. Only when agricultural productivity was increased could labor be released from farming for work in an industrial sector.[5]

Relative capital scarcity posed more difficulties. A major problem facing promoters of American manufactures was to attract mercantile capital away from the rival investments of land and trade. Speculation in land could be highly remunerative. Although they were not always realized, thirty-fold returns were expected from Pennsylvania land speculations in the early 1790s. Profits could be as high, or higher, in trade, especially in new markets opening up, such as the China trade. When the first pepper cargo from the East Indies reached Salem, Massachusetts, in 1793, it sold at a 700 percent profit. In trade, however, high profits were accompanied by the risk of equally severe losses.[6] The scarcity of capital pushed up its cost. There were maximum legal interest rates, which varied from state to state, but generally these understated the actual rates charged. In New York State, for example, the legal rate was 7 percent but 8 to 10 percent was commonly earned in the New York money market in the early 1790s.[7] Manufacturers therefore had to demonstrate profits of at least 10 percent with greater security of return than was available from investments in land or trade.

In Britain textile manufacturing regularly exhibited a minimal return of 10 percent or more. In the woolen industry one clothier in the West of England reckoned that he would make a 10 percent net saving in labor costs as a result of installing new spinning machinery, an estimate that he published in 1791. In the cotton industry, profits were much higher, though levels were not exactly known. When Richard Arkwright, the pioneer of factory cotton spinning, died in 1792, the *Gentleman's Magazine* reported that he left property worth £500,000. Knowledge of manufacturing profits also circulated privately and so might filter to the United States by informal channels. In Manchester, for example, it was commonly reported that Samuel Oldknow, a muslin manufacturer, earned £17,000 per year two years running in the early 1790s.[8]

Certainly American proponents of manufacturing were cognizant of the high profits Britain's textile entrepreneurs enjoyed. At one of the early meetings of the Pennsylvania Society for the Encouragement of Manufactures and the Useful Arts, Tench Coxe described a European factory that employed "a few hundreds of women and children, and performs the work of twelve thousand carders, spinners and winders."[9] One Philadelphia merchant observed that Britain's consumption of 22.6 million pounds of cotton in 1787 cost £2,230,000 but secured a value added by manufacturing (which he calculated by deducting cost of raw materials from selling prices of yarn and cloth) of £5,270,000.[10] Whether profit levels attained in British cotton and woolen manufacturing could be attained in the United States if the new technologies were to be transplanted there was unknown in 1790.

Recruiting a factory work force posed problems. There was the matter of a supply of unskilled labor. In America in 1790, the lowest-paid adult white and free male workers were farm laborers. With room and board free, they received about £17 in wages each year. Unskilled urban workers earned similar wages. Therefore wage levels of five to eleven shillings a week had to be exceeded if free white adult males were to be recruited by mill owners.[11] Indentured labor was another possibility but an unlikely one. In 1788 three-year indentured servants in Pennsylvania required an initial outlay of £12–£15 for adult males and £6–£8 for children aged nine to twelve years. Further, they had to be kept for three years regardless of business fluctuations, and their work, not rewarded with wages, might be very indifferent unless closely supervised.[12] For unskilled work, two other sources might be tapped: children and immigrants. America's population was very young. The federal census of 1790 showed that nearly half of the white males were under the age of fifteen. Apart from wide geographic dispersion, obstacles to employing children sprang from a strong American suspicion that manufactures had a morally and politically corrupting influence. To this prejudice two of the most venerated founding fathers, Franklin and Jefferson, lent their weight. Without safeguards against the exploitation of child labor the widespread use of children as in many English factories was not immediately likely.[13] Immigration, small by subsequent standards, offered another source of unskilled labor. Over twenty-five thousand Scottish and Irish immigrants arrived in the ports of the Delaware from 1783 to 1789, according to the British consul in Philadelphia.[14] Their entry into agriculture would be postponed if the differential between land prices and agricultural wages on one hand, and manufacturing wages on the other, sufficiently favored working in industry.

Another problem was the supply of skilled labor. Only European immigration could provide the textile mill managers, skilled operatives, and machine makers needed for the new technology. Would enough men with appropriate experience and reliability cross the Atlantic? If British artisans came, would they not be carrying preindustrial skills, their possessors in flight from technological unemployment? Would wages and profits be sufficiently high in America's manufacturing sector to attract and keep any industrial immigrants? Precise answers to these questions are difficult to give. A subsequent chapter in this book examines the data on the immigration of British textile workers. As for wage levels, most craftsmen and artisans in American cities received around £50 per year in the late 1780s.[15] Skilled factory wages would have to be higher than this, or else would have to be supplemented with fringe benefits like free housing, or with expectations of other economic or noneconomic opportunities.

Serious as were the U.S. labor and capital supply problems, far greater disabilities loomed on the demand side for promoters of an American textile factory system. One powerful deterrent was the tiny size of the potential domestic market. Of the 3.9 million people recorded in the first census in 1790, 18 percent were slaves who fell into the 30 to 40 percent of the population with little or no personal income and property. Small farmers were the most numerous element, composing 40 percent of the whites and a third of the whole population. In the whole of the northeastern states, they worked mainly subsistence farms. Larger farmers, chiefly in the middle Atlantic and southern states, who worked commercial farms, were less than 10 percent of the whites. And at the top of the economic scale, 10 percent of the population controlled half of the country's wealth. In short, two-thirds of the people lived at subsistence level, growing their own food and making their own clothing, and offered slim markets for industrial goods in 1790.[16] Furthermore, nearly all the population was widely scattered through rural territory. Some 95 percent of the people lived in settlements of under twenty-five hundred inhabitants. The majority of these were in the Atlantic states stretching thirteen hundred miles north-south and two hundred or three hundred miles west from the coast into the Allegheny Mountains, which about two hundred thousand people had crossed by 1790.

For manufacturers, the scattered nature of American markets was compounded by inadequate transportation networks. In 1790 roads were in a deplorable condition, having been neglected by the states and Congress during the Revolution and after. Dirt roads, subject to muddy conditions or flooding at fords, meant that only small freight loads, on pack horses or in small carts, could be shifted with any facility, and then only slowly. A stagecoach in the early 1790s took three days to travel from Boston to New York, a distance of 250 miles, and a wagon crossing the Alleghenies from Fredericksburg, Virginia, through the Shenandoah valley into Kentucky, could make only sixteen to twenty miles a day. Consequently, overland freight rates were high. In 1793, for example, wagon transportation charges made up 29 percent of the selling price of flour produced from wheat grown in the Shenandoah Valley and sold in Alexandria, Virginia, eighty miles away.[17] Water transportation was inhibited by two factors: All of the rivers in the hinterland flowed eastward from the Alleghenies, and many tracts of land lay between river valleys and so were inaccessible by water. Thus, lumber and farm produce moved swiftly downstream on rafts toward the seaport cities, but manufactured goods moved inland by water only through arduous poling or towpath towing.[18] Manufacturers therefore were confined to small, local markets. While affording monopoly conditions to workshop producers, the limited size of such markets denied entrepreneurs access to internal and external economies of scale without which the advantages of specialization could not be realized. Factory production was subject to the law of diminishing costs: only large-scale production and mass markets made mechanization worthwhile.

Concentrated populations were found in cities; however, there were only two dozen urban settlements with over twenty-five hundred inhabitants in 1790. New York was the largest, with thirty-three thousand, followed by Philadelphia with twenty-eight thousand. Then came Boston, Baltimore, and Charleston with between ten thousand and twenty-five thousand. Seven towns had populations between five thousand and ten thousand, and twelve more had under five thousand people. In the cities, upper and lower economic classes tended to be larger than in rural areas. Both classes were unlikely consumers of American manufactured goods in 1790, the poor because of their low standard of living and the upper groups, who derived their wealth from trade, because of their decidedly European tastes. Opinion leaders like the Binghams of Philadelphia displayed London furniture and carpets and French tapestries and wallpaper. Clothing followed European styles among the upper classes in America, despite the democratic influence of the Revolution. One traveler attending the Philadelphia theater in 1794 remarked on the ascendancy of English and French dress styles among the audience.[19] And styles were closely linked to cloth quality and origin, as well as cut.

Against these massive economic obstacles to the growth of a capitalist manufacturing sector, however, America possessed some advantages in 1790. Its natural resources, still largely unexploited and unknown, afforded innumerable streams for mill seats, plentiful supplies of timber for the construction of machinery and factories, and minerals (such as coal, iron, copper, and tin) invaluable in machine-making and power-transmission equipment. In the South, farmers grew the prime raw material for industrial textile manufacturing, although there were difficulties in producing enough cotton of the right quality. Short-staple or green-seed cotton was full of seeds that had to be removed manually; one person could clean only one pound of cotton per day. Long-staple or Sea Island cotton was cleaned more easily between rollers, but it could be grown profitably only along the coasts and on the islands off South Carolina and Georgia. Despite these limitations, the United States exported 3,135 bales or about 1,130,000 pounds of cotton in 1790, signaling America's international competitive advantage with respect to raw materials.[20]

Available labor had two assets. The basic principles of fiber processing were well known among the lower economic classes, some of whom would ultimately filter into a factory system. During the first stages of industrial growth, this labor supply theoretically reduced the number of difficulties posed in training operatives and instructing machine makers. Second, the United States in 1790 boasted a range of preindustrial craftsmen, skilled in working wood and metals and in traditional textile-machine-making, comparable to that found in England on the eve of industrialization. With this gamut of craft skills, the new machinery could be built without difficulty once design data were supplied from British or European sources.[21] Woodworkers constructed machinery framing and card cylinders. In 1787 and 1788, for example, Enoch Richardson, a Philadelphia carpenter, custom built three jennies for the Pennsylvania Society.[22] A variety of workers in iron, steel, brass, and tin participated in the construction of the gearing, cams, weights, levers, blades, plates, rollers, and spindles found in the whole range of early textile-processing machinery. Iron furnaces were the obvious source for cast-iron pieces. Such a furnace provided the Beverly company with a singeing plate in 1789. Wrought-iron work could be shaped by blacksmiths, and turners and whitesmiths were commissioned to make spindles and rollers in Philadelphia in the 1790s. A clockmaker, John Bailey, added two spinning heads (of four sets of rollers and spindles each) to the twenty-four-spindle water frame that Moses Brown bought from John Reynolds in 1789.[23]

In addition to these craftsmen, the early organizers of American manufactories relied on makers of traditional textile accessories. At Worcester in 1789, a Mr. Grout supplied a cane reed of twenty-four biers at 2¼d. a bier. Philadelphia, the federal capital, in 1793 boasted two reed makers, one comb maker, three spinning wheel makers (who built reels as well), two millwrights, and eight card makers.[24] Apart from the millwrights, whose work in installing power sources and transmission systems falls outside the scope of this study, the most important of these were the card makers.

Card clothing was critical in both preindustrial and industrial cotton and woolen manufacturing. It consisted of sheets of leather pierced by long wire staples, bent at an angle, to give a brush effect. Mounted on hand cards or in the carding machine on drum, rollers, or flats, the clothing brushed through the raw cotton or wool, getting rid of dirt and placing the fibers in a roughly parallel alignment in preparation for spinning. Attempts to improve the equipment that made card clothing were initiated by several American inventors before 1790.[25] In July 1785 it was reported to the Board of Trade in England that "they have got at Boston and in the Northern Colonies the Engine by which these Cards can be made more expeditiously."[26] Two devices were involved; both were said to have derived from an English inventor, John Kay. One bent the wire teeth, and the other pricked the leather. It is known that Oliver Evans, the millwright, developed a wire-cutting machine during the Revolution that cut teeth at the rate of fifteen hundred a minute. And in 1784 a New Haven goldsmith, Ebenezer Chittendon, was credited with a similar machine that cut teeth for cards at the rate of eighty-six thousand an hour. Another builder of card clothing machines was Nathan Cobb of Norwich, Connecticut, whose "Machines for Cutting, doubling & Crooking Card wire" were in use in Boston and Philadelphia in 1790.[27]

Wire-cutting and bending machines were employed in a number of card clothing manufactories in 1790 in Philadelphia, Boston, and Leicester, just west of Worcester, Massachusetts. All of them relied on household workers to set the teeth in the leather. Giles Richards's machines and his nine hundred hand setters at Boston reportedly produced sixty-three thousand pairs of hand cards annually in 1789.[28]

Infusing this reservoir of skill was a keen awareness of the possibilities of the machine. Whether it derived from educational provisions, or from the utilitarian values of the colonial experience, or from other directions is unclear. What is certain is that in the 1780s throughout the middle and northern states, from Philadelphia scientists like Franklin and Rittenhouse to New England mechanics like Abel Buell and Jacob Perkins, there was an intense interest in machine technologies.[29] Solow regards this as the basic element in the structure of cognition that he believes is a precondition of the capacity to assimilate a new technology.[30]

American trade difficulties, albeit temporary, lent special credence to the case for domestic manufactures. Foreign trade, as measured by exports to Britain, throughout the 1780s failed to reach two-thirds of its pre-Revolutionary level. In the carrying trade, Americans were excluded from the British empire, a severe blow in the case of the lucrative West Indies trade, and were confined by European countries to carrying American exports directly to Europe. At the same time, demand for America's primary products—tobacco, rice, wheat and flour, and naval stores—was not expanding. Secession from the British empire largely explained the American predicament. American goods were placed under heavy import duties by all European states, with the intention of protecting their own domestic or imperial interests.[31] Congress began to retaliate in 1789, passing a tonnage act giving preference to imports carried in American ships.[32] Even so the U.S. balance-of-payments deficit (calculated by one economic historian using five-year moving averages to be $70 million for 1790–1794) disturbed American merchants and commercial farmers, who suffered most immediately from this dismal situation.[33] While merchants probed new Asian markets and some American leaders like Jefferson advocated adherence to primary production and loosening foreign commercial restrictions through diplomacy, others argued that manufactures were the optimum economic and political solution. Secretary of the Treasury Alexander Hamilton declared in 1791 that a manufacturing sector "founded on the basis of a thriving Agriculture" would always lead to "a higher probability of a favourable balance of Trade."[34] A vigorous pressure group that shared Hamilton's economic philosophy worked in the 1780s and 1790s through press and political connection to mobilize the nation's industrial consciousness.[35] As much as profit, their motivation was patriotism. Economic independence seemed the logical complement to political independence. To this end, they were instrumental in bringing a number of skilled textile workers, and with them knowledge of the new textile technology, to America from Britain in the late 1780s.

But in 1790 America's most prominent assets—derived culture, cheap raw materials and motive power, traditional craftsmen, ingenious artisans, and patriotic entrepreneurs—looked insignificant against the harsh realities of high capital and labor costs and small, inaccessible markets. If this was America's position in 1790, what was the relative situation of British and American textile manufacturing? And what constituted the novel elements in Britain's new technologies?

During the two decades prior to 1790, Britain's factory-based cotton industry had experienced phenomenal growth. At this date Hargreaves's jenny, patented in 1770 and applicable to home, work-shop, and factory, accounted for 1.4 million spin-dles. Arkwright's water frame, patented in 1769 and confined to factory production because of its power needs, provided 310,000 spindles. Crompton's mule, developed in the 1770s and used in work-shop and factory, contributed 700,000 spindles.[36] Factories housing water frame and mule spindles numbered around 200 in 1790.

A recent typological analysis suggests that three basic cotton factory types emerged in Britain be-tween the 1770s and 1790s. The small factory, with horse capstans to drive carding machines and con-taining hand-operated jennies and mules, was val-ued for insurance purposes at around £1,000, or as much as £2,000 if a steam engine substituted the horse capstan. Second there was the medium-sized spinning mill, three or four stories high, seventy to eighty feet long by twenty-five to thirty feet wide, designed for a thousand spindles and powered solely by water. Valued at around £3,000, this type was modeled on Arkwright's water-frame mills. And, in the 1790s steam-powered spinning mills began to appear; containing about three thousand spindles, they cost a minimum of £10,000.[37]

In the early 1790s Britain's cotton industry was found chiefly in three regions: the Midlands, the North of England, and Scotland's Clyde Valley. Fifty-five cotton mills were in the Midlands; by 1795 another seventy mills were in Lancashire in the the North, and fifty more were in Scotland. At this point fixed capital in Britain's cotton industry to-taled between £2 million and £2.5 million.[38] In contrast the workshop and factory cotton industry in the United States looked diminutive, but not

for lack of entrepreneurial effort to break out of household manufacturing by acquiring Britain's new technology.

Two twenty-four-spindle jennies, the first known in America, were made at Philadelphia over the winter of 1774–1775. One was built for an immi-grant weaver manufacturer from Derbyshire, Joseph or John Hague, who was prompted to com-mission and supervise its construction because of the high labor costs of hand spinning in America. The other was built by Christopher Tully, who pub-lished its drawing and specification in *Pennsylvania Magazine* in 1775. Both showed a number of im-provements over Hargreaves's design of 1770.[39] Whether the modifications were added in Britain and then carried to America or made in America is unknown. The fact that the jenny was used in America within five years of its British patent date implies a rapid transfer.

Perhaps because of the dislocations of the Revo-lution, no carding machine appeared in America before 1783. A carding engine of unknown design, along with other machines, was smuggled into Philadelphia in 1783 and was set up by a Joseph Hague, probably the jenny builder. After four years of financial and technical difficulties, Hague sold this equipment (including two jennies and a sixty-four-spindle mule), which then fell into the hands of an English cotton merchant, Thomas Edensor, who promptly packed the lot up and in May 1787 sent them back to Liverpool. When the Manufacturing Committee of the Pennsylvania Society for the Encouragement of Manufactures and the Useful Arts searched for carding and spin-ning machinery in the winter of 1787–1788, Joseph Hague came forward to build the card.[40] It is quite

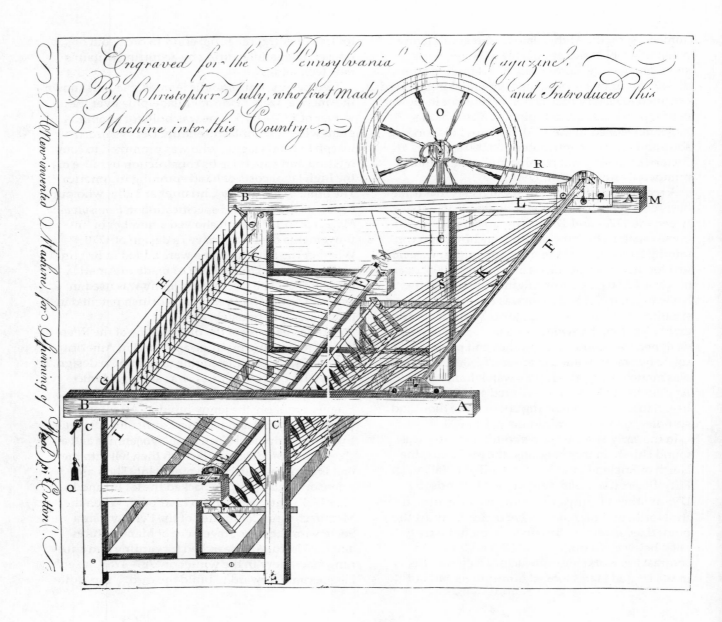

likely that this was the "John Hage" who emigrated from Derbyshire in 1774 and the following year, as Joseph Hague, petitioned the Pennsylvania Assembly for a premium for his jenny, not least because in November 1787 the British consul in Philadelphia, Phineas Bond, reported to his foreign secretary the rumor that Hague could be arrested at "a Place called Simonthly, near Hayfield in Derbyshire."[41] This time, the Hague in question succeeded in obtaining the design of Arkwright's rotary carding machine. On January 19, 1788, Thomas Wood, a member of the Pennsylvania Society's Manufacturing Committee, reported that he had "ingaged with John Hague of Alexandria in Virginia to make [a carding machine] on their joint Acct. & Resque."[42] By March 12 the machine was completed. Carding cotton at the rate of fifty pounds a day, it was displayed in the grand federal procession that summer.

New Englanders also exhibited a strong interest in procuring Britain's new labor-saving textile machinery during the 1780s. One prime mover in Massachusetts was the immigrant Hugh Orr.

1.1 One of the first American spinning jennies, built in 1775 by Christopher Tully. Compared with Hargreaves's patented English jenny of 1770, Tully's machine incorporated a number of improved features: a driving wheel on a horizontal axle (O); power transmission to spindles by a tin layshaft (E); wheels and an opening mechanism on the clasp (F,K); activation of the faller wire from the carriage clasp; friction-reducing glass bases under the spindles; tensioning devices for both the driving wheel and the layshaft; and an adjustable roving box. (Source: *Pennsylvania Magazine,* 1775. Courtesy American Antiquarian Society.)

Trained as a gunsmith in Scotland, he continued his trade after emigrating in 1740 and during the Revolution set up a cannon foundry at East Bridgewater. Through his energies, the brothers Robert and Alexander Barr came to Bridgewater from Scotland in 1786 to set up cotton carding, roving, and spinning machinery, including, evidently, Arkwright's water frame. The sponsorship of the Massachusetts legislature, in which Orr was a senator, brought the three or four machines into the public domain; deposited in Orr's house, they stood open to public inspection and became known as "the State Models."[43]

That was in May 1787. Shortly before, a weaver, Thomas Somers, had arrived in Massachusetts with descriptions and models of carding and spinning machines he had collected in England the previous summer. Somers emigrated to America before 1785 because in that year an association of Baltimore tradesmen and manufacturers had dispatched him on a machine-collecting expedition to England. On his return, he found his promoters unenthusiastic, so he headed north for Boston, presumably because its manufacturers' circular of 1785 had inspired his sponsorship by the Baltimore group. Under Orr's eye, Somers performed some trial work, which made a group of Boston merchants sufficiently confident to sponsor the famous Beverly Cotton Manufactory, incorporated by the state of Massachusetts in 1789. The technical side of the manufactory was in the hands of Somers and another British immigrant, James Leonard. Visitors to the factory between 1788 and 1790, including George Washington, observed one Arkwright roller card (carding fifty pounds of cotton a day), four jennies (one of eighty-four spindles), a twisting mill, a warping mill, and fly shuttle looms with the shuttles driven by springs.[44]

The exhibition of state models at Bridgewater attracted several Providence merchants, including Moses Brown, who sent a clockmaker friend to examine them and report back to him in spring 1788. That fall Brown went to East Bridgewater himself and then toured the factories at Beverly, Worcester, and Hartford (woolen). With the encouragement of several Beverly promoters and the technical advice of immigrant English woolen workers Joseph Ashton and Thomas Ashton at Hartford, Brown decided to start textile manufacturing at Providence with his son-in-law William Almy in spring 1789. They purchased a carding machine and a twenty-eight-spindle jenny in May (copied from Beverly machines for another group of Providence promoters led by Daniel Anthony). By June they had hired a Scottish fly shuttle weaver, Joseph Alexander, to supervise cloth manufacture. His acquaintance Thomas Kenworthy, another immigrant, was put in charge of yarn spinning. James Leonard, lured from Beverly to act as a consulting engineer, was then dispatched to New York to acquire knowledge of finishing techniques and also to secure cutting knives and guides for velvet work. Brown, having organized a production line consisting of hand picking, machine carding, jenny spinning, handloom weaving, and hand finishing, turned his attention to spinning hard warp yarns with water power. For twenty-two pounds ten shillings, he purchased from a fellow Quaker, John Reynolds of East Greenwich, a twenty-four-spindle copy of the state model spinning frame (which Reynolds had unsuccessfully tried to operate), and on August 8, 1789, the *Providence Gazette* announced that Brown

and Almy had a thirty-two-spindle "mill after Arkwright's construction." By fall 1789 they had constructed a second frame, this one with twenty-four spindles, in a building at Pawtucket. But successful operation eluded their managers. Despite Brown's visits to New York and Philadelphia, they were unable to recruit a competent water-frame spinner, and a disappointed Moses Brown withdrew from the partnership. His place was taken by Smith Brown, Moses' cousin, who ran a dry goods store in Providence and was already investing in stocking machinery. The textile manufactory continued as Almy & Brown. Such was the position when Samuel Slater, newly arrived at New York after finishing a six-and-a-half-year apprenticeship in Jedediah Strutt's Derbyshire mills, wrote to Moses Brown on December 2, 1789, offering to install the Arkwright spinning system.[45]

A handful of smaller manufacturing enterprises formed the remainder of America's cotton industry in 1790. At Worcester, Massachusetts, a cotton manufactory was established in February 1789. It had a carding machine and spinning jenny and two hand looms running by that fall.[46] In New York the cotton manufactory belonging to the city's Manufacturing Society was technologically no more advanced: a card and two jennies, which Slater thought "not worth using," were the most advanced machinery in the Vesey Street factory.[47] At Portsmouth, New Hampshire, Wilmington, Delaware, Baltimore, Maryland, and elsewhere, patriotic merchants and tradesmen also promoted textile mills based on English models. By 1790, however, these efforts had advanced little further than the advertisement of aspirations in local newspapers.[48]

American methods and standards of cotton weaving (on the household hand loom) and cotton finishing (apart from chlorine bleaching and roller printing conducted in workshop premises in Britain) were much the same as those carried on in England at this time. Several American cotton finishers came from England. In Philadelphia in the 1780s, for example, the calico and linen printers or engravers originating from England included John Norman, Thomas Bedwell, Henry Royle, and, most famed for his skill and longevity in the printing trade, John Hewson, who emigrated in 1773 and retired in 1810.[49]

The U.S. cotton spinning industry lay far behind that of Britain in 1790. Compared with Britain's estimated 2.41 million machine spindles and £0.7 million of fixed capital investment, the United States operated fewer than two thousand jenny spindles in about ten cotton manufactories (workshops or mills in which one or two stages of production were performed by hand- or water-powered machines), whose fixed capital investment was under £1,500 each. Successful commercial operation of Arkwright water-frame technology or mule-spinning technology still eluded American entrepreneurs, despite their strenuous efforts to obtain these new technologies in the 1780s.

The most advanced machine technology in Britain's woolen industry in 1790 was found in Yorkshire. Here small capitalists used low grades of wool and a machine technology derived from the cotton industry to capture domestic and overseas markets in coarse qualities of broad and narrow cloths. In the two previous decades, at least fifty-one scribbling and carding mills, the vast majority powered by water, came into operation in the West

Riding. Such mills averaged £1,000 in fixed capital. About £250 was needed for machinery for a typical mill that ran six to eight scribbling and carding machines, costing £20 to £30 each, and two or three slubbing billies, costing around £15. Jennies of forty or more spindles were usually worked in clothiers' homes, though some West Riding carding mills installed them in the early 1790s. Hand loom weaving, as in the cotton industry, took place in weavers' homes. Wet finishing was concentrated in the ancient fulling mill. In the 1780s the West Riding boasted over a hundred fulling mills, and by 1790 many provided premises for scribbling and carding machines. Cloth dressing was carried out in raising and cropping shops, and gig mills, for raising, were slowly being adopted.[50]

In the West of England, larger capitalist manufacturers were aided against Yorkshire competition by the country's population growth, a rise in the standard of living, and new cloth types (chiefly Francis Yerbury's cassimere). Partly because the region concentrated on quality production, the new machine technology arrived more slowly. In 1790 western clothiers started to install carding machines and jennies, but the fly shuttle was still unknown. In addition to fulling mills, gigs were used, especially in Gloucestershire where they first appeared in the sixteenth century. No estimate of the region's fixed capital investment in woolen manufacturing has been made; in any case the industry's manufacturing assets chiefly lay in the unrivaled reservoir of skills and refined techniques, developed over the centuries, that placed it ahead of Yorkshire and most, if not all, European rivals in the production of fine-quality woolen cloth.[51]

Apart from the medieval fulling mill, which was

introduced to America in colonial days, very few exceptions to the predominant household organization of woolen manufacturing appeared in the United States before 1790. Hamilton, in his *Report on Manufactures* (1791), singled out the woolen manufactory at Hartford, Connecticut, as the most "promising essay" but believed "difficult and uncertain" the endeavor to "cherish and bring to maturity this precious embryo."[52] Projected by an unincorporated joint stock company in 1788, its original capital of £1,280 was increased to £2,800 a year later in order to purchase broad looms, fulling stocks, and shearing frames. Relying on English army deserters and ex-prisoners of war with textile trades experience, the manufactory was organized on the putting-out system, with a workshop for wool sorting and cloth finishing, and a fulling mill. Its only piece of new textile technology in late 1791 was one willow or picker.[53] This is strange because thirty miles away at Norwich, carding machines and spinning jennies were currently being built by a Mr. Lathrop.[54] Though its cloths were of fair quality, Colonel Jeremiah Wadsworth's enterprise was not yet abreast of technical developments in American cotton manufacturing, to say nothing of British woolen manufacturing. Washington, who visited the factory in 1789, commented in his diary: "Their Broadcloths are not of the first quality, as yet, but they are good; as are their Coatings, Cassimeres, Serges and Everlastings."[55] Patriotically, Washington wore a suit of Hartford broadcloth at the opening of the second session of the first Congress.[56]

In household manufacturing, the American bias was obviously toward coarse woolen fabrics whose production required lower labor inputs of skill and time than did finer cloths. For sorting wool, Wily in his *Treatise* defined three qualities, compared with between ten and seventeen grades distinguished by English wool staplers. In cloth raising Wily suggested only one course (treatment) with the teazel handles, as opposed to as many as twenty-nine in the West of England.[57]

In assessing American chances of acquiring the new British technologies, novel technological elements, unfamiliar to Americans, must be identified. From their efforts simply to establish working machines, American entrepreneurs apparently found a lack of labor skilled in the new industrial technologies the major barrier to technical success. Philadelphia merchants had to abandon a spinning mule and New England merchants a water frame because the machines lacked experienced men competent to build, run, repair, and manage the equipment. What was the nature of these new operative, managerial, and machine-making skills, and how did their nature limit the manner in which the skills could be taken to America?

Two distinct cotton-spinning technologies, the Arkwright or water frame and the Crompton or mule technologies, evolved in Britain.[58] Both were unfamiliar to Americans. In Arkwright mills, which made coarse, hard-spun yarns, operative labor required no skill. The physical labor of cotton picking (cleaning) needed no more than the strength of

1.2 English flat-top cotton carding machine, as depicted in 1812. The carding action (fig. 4) occurs between the card teeth on the main cylinder and the teeth on the underside of the flats. (Source: Rees, *Cyclopaedia*. Courtesy Merrimack Valley Textile Museum.)

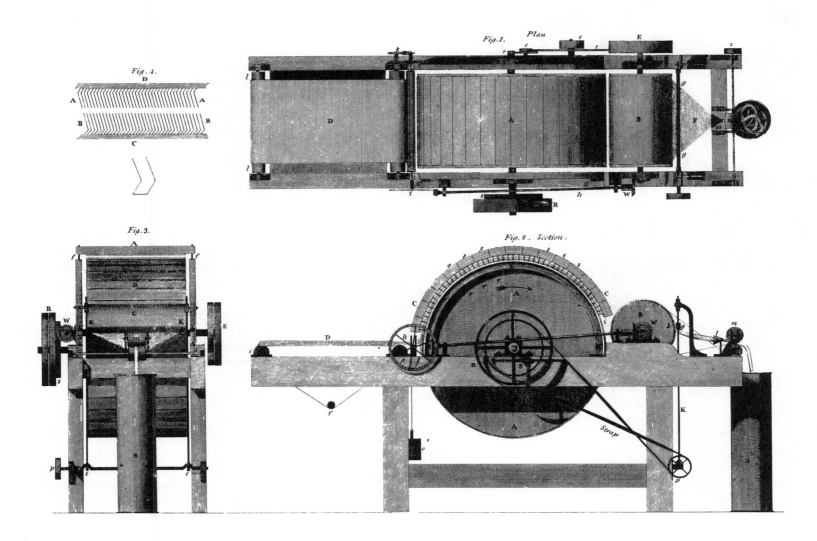

Fig. 4.

Fig. 1. Plan

Fig. 3.

Fig. 2. Section.

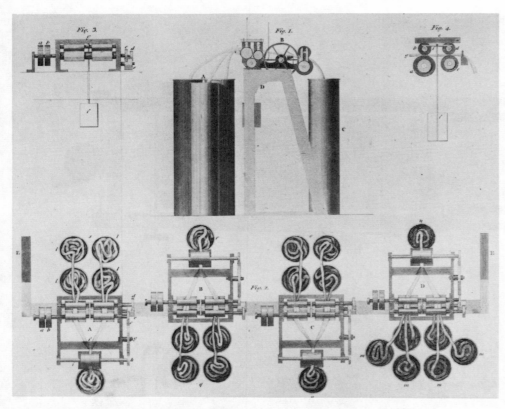

1.3 English drawing frame of four heads, as depicted in 1812. Four ends are being doubled at each head in fig. 2. Drawing rollers, the key element of the machine, are shown in fig. 4. (Source: Rees, *Cyclopaedia*. Courtesy Merrimack Valley Textile Museum.)

healthy adult women; the portering chores of feeding and discharging carding machines, the strength of children; and the carrying and monitoring routines of doffing and piecing in drawing, roving, and spinning frames, the strength and alertness of children. No operative task required more than a few weeks' learning time. The Arkwright system substituted capital for labor, machines for skill, factory for home, and mill discipline for family work routines. Fixed hours, punctuality, regular attendance, constant attention, cleanliness, sobriety, and, above all, submission to the pace and work flows of mechanized production were demanded in the water-frame factory.[59] Mill management, however, demanded new skills, especially in the technical and labor relations ambits. In replacing the old batch methods with flow production, Arkwright managers handled more complex production engineering problems than had previous manufacturers.[60] Flow production, as opposed to putting-out or workshop organization, additionally meant attention to the design and development of the whole production line and prevention of imbalances between the various processing stages, both in capital and in running speeds.

At each processing stage, different technical and managerial problems needed solution. The first set of production decisions facing mill managers related to the quality of cotton available. On this depended the fineness of the final yarn and also machinery adjustments at subsequent stages. To effect cost savings, bales of different cottons were frequently mixed together. Because cotton has many properties, the selection and proportioning of cottons required informed judgment. Lacking sophisticated textile testing equipment, managers were forced to rely on viewing, handling, and experience in the trade in making the crucial mixing decisions. And different cotton types were entering Britain in the 1790s as the sources of cotton imports changed.[61]

It was also the manager's responsibility to match his raw materials to his capital equipment and his manufacturing techniques to his raw materials. This began at the carding stage, which James Montgomery regarded as the most important subprocess in yarn production; decisions and practice at subsequent processing levels were predicated on those of the carding stage.[62] At this point a manager directed his overseer to produce a size of card sliver suited to the drafting limits of his drawing, roving, and spinning frames. From sample weighings of the card sliver, the overseer estimated its grist or fineness. Change gears in the card were then replaced to effect a reduction or increase in the sliver grist. Overseers were responsible for fitting and setting up their carding machines and for keeping them in good condition. The former work included remedying warping in cylinders or flats, a problem with new machines unseasoned by warm mill temperatures; ensuring that the card clothing was securely fixed to cylinders and flats; and setting the card teeth (the tolerance between working surfaces), which some overseers did by eye and ear, others by metal gauges. Maintenance routines included oiling axles and gears, tensioning belting, brushing card teeth daily, sharpening teeth every nine months with a grinder, and replacing worn card clothing or changing it to suit a different cotton batch. Under the mill manager's executive direction, overseers also made several adjustments to drawing, roving, and spinning frames. Changes in

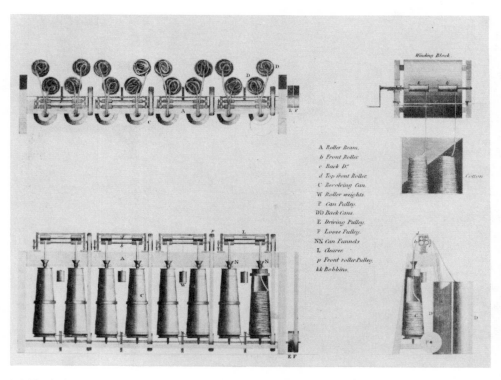

1.4 English roving can frame, as drawn in 1812. Rotation of the cans (C in the lower right diagram) inserted twist in the drafted sliver. (Source: Rees, *Cyclopaedia*. Courtesy Merrimack Valley Textile Museum.)

1.5 English water frame, as drawn ca. 1812 in Messrs. Strutts's mill, Belper, Derbyshire. Figure 4 illustrates the core elements: the drafting rollers, flyer, and bobbin. (Source: Rees, *Cyclopaedia*. Courtesy Merrimack Valley Textile Museum.)

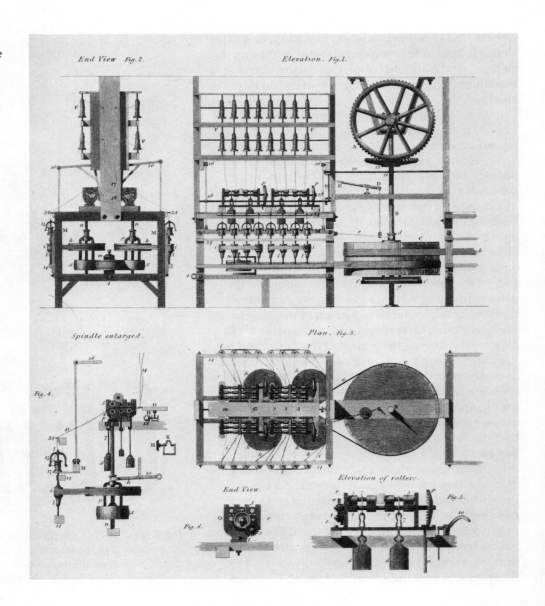

roller settings, speeds, and weights controlled draft or yarn fineness; spindle speeds altered twists per inch or yarn hardness. In 1790 only several years' experience or an older manager's or overseer's notebook imparted an awareness of these basic maintenance problems and the mechanical adjustments necessary for a range of fiber and yarn qualities.

A practical knowledge of cotton and the adaptation of a production line to manufacture its various grades was not the only area in the early factory where the management functions of planning, organizing, and controlling were exercised. Others were the recruitment, training, and disciplining of the work force; and accounting, in which departmental bookkeeping was usually adequately performed but overall cost-and-revenue statements eluded definition.[63] But because the technology was new and unfamiliar, technical knowledge of Arkwright spinning was the prime requisite in a mill master.

The new skills of the Arkwright mill manager and overseer were acquired, as the system spread through the British cotton manufacturing districts in the 1770s and 1780s, by men who learned by doing. Americans faced the choice of recruiting such men or of learning by doing themselves. British mill managers working in America would need at least their British salary levels, around £200 or £300 a year. Overseers earned much less. The card-room overseer at Styal, Cheshire, in 1790 received fifteen shillings and his spinning room counterpart twelve shillings sixpence to thirteen shillings a week, on top of which there were fringe benefits like cottage accommodation and a large garden. Perhaps an annual wage of £50 would se-

1.6 Close-up of roller beam in section of full-size English water frame dated to the late eighteenth century. Notable are the metallic components: pigtail guides, rollers, weights, gears, and the arm operating the clutch on the roller drive. (N-W Museum of Science and Industry, Manchester; photo by author.)

cure a British mill-room overseer.[64] The alternative of learning by doing would take several years and involve unknown costs; hiring experienced men was preferable.

A cursory analysis of machine making in an Arkwright mill suggests three broad areas where new skills were needed. First was machine design. Without a knowledge of the structure of the new carding, drawing, roving, and spinning machines, machine makers could not begin work. Next was the production of new components, in particular carding machine cylinders and fillet card clothing, spindles, and rollers. Traditional spinning wheels used forged spindles, but for the water frame they had to be ground more accurately and set true in order to reduce spindle wobble at high speeds.[65] High elasticity constituted another valued property of the best spindles.[66] Reflecting the technical skill absorbed in their production, spindles and rollers represented a fifth of the cost of a spinning machine in the 1820s.[67] Lastly, the static and dynamic interrelationships of the machines' crucial parts had to be established, as with dimensions, speeds, settings, or weights. Machine makers in Britain could earn as much as overseers. At Styal in 1790, the machine makers consisted of two roller coverers, three smiths, two joiners, seven clockmakers, and two turners, each earning weekly wages ranging between twelve and twenty-five shillings.[68] Again recruitment of such workers involved less risk than learning by doing, which in Arkwright's own case took six to eight years.[69]

In Crompton's system, high degrees of skill and physical stamina were the salient labor requirements of the spinning mule, which spun yarns of all counts and qualities. In the spinning cycle of the

hand mule in 1790, both movements—the outward draw and the inward run of the carriage (putting up)—were hand powered and hand controlled, demanding the physique, coordination of hand and eye, and average intelligence of a strong adult, usually a male. His physical strength propelled the carriage and the spindles. His dexterity maintained a correct balance between drafting and twisting. If the carriage came out too fast, the ends broke; if it came out too slowly, they were overtwisted, and yarns coiled up on spindle tips. Putting up especially tested his coordination. Carriage and spindle speeds had to be so regulated that the yarn moved onto the spindles without snapping or coiling up. Simultaneously the spinner had to manipulate the faller wire to form conically shaped cops, built by spiraling the yarn up and down bare spindles to make light and compact yarn packages from which yarn could be withdrawn without cop rotation and therefore were ideal for shuttle bobbins. The mule spinner was also responsible for changing the yarn count as directed by his customers or manager. Since the distance traveled by the carriage was fixed, the yarn count could be altered only by changing the speeds of draft rolls on the roller beam and/or the length of roving entering the spinning zone between the draft rolls and carriage spindles. For a new count, the spinner would change his draft roller gears and choose a fresh position for the stud in the count wheel, which governed the time taken to activate the lever disengaging power from the drafting rollers, and thus the length of roving in the spinning zone.[70] Clearly mule spinning was a very skilled job; even after the draw was powered, cop formation in winding and tuning the other movements of the machine remained in the hands of the

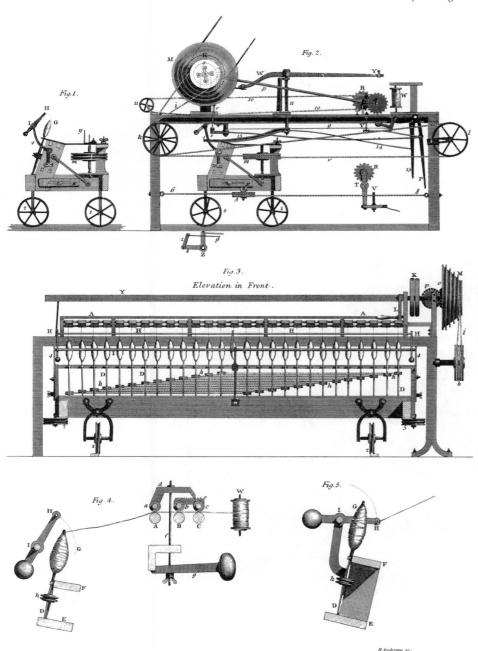

1.7 English cotton mule, as depicted in 1812. In the mule, roving underwent drafting twice: in the rollers (near *R* in fig. 2) and between the front rollers and the spindles of the carriage as the carriage moved outward. Figures 4 and 5 show the two positions of the faller wire during drafting and winding respectively. (Source: Rees, *Cyclopaedia*. Courtesy Merrimack Valley Textile Museum.)

1.8 English roller woolen carding machine, as drawn in 1818. Notable are the wooden framing and cylinders, the positive chain drive to the worker rollers, the fluted wooden delivery cylinder, and the crank-driven windlass by which the delivery cloth moved. (Source: Rees, *Cyclopaedia*. Courtesy Merrimack Valley Textile Museum.)

1.9 Close-up of teeth on a double woolen carding machine built by Artemas Dryden, Jr., of Holden, Massachusetts, pre-1830. Teeth on the worker cylinders (upper right and upper left) act against teeth on the main cylinder. Fibers caught on the worker are removed and returned to the main cylinder by a clearer roller (the smaller cylinder next to each worker, one being visible here). Adjustments to the poppet heads supporting the worker and clearer axles altered the tolerances between the card teeth, to suit the stock being carded. (At Merrimack Valley Textile Museum; photo by author.)

spinner, who earned about thirty shillings a week in the early 1790s, approximately three times a farm laborer's wage, though out of this he paid an assistant or piecer a few shillings.[71]

Learning times for mule operatives in the late eighteenth century are unknown. However, some guesses can be made on the basis of subsequent mule spinners' training times. Dr. Harold Catling, who himself worked in the mule gate before moving into textile engineering, estimates that the operative skills of spinning on a hand or automatic mule required no more than three months' learning time.[72] However, the mule spinner also had to learn maintenance skills. This maintenance work needed three to four years to learn on the hand mule and, certainly, a minimum of three years and generally seven years on the automatic mule. The surprisingly short operative learning times, Dr. Catling emphasizes, applied only to people brought up in the mule spinning room. They developed an instinctive knowledge of fiber behavior and machine operation in their formative years and in their early teens practiced operative skills like piecing before they took charge of a mule. The desirability of starting operative training early was emphasized in Arkwright mills, where ten was regarded as an optimum starting age, though this may have been an argument to buttress the case for child labor.[73] Unquestionably, adults unfamiliar with mule spinning equipment would need longer than three months to learn to operate, and more than three years to learn to maintain, a hand mule.

Managing and machine-making skills in the Crompton system were identical to those in Arkwright mills. Although data have not survived for the 1780s, it may be assumed that a mule work-shop manager's salary was not much different from the Arkwright mill manager's. Since mule spinners were their own bosses to a great extent, spinning overseers did not enter the picture before mills reached tens of thousands of spindles in size. Carding-room overseers would have received wages parallel to those in water-frame mills. Most likely, Americans found it less costly to hire these skills than to set about learning by doing.

The carding machine (but not its cruder variant, the scribbler), the billy, the jenny, the gig mill, and the shearing frame made up the new mechanical technology in Britain's woolen industry unfamiliar to American woolen manufacturers in 1790. Unlike operative tasks in carding, the billy and jenny required much skill, not least because the imbricated woolen fiber was more difficult to spin than the convoluted cotton fiber.[74] The failure of early woolen cards to produce a continuous sliver necessitated an additional processing stage in which short rolls, or slubbings, two or three feet long were pieced together. The slubbing billy performed this work. As in the mule, the slubber controlled twisting and drafting. Children or piecers were employed in the unskilled work of rubbing the card rolls onto the ends of the slubbings. In a sixty-spindle billy the slubber required the assistance of four piecers, each responsible for fifteen ends. Children who joined cardings unevenly or rubbed ends too weakly so that they broke occasioned the proverbial cruelty of the slubber and were spurred with a strap or the more handy billy roller.

The jenny, like the mule, had two basic movements: the outward draw and the winding-on motions. But with the jenny more than with the semipowered mule, the spinner controlled yarn

1.10 English slubbing billy, as drawn in 1818. From the feeding cloth on the left slubbings were drafted between the billy roller *C* and the spindle tips on the moving carriage (on right). (Source: Rees, *Cyclopaedia*. Courtesy Merrimack Valley Textile Museum.)

1.11 English spinning jenny, as depicted in 1818. Slubbings from the spindles in the sloping roving box pass up
to the wheeled clasp on top of the machine frame and are
stretched as the spinner moves the clasp away from the
yarn spindles (on left). (Source: Rees, *Cyclopaedia.* Courtesy Merrimack Valley Textile Museum.)

1.12 English gig mill, as drawn in 1818. Teasels carried on a cylindrical frame (*FF*) brushed the woolen cloth, raising the surface fibers. Jets of water (from pipe *N*) moderated the harshness of the raising action. (Source: Rees, *Cyclopaedia*. Courtesy Merrimack Valley Textile Museum.)

quality; his strength, coordination, and judgment regulated drafting, as well as twisting and winding. No assistants were needed in jenny spinning, which could be performed at home or in a workshop by an adult male or female or a strong teenager.

Both the gig mill and Harmar's mechanized hand shears (which were still being perfected in 1790) presupposed that their operatives were experienced in cloth finishing. In the former, for example, the operative had to know how long to work the cloth; in the latter, how close to run the shears to the cloth and how soon to move the shear carriage forward.

The prevalence of piecework and part-time work makes any assessment of earnings, and the cost of skills, in the woolen industry very difficult. Spinning rates halved after the introduction of the jenny, dropping from a shilling a score to six and a half pence a score. A score of skeins, each 320 yards long, could be spun on a spinning wheel in a day, giving weekly earnings of six shillings in the 1770s. If a forty-spindle jenny in the 1790s had only ten times this output, the spinner could earn thirty shillings a week. But this assumes full work and steady piece rates, which were not always the case. If jenny operatives received thirty shillings a week, their earnings were more than those of woolen weavers, who made a pound a week on a double loom or a fly shuttle loom, and up to twice those of shearmen, who earned fourteen to twenty shillings a week in the West of England. Much less is known about wage rates in Yorkshire, but by 1808 slubbers in one West Riding mill averaged twenty-two shillings sixpence a week.[75] Learning

times on the woolen jenny must have been comparable to those on the cotton hand mule.

None of this new woolen equipment impelled factory organization. All could be operated in the home, like the jenny, or in a small-scale carding mill or finishing workshop. Unlike Arkwright's system, the new woolen technology might be deployed to reinforce household manufactures by putting out the spinning and weaving while concentrating preparatory and finishing processes in carding and fulling mill complexes. This preserved the older managerial skills of the manufacturer in organizing a hybrid putting-out and workshop system and exercising control over widespread household spinners and weavers.

Preindustrial skills met nearly all the machine-making requirements of the new woolen technology, except in the area of design and the case of the carding engine. Clearly the designs of unfamiliar machines (card, billy, jenny, gig) had to be known before construction began. But apart from the card, none of the woolen equipment involved close mechanical adjustments to suit fiber processing, as were necessary with cotton rollers and spindles. However, the carding engine needed the new components of roller card cylinders and fillet card clothing, together with new skill in establishing the static and dynamic interrelationships of its parts.

In 1790 scattered groups of American merchants and patriots, eager to establish the new British cotton and woolen technologies in the United States, possessed few actual economic advantages. Besides a shared cultural background, they had only the potential of raw materials, motive power, and a reservoir of traditional skilled craftsmen. These were hardly sufficient in the face of high capital and labor costs and looked negligible against the disability of small, local markets, which threatened to contain demand to uneconomic proportions. First attempts to transplant the new British technology foundered for several reasons in the 1780s, not least of which was the nature of the technology itself. An examination of the novel aspects of the textile technology that eluded the grasp of American entrepreneurs showed that it would have to be accompanied by skilled managers and machine makers and, in the case of mule spinning and woolen technology, skilled operatives if an expensive learning-by-doing period of one or two decades was to be avoided.

2
Private and Public Protection in Britain: Secretiveness, Prohibitory Laws, and Patent Practices

While enterprising Americans and Europeans evinced intense interest in Britain's new industrial technologies, their concern encountered varying degrees of hostility at all levels in Britain, from the individual worker up to the national government. The extent to which American imitators could penetrate the barrier of British protectionism depended on the nature and effectiveness of the security measures imposed at points of possible information leakage: the factory or workshop, technical publications, the ports, and the patent offices. Complicating any assessment of these measures is the possibility of variations in application over time and between places in the textile districts.

Secretiveness, a preindustrial posture deriving from the apprenticeship system in which the novitiate craftsman swore to keep his master's secrets, widened in Britain with the advent of radically new technologies. With good reason in the late eighteenth century, inventors feared machine-breaking mobs, the loss of patent eligibility, and the depredations of domestic or foreign rivals should their devices or machines be learned about abroad. Similarly, manufacturers' stances of suspicion and distrust were not based on groundless paranoia; labor piracy, loss of technical information to British or foreign rivals, even the discovery of patent infringements, awaited open managements.[1]

Secrecy was preserved in several ways. Factories assumed the defensive features of a medieval castle: main shops built around a quadrangle yard, small windows, and narrow gateways, as in Benjamin Gott's Bean Ing Mills at Leeds.[2] Workers were sworn to secrecy. Robert Pilkington, who claimed to have invented the spiral application of fillet or garter card clothing to produce a continuous card sliver, told Arkwright in 1775 that he and his partner "proposed swearing the hands we employed, that they should keep it a secret."[3] Oldknow, who developed a method of improving fine muslin finishing in 1783, told a correspondent that his workers were "under an obligation not to disclose the secret."[4] Mills were closed to strangers. Pilkington shut his doors to those not under oath to guard his invention. In the 1780s and 1790s, Robert Owen recalled much later, "cotton mills were closed against all strangers, and no one was admitted. They were kept with great jealousy against all intruders; the outer doors being always locked. Mr. Drinkwater [Owen's capitalist partner] himself . . . came only three times during the four years I retained the management of it [Manchester fine cotton spinning mill, 1791–1795]."[5] The visitors Drinkwater brought "had influence with him" and noticeably were no rivals: an astronomer, a lawyer, and a cotton lord. One of the mill rules at Taylor & Co's. cotton mill at Halliwell, Lancashire, in 1804 read, "If a stranger comes into the Factory, the spinner who spins next to the Door to send for the Overlooker but if he cannot be found, then the spinner who came last in that room to order him out."[6]

Several techniques of industrial concealment persisted after 1815. Some factories shut their doors to all visitors. Johann Georg Bodmer, the Swiss engineer and inventor who toured British industrial districts in 1816 and 1817, was unable to gain admittance to a few mills, including Pilling & Wood's cotton spinning mill in Manchester and a spindle factory in Sheffield. Other mills selectively exhibited their operations to visitors. In his Manchester cotton spinning factory, a Mr. Holt allowed Bodmer "to see everything except the rag cutting and the twisting frames."[7] Intentionally or by accident, Bodmer was not shown the blowing machine or packing department in McConnell & Kennedy's Manchester mill. Edward Temple Booth, a Norwich worsted manufacturer, idiosyncratically allowed foreigners into his factory only "with caution": "When the machinery is peculiarly complicated, you may show it with good effect, I think, because it makes the difficulty of imitation appear greater."[8] Such arrogance was unwarranted, as technically competent Continental and American visitors proved. Other forms of secretiveness included the use of English erectors with machinery exported abroad, as was the practice of London engineers (in some cases too because local engineers could not do the job), and the refusal to instruct foreigners.[9]

For a large number of reasons, secretiveness became less defensible in the early nineteenth century. There was little point in concealing common knowledge except for a leading firm anxious about an increasingly hollow reputation. Boulton & Watt's "uncommon degree of mystery" was explained in 1824 by Alexander Galloway, a London engineer, as something of the sort: "They have nothing to show beyond what is well known in other places; they continue from pride that exclusion which before was dictated by interest."[10] The reported openness of Glasgow manufacturers stemmed from their use of widely known technology in spinning coarse and medium yarns.[11] In 1824, however, a Glasgow cotton spinner asserted that "immediately after the peace [after Napoleon's defeat in 1815] we were more liberal than at present" in admitting strangers to mills.[12]

Manufacturers in the West of England who relied on manual techniques more than machines to produce their fine cloth finishes had nothing to lose in showing foreigners their machinery, for the essence of their success was invisible to a transient visitor. One told two visiting American manufacturers, Zachariah Allen and Abraham Schenk, who arrived without letters of introduction in summer 1825 and openly declared their trades and nationality, "that the world was wide enough for us all, and that he would show us his mill with pleasure."[13]

Patentees, owners of patent rights, and builders of machinery were, naturally, far more anxious than manufacturers to advertise the new capital equipment. John Jones of Leeds, who made patent steam brushing machines, and Samuel Stead of Gomersal, who made woolen mules and gigs, willingly allowed Zachariah Allen to inspect their products and note dimensions, materials, and costs, in hopes of attracting a customer, despite the illegality of machinery export.[14] For the same reason manufacturer-machinemakers like William Hirst of Leeds, who had patented steam fulling, showed their premises off to foreign visitors.[15] And also, from the inventors' viewpoint, those refusing to disclose their inventions prevented further improvements and new applications and thereby limited their own financial rewards.[16]

Subtle influences played a part in opening the oyster of English technical achievement to the gaze and grasp of foreign visitors. One was the network of commercial and social obligations stretching from provincial manufacturers to London merchants and then, by foreign diplomats and commercial agents, to foreign importers and manufacturers. Such a mechanism aided the movements of Bodmer through British manufacturing districts in 1816 and 1817. By using a chain of letters of introduction, he was admitted to a number of important factories in Birmingham and the Black Country, in Manchester, in Glasgow, and in Leeds. Besides collecting letters of introduction from one district to open doors in the next on his itinerary, Bodmer also relied on London men to exert leverage on provincial manufacturers. His own brother, settled in London as a timber merchant and cloth cutter, gave Bodmer letters of introduction to "Birmingham and Wedgwood." By being "less than truthful in my replies to their questions," Bodmer convinced the Manchester Quakers that he was not "an expert on machinery" and gained admittance to their cotton spinning factories.[17]

The conflict of interests between trade and manufacturing, like that between machine making and manufacturing, posed another subtle influence. For example, a merchant manufacturer like Benjamin Gott was almost obliged to open his mills and offer hospitality to American visitors, as he did to Zachariah Allen in April 1825, because he directly and heavily engaged in the American trade.[18]

Two more factors undermining industrial secretiveness can be identified in this period. One was the foreign community living in Britain. Its members provided routes into merchanting-manufacturing-machinemaking circles, could act as foreign correspondents, or on eventual repatriation might export British technology. Among Americans domiciled in Britain after 1812 were Joseph Chesseborough Dyer, who settled at Manchester in 1811; Jephtha Avery Wilkinson, who worked with Sharp

& Roberts at Manchester between 1817 and 1823; and William Davy, the American consul at Leeds for a number of years before his death in 1827, and his son Albert.[19] Along with Continental merchants, they composed what Radcliffe in old age bitterly called "the Foreign Anglo Junto," a fifth column whom he suspected of betraying the British manufacturing interest. More than this, his diatribe accused this junto of promoting free-trade measures, the other undermining trend at work in this period.[20] Free trade envisaged the free export of machinery, in which case the boundaries of industrial secrecy logically shrank to the inventor's workshop and mill trial machines. Henry Houldsworth, Sr., the Glasgow cotton spinner and machine maker, predicted in 1824 that free machinery export would bring "a greater number [of foreigners] to this country, with a view to ordering machinery; and besides they would, through their machine makers or friends, have greater opportunities of entering our mills than they at present have."[21]

The publication of technical information was nevertheless stifled by the prevailing atmosphere of secretiveness. James Ogden, author of *A Description of Manchester* (1783), confessed that he omitted technical information from his volume for fear that the French would benefit.[22] Even the publication of commonly known hand loom weaving techniques was resisted. John Duncan, who wrote a hand loom weaving treatise published at Glasgow in 1807, met the objection that his work might "operate to the prejudice of those who practise weaving" and that a knowledge of it "may be acquired out of this country, and, consequently, the manufacture may become less productive."[23]

After the turn of the century a breakthrough in publishing technical knowledge came with the publication of *The Cyclopaedia: or Universal Dictionary of Arts, Sciences, and Literature* edited by Dr. Abraham Rees.[24] Issued in seventy-nine parts between 1802 and 1820, it was the first publication to describe systematically Britain's new industrial technologies. However, restraints on the propagation of technical information lingered beyond the completion of Rees's *Cyclopaedia*. James Butterworth, when compiling his *Complete History of the Cotton Trade* in the early 1820s, encountered determined opposition to this sort of disclosure on the part of "some of the great trading people of Manchester."[25] Operating rules, which Butterworth intended to publish, were a final step in completing the reduction of the new technology to a verbal state, complementing the machine designs and basic operating principles available in Rees's *Cyclopaedia*. Patriotism and the northern climate of protectionism evidently compelled Butterworth to suppress such information.

Hesitation to publish technical data virtually evaporated in Scotland, if not Lancashire, by the early 1830s. In the preface to his *Carding and Spinning Master's Assistant* (1832), which at last reduced practical experience to a series of operating rules by which a spinning factory could be managed, James Montgomery gave no hint of opposition toward his publication. On the contrary, at the request of "some particular friends" he published these notes, which he had compiled over several years for his own personal use. Montgomery argued in his preface that his explication of cotton spinning would rapidly educate young mill managers, serve as a corrective to managers and overseers who were disposed toward secretiveness, and place the business on sound theoretical principles, which was rarely done when experience was wholly practical and communication purely verbal and imitative.[26]

At Britain's ports, the outward movement of artisans and machinery came under the restraints of the prohibitory laws.[27] Their effectiveness in sealing off Britain's reservoir of technology from the rest of the world cannot be measured exactly; however, the weaknesses in the administration of these laws leave no room for doubt. Administration of the laws was in the hands of the Treasury (through its subordinate department, the Board of Trade) and, in wartime, the Privy Council. In the early 1780s skilled workers convicted of emigrating lost their nationality and property if they did not return within six months of a consular warning. Recruiting agents were subject to a £500 fine for each worker enticed abroad and twelve months' imprisonment. The export of textile, metalworking, clockmaking, leatherworking, papermaking, and glass-manufacturing equipment was prohibited. Penalties against exporting the dead instruments of trade (the implements and machines) were lower than for transplanting the living instruments (the artisans): a £200 fine (£500 for printing equipment), forfeiture of equipment seized, and twelve months' imprisonment for illegal export or attempted export. Since steam engines lay outside the prohibitory laws, steam engine erectors, millwrights, and mechanics were allowed abroad, but under license.

As part of the war effort, the Privy Council determined to tighten the application of the laws. It issued an order on April 8, 1795, compelling captains of foreign vessels to submit to the port officer before clearing a list of passengers' names, ages, oc-

cupations, and nationalities. Any British artificers, manufacturers, or seamen discovered were to be arrested. The measure apparently owed much to the indefatigable reporting of Phineas Bond, consul general at Philadelphia. Patterned on Bond's proposals, the Passenger Act of 1803 also supplemented the prohibitory laws in limiting the number of passengers British vessels could carry for every two tons of a ship's unladen capacity. The act not only relieved conditions of transatlantic travel but also struck a blow at emigration traffic. Castlereagh, president of the Board of Control in 1803, told John Quincy Adams that this was the real purpose of the act.

Skilled emigrants nevertheless continued to pass eastward and westward through the customs net, detected only by accident. The greatest loophole in the laws related to the difficulties of identifying skilled workers. The discovery of a marriage certificate and letters of introduction to a Philadelphia inhabitant found in the sea chest of a Manchester spinner during a muster of passengers at Liverpool in 1811 inspired the proposal of one radical solution. James Kipping, the surveyor on special service at Liverpool, suggested that all aspiring emigrants produce a certificate, attested by their local magistrate or parish officer, stating that the holder was not a prohibited emigrant, but this plan meant a passport system that threatened the rights of freeborn Englishmen, so it was shelved.

Reversing the tide of emigration was first tried in 1797 when Phineas Bond persuaded the Privy Council to pay the return travel expenses of a British immigrant in Philadelphia. Ironically the artisan, James Douglas, later moved to France, where he introduced Britain's latest woolen carding and spinning machinery. Two weeks before the signing of the Treaty of Amiens in 1802, the government considered a bill to offer amnesty to returning artisans. Nothing more was heard of this, though the postmaster general instructed his packet captains to grant free passages to returning artisans who held consular identification certificates. Other proposals to reward returning artisans, inspired by British involvement in Austrian cotton manufactures, were made in 1805. But neither these positive measures nor the negative threat of Algerine corsairs in the Mediterranean or the British army's invasion of the United States after 1812 seem to have had much effect in flushing out immigrant workers.

Customs found it almost impossible to halt the illicit export of machinery. They complained that their staffs were insufficient for exhaustive and close searching; wharf facilities for unpacking and inspecting were often lacking; large cargoes could not be examined in detail; and when parts of different machines were mixed up and packed together, it was impossible for searchers to identify the original equipment. Few customs officers were both legal and technical experts.

Despite manufacturers' and Customs' complaints, the Board of Trade shied away from tougher statutory measures against the export of machinery. Instead it urged customs officers to greater vigilance and modified the laws to accommodate changes in technology and the positive interests of British manufacturers. Simple preindustrial tools were virtually removed from the prohibitions in 1785. A second protectionist exemption, established in 1797, allowed the licensed export of primary processing machinery. Between 1797 and 1806, the Privy Council passed five applications to export cotton packing or baling presses to India, the East and West Indies, and the United States. Mechanized cotton packaging opened the distribution bottleneck between Whitney's recently patented American cotton gin (1794) at the harvesting level and the new manufacturing technology at the processing stage.

The cessation of hostilities in 1814–1815 shifted the responsibility for implementing the prohibitory laws from the Privy Council to the Board of Trade. The change scarcely altered either the interpretation or the effectiveness of the laws, at least until the early 1820s. However, peacetime brought a new problem: the legal definition of the artisan. At Liverpool, after customs officers argued with magistrates about the meaning of the term, the Liverpool collector wrote to the customs commissioners in 1817 asking "whether common workmen, such as Shipwrights, Coopers, Ropers, Joiners or House Carpenters, Brick Makers &c. [are covered by the laws]."[28] Despite the Board of Trade's enquiries to the customs secretary, the attorney general, and the solicitor general, no agreed definition was recorded in the board's minutes.

With regard to machinery export, the exemptions allowed before 1812 continued thereafter with the addition of one more, imperial preference. Thus, whereas Customs was ordered to seize ironworking rollers destined for France, Alexander Galloway was licensed to export copperworking rollers to Egypt and Peter Fary was allowed to take worsted-spinning machinery to Tasmania.

Changes in the postwar political and economic climate eventually brought major modifications to the prohibitory laws. In 1820, the Privy Council's wartime order in council of 1795 was suspended.

And in 1824, all restrictions on emigration were lifted, not least because the issue divided the manufacturers and machine makers. On the other hand, the proposed repeal of the prohibitions on machinery export aroused nearly solid opposition from the textile manufacturers and machine makers, apart from the London machinists. They feared that a loss of overseas markets in the long term, and, more immediately, curbs on the current expansion of business activity, would more certainly arise from free machinery export than from free emigration. The depression of 1825–1826 strengthened the manufacturers' hand; until 1843, machinery export remained under a licensing system operated by the Board of Trade.

In the first years of the system, the Board of Trade frequently had difficulty in deciding what sorts of machinery could be exported under license, either because the law was ambiguous or because it was silent. From the start, Huskisson, the board's president, established the overarching principle that licenses be granted in the interests of British manufacturers. Three sorts of textile manufacturing equipment were generally released for export under license between 1825 and 1843. Preindustrial equipment, like wool combs, spinning wheels, and textile printing blocks, or obsolete industrial machinery, like the "old defective machinery" for spinning worsted sent to Rotterdam, passed without difficulty. Industrial machinery for preparing fibers for manufacture was allowed to leave the country, but machinery that divided a mass of fibers was not. This principle was formulated by the board and relayed to customs collectors by the customs commissioners in a general order dated November 14, 1838. In cotton and woolen man-

ufacturing, the prohibition covered equipment for carding and subsequent processing stages. Third, various finishing machines received export licenses. Engraved copper rollers, calico printing machines, and calenders for cotton finishing were released. So were woolen cloth washing, drying, and pressing machines, but not fulling, raising, or shearing equipment. In the case of calico printing machines, the board prohibited their export to the United States until December 17, 1830, possibly in the hope of retarding industrial calico printing, to which New England manufacturers were turning in the late 1820s.[29]

Since the Board of Trade developed its licensing policy primarily as a reaction to pressure from British manufacturers, the absence of such pressure from machine-tool makers must have been responsible for inattention to the outflow of machine-making equipment, though one of the Manchester engineers voiced his concern in 1824. Export of rollers, slitters, presses having screws over one and a half inches in diameter, cutting-out presses, and lathes for "plain, round and engine turning" was prohibited under the 1825 and 1833 acts. Yet between 1830 and 1840, over thirteen hundred metal-working rollers for iron, copper, tinplate, and zinc; forty-one lathes, including planing, boring, and screw-cutting lathes; and twenty cutting engines, including gear cutters and punching presses, were licensed for export. Inasmuch as these tools would have incorporated screws larger than one and a half inches in diameter, the total number of machine tools exported must have been greater still.

Loopholes soon appeared in the machine-licensing system, and a variety of smuggling routes, some old, some new, were soon firmly established. For example, machinery might be misrepresented or inadequately described in an application (until 1835, that is, when the Board of Trade rejected such applications). Or a piece of machinery might be broken up and exported as separate components in hopes of confounding customs officers. Nearly 20 percent of the machinery seizures made at Liverpool from 1830 through 1839 were of unidentified machine parts; at other ports the proportion was much less.

Smuggling or totally bypassing the licensing system was facilitated by the coming of cross-Channel steamboat services, operating by published timetables, in the late 1830s. Smugglers would bring cases of machinery to the dock at the last moment with the intention of avoiding customs inspection. Small machinery parts might be taken aboard in passengers' baggage, which Customs never, reportedly, examined. Sometimes machinery was concealed in bales of cotton. Machinery bound for the Continent might also go by the coastal trade. In 1839, there were only forty-nine revenue cruisers and twenty-one tenders, manned by 6,183 sailors, to monitor the seven-thousand-mile coastline of England, Scotland, and Ireland. Yet the nature of some of the machinery required wharves, cranes, tackle, and experienced dockers. Customs witnesses before the 1841 Select Committee on the laws affecting the exportation of machinery confirmed that illegal exports were boldly made through major ports rather than on remote moonlit beaches, the romanticized setting for smuggling. And customs seizures included machine frames, as well as boxes and chests of parts.[30]

If machinery or machine parts were not themselves taken abroad by any of these means, statutory prohibitions might be avoided by transmuting them into other forms. In 1833 the Board of Trade's minutes recorded "the transmission of drawings cannot be prevented."[31] Eight years later, even verbal descriptions were reckoned sufficient for a machine maker to copy a design. Artisans, of course, constantly represented a threat to the retention of technology. Either they were recruited from a British mill or machine shop, or else foreign apprentices spent part of their time with British firms. Against these numerous possibilities and the metamorphotic nature of the technology, customs men were almost powerless.

On the other hand, there are pointers toward some success for customs in apprehending illegal machinery exports. Availability of insurance against detection at first sight suggests a low regard on the part of unscrupulous exporters for the efficiency of the customs officers, who made 289 machinery seizures in the decade 1830–1839. For the hazard of customs detection, insurers charged a premium of 10 to 25 percent in 1841, and this was for small machine parts; larger pieces were charged more. This suggests a not unreasonable chance of customs detection, though a diminishing one since smugglers' insurance premiums were quoted at 30 to 45 percent in 1825.[32]

Official machinery exports, as recorded in the 2,098 applications made to the Board of Trade from 1825 through 1843, predominantly involved textile machinery (table 2.1). It is evident that textile machinery figured heavily in those applications denied export licenses and that exporters progressively learned what might be sent abroad under license.

2.1. Applications Made to the Board of Trade for Machinery Export Licenses and Proportion of Refusals, 1825–1843

	Total Applications	Applications for Textile Machinery		Applications Refused		Textile Refusals	Textile Refusals as % of Total Refusals	Textile Refusals as % of All Textile Applications
		No.	%	No.	%			
1825	10	2	20	1	10			
1826	24	12	50	2	8.3	1	50	8.3
1827	37	27	72.9	13	35.1	9	69	33.3
1828	52	29	55.8	10.5	20.2	9.5	90.5	32.7
1829	48	29	60.4	13	27.1	12	92	41.4
1830	67	34	50.75	14.5	21.6	13.5	93	39.7
1831	52	31	59.6	10	19.2	9	90	29
1832	69	30	43.5	8	11.6	8	100	26.6
1833	85	46	54	11	12.9	11	100	23.9
1834	124	78	63	26.5	21.4	24.5	92.5	31.4
1835	124	76	61.3	15.5	12.5	14	90	18.4
1836	124	77	62	14	11.3	14	100	18.2
1837	135	88	65.2	9	6.6	8	89	9.1
1838	204	151	74	19.5	9.5	19.5	100	12.9
1839	178	134	75.3	10	5.6	10	100	7.5
1840	186	133	71.5	11.5	6.2	11.5	100	8.6
1841	200	135	67.5	7.5	3.7	7.5	100	5.5
1842 to 16 August	105 ⎫ 244	73 ⎫ 193	69.5 ⎫ 79	7.5	7.1	7.5	100	10.3
Remainder of 1842	139 ⎭	120 ⎭	86.3 ⎭	10	7.2	8	80	6.7
1843	135	104	77	4	2.96	4	100	3.8
Total	2,098	1,409	67.1	218	10.4	201.5	92.4	14.3

Source: PRO, BT 6/151, 152 (Machinery Books, 1825–1843).
Note: Figures for 1842 break in mid-August, when the Board made sweeping relaxations in its administration of the laws. The system ended in August 1843.

Few applications were made to export machinery to the United States; most involved northern Europe. American interest is disclosed in table 2.2. Furthermore a slightly higher proportion of American applications were denied export licenses: 13 percent, compared with 10 percent of all applications. And a lower percentage of these American refusals involved textile machinery (80 percent) than did all refusals (92 percent). Nearly two-thirds of the 158 American applications included textile machinery, chiefly for calico printing. A few printing machines were officially exported after December 1830, but far more numerous were calico printing rollers, plain and engraved, and doctor blades.

Other equipment refused export licenses to the United States included iron rolls for rolling iron (prior to 1834), eighty rolls for rolling copper (1829–1835), copper engraving machines (in 1827), spinning machines, some patented (1831, 1836, 1840), carding engines (1833), jacquard looms (1835, 1838), cotton rollers and spindles (1838), flax spinning machinery (1841), and a stocking frame (1840). After the relaxation of the export law in August 1842, leading English machine makers like Parr, Curtis & Madeley, Higgins & Sons, and J. Hetherington sent whole lines of cotton spinning and weaving machinery to the United States. Taken together, these applications suggest a strong American interest in cotton finishing, especially calico printing, and in metalworking.

Patents potentially represented a major vehicle of technology transfer: "Our Magazine of Secrets at the Patent Office is exposed to all Foreigners," a Birmingham industrialist told Matthew Boulton in 1789.[33] Whether this was true depended on the sort of information conveyed in a patent of invention and the extent to which foreigners had access to British patents.

After 1778, patent-case law required that the specification contain a full description of the invention. The question of obscure specification was raised in the trial over Arkwright's 1775 patent. It was acknowledged that specifications should be understood by "a mechanic of reasonable general knowledge in his profession and thoroughly acquainted with the machines before in use" so that he could build the invention from the specification.[34] By their very nature, British patents therefore presented some difficulty to American borrowers; they were comprehensible only to artisans already familiar with the technology. This implied that as the technology developed, the patent specifications became harder to understand because the up-to-date knowledge that they assumed was increasing.[35]

Since attempts to alter the law governing patent specifications and efforts in 1785, 1801, and 1820 to introduce a system of secret patents (like the French used) failed, the chaos of administrative procedures also guarded British patents from interested foreigners.[36] Their plagiarism was, of course, beyond the powers of English law to check. Patent rolls were scattered among three offices: the Petty Bag Office, the Rolls Chapel Office, and the Enrolment Office. Only the more recent patents were located at the Enrolment Office. On top of the admittance fee of a shilling at each office, the fee for reading was two shillings and sixpence per patent. The patents, written and drawn on vellum, were sewn end to end to form rolls, each containing several dozen patents, the number depending on their length. Long patents before 1829 ran to twenty or thirty

2.2. Applications for Machinery Export Licenses to the United States, 1825–1843

	Total Applications	Applications Citing U.S.A. as Destination	Applications to U.S.A. Entailing Textile Machinery	Applications to U.S.A. Refused Licenses	Applications to U.S.A. Entailing Textile Machinery Refused
1825	10				
1826	24	1	1		
1827	37	4	2	1	1
1828	52	3	1	2	1
1829	48	6	2	2	1
1830	67	8	4	3	2
1831	52	13	5	3	2
1832	69	7	1		
1833	85	7	3	1	1
1834	124	9	4		
1835	124	15	10	2	2
1836	124	15	10	1	1
1837	135	7	5		
1838	204	11	10	2	2
1839	178	8	6		
1840	186	13	10	2	2
1841	200	9	7	1	1
1842 to 16 August	105 ⎫ 244	5	4		
Remainder of 1842	139 ⎭	8	7		
1843	135	9	7		
Total	2,098	158	99	20	16

Source: PRO, BT 6/151,152 (Machinery Books, 1825–1843).

skins of descriptive matter, with perhaps fifteen or twenty skins of drawings. Because the patents were filed in no particular order at these offices, searches could occupy hours. By the 1820s Moses Poole of Lincoln's Inn, a patent agent in the attorney general's Patent Office, had "for his own amusement" compiled a classified index, but it was subject to error and under Poole's personal control.[37] Meanwhile patents of invention continued to multiply. In 1790 they totaled seventeen hundred; over the next forty years, another forty-three hundred were added. Having located a particular patent, the reader was allowed official copies only and at some considerable expense. The 1829 rate was three shillings fourpence for a folio of seventy-two words, plus that same amount for the signature and a stamp duty for authentication. Drawings further pushed up the price of patent copies, which could cost between two and forty guineas by 1829.[38]

Patent journals seemed to offer a shortcut to reading British patents. The first trade journal of this sort, the *Repertory of Arts, Manufactures and Agriculture*, began in 1794. Its publishing record of patents for cotton and woolen cloth manufacturing inventions is shown in table 2.3. (Appendix A contains a list of published patents.) This record corresponds closely with the claim made in 1829 that the trade journals published no more than a quarter of the patents granted each year.

More important than numbers of patents was the technological significance of patents published. The commercial value of an invention is wholly unknown to a patentee; he can but guess at its profitability in the light of probable demand. Only with hindsight is it possible to identify significant inventions in terms of profitability or the extent of

2.3. Textile Patents Published in the *Repertory of Arts*, 1794–1830

Period	Textile Patents Awarded[a]	Number Published in *Repertory*	Percent Published
1794–1812	106	22	20.7
1813–1824	106	33	31.13
1825–1830	91	20[b]	21.9

[a]These patents are only those for cotton and woolen preparatory, spinning, weaving, and finishing mechanical inventions. Omitted are other technologies like worsted or knitting, techniques like dye recipes, and other fibers like silk.
[b]Eleven in abridged form, eleven in full, and two in both abridged and full forms.

usage. In this early period, an impression of adoption can be derived from contemporary descriptive sources.[39] From these it is evident that a number of the published specifications included machines that would become popular in their respective industries. For the years 1794–1812, specifications were published for several cotton batting or picking devices, Robert Miller's wiper power loom of 1796, Whittemore's American card clothing machine, and several cloth raising and shearing devices. Among the published cotton manufacturing patents after 1813 were Buchanan's self-stripping top flat card, Raynor's fly frame differential, Bradbury's twisting stop motion, Horrocks's variable batten speed motion, Taylor's dobby head, and Wilson's jacquard head. In woolen manufacturing, where inventive interest focused on cloth finishing, they included the rotary cloth shear. Of the twenty patents published from 1825 through 1830, only one was of major and enduring value, and this, Thorp's ring spindle, was imported from America.

On balance, rather more inventions which proved of lasting value were left unpublished. The major omission before 1813 was the Radcliffe and Johnson dresser. After 1813 the omissions in cotton included Bodmer's double lapper, sliver lapper, and railway carding head; Roberts's six-harness cam power loom; Potter's power loom dobby; Wilkinson's reed-making machine; and Ridgway's interchangeable shells for calico printing rollers. In woolen manufacturing, Collier's helical shears, Bayliss's rotary cloth washer, Flint's rotary scourer, and Daniell's steam boiling system were the major omissions. Chief among the specifications left out of the *Repertory* from 1825 through 1830 were those for the Taunton speeder and Arnold's differential,

both from America; Lane's roving frame builder motion; Danforth's cap spindle, also American; and Roberts's mule winding-on motion, in cotton manufacturing. In wool none of the major condensing and piecing machine patents and few of the numerous fulling, raising, and shearing patents (which encompassed many incremental improvements and experiments) were published.

Trade journals in fact afforded several forms of uneasy protection to the British patentee. Walter Henry Wyatt, editor of the *Repertory*, asserted in 1829 that his choice of patents for publication was guided by the length of the specification and copying costs rather than by the importance of the patent. And an inventor might prevent a trade journal from publishing his patent specifications, as the experience of William Radcliffe, in trying to confine his weaving patents to British manufacturers in 1802–1803, illustrates. Radcliffe's solicitor initially recommended that he take out a secret patent, but John Peel, Robert's brother, advised against this course as being too troublesome and expensive. Radcliffe's solicitor then suggested that the patents be taken out in the name of an ordinary workman. Radcliffe did this, knowing that his own name, or that of his partner in the published list of patents, would certainly attract a foreign rival's attention. Additionally—and because his partner had found a Berlin manufacturer building a version of Anthony Bowden's batting machine from the specifications in the latest issue of the *Repertory of Arts*—Radcliffe went to Wyatt and persuaded him not to print his patent specifications.[40]

Besides inappropriate editorial criteria and the possibility of suppression of publication, a patentee could secure protection by the delay between his patent award and the journal's publication of the specification. The length of time between patent enrollment and publication haphazardly varied between five months and five years in the period 1790–1830. Here a patentee's perception of his commercial interests obviously played a large part.[41]

The *Repertory* nevertheless provided a few investigative tools for foreign observers. One was a correct list of British patents of invention most recently published. From 1796 on, it accompanied all issues, which by 1802 were appearing monthly. This regular listing of patents by date, patentee's name, location and occupation, and patent subject gave enough clues to informed readers to guess which patents were worth pursuing or which towns and villages worth exploring on an English visit. The proliferation of patents led in the 1820s to other aids. Besides printing patents in an abridged form, from 1827 on the *Repertory*'s editor added a comprehensive list of expired patents to lists of those currently awarded. Three years later he printed a list of numbers of patents annually awarded between 1675 and 1829. But no vital subject index to British patents was published until the mid-nineteenth century, and in 1829 none existed apart from Moses Poole's, according to John Millington, professor of mechanics at the Royal Institution.[42]

Patent journals were therefore an inadequate source of patent information, apart from the patent lists. For important pre-1794 patents and for three-quarters of those after that date, foreign imitators had to have the resources of time and money to visit London. Their only alternatives were to tour provincial manufacturing districts or recruit British artisans.

Like the novel aspects of the new technologies, private-level secretiveness and government-level protection reinforced the artisan's role as the prime international carrier of industrial technology before the 1820s. Because technical knowledge was incompletely or imperfectly published before this time, the experienced artisan could most easily break through a firm's secrecy and relay an understanding of the latest manufacturing equipment and methods to foreign rivals. Only a persistent, wealthy, and technically intelligent foreign visitor was likely to penetrate British firms and industries. Even if the visitor could return home to command adequate capital and labor resources, he might find it less risky to recruit British managers, operatives, and mechanics than to rely on his own powers of observation and memory, on patent specifications, or on information picked up in Britain's manufacturing districts. Despite the prohibitory laws, recruitment of British artisans was not difficult, compared with contemporary France and Prussia where workers' passport systems operated.

3
Problems for Diffusion Arising from the Structure and Growth of the Technologies

A torrent of technical changes characterized Britain's cotton and woolen industries between 1790 and 1830. Most visible was the spread of mechanized and powered production to weaving and finishing, but alongside all the new technology, traditional skills and methods persisted and sometimes came into conflict with the new industrial forms. In such a period of transition, both old and new technical patterns and their points of friction held implications for technology diffusion.

Probably the most threatening feature of traditional technical patterns in the British textile trades derived from the regional diversity of technical systems untouched by mechanization—in particular, mensuration systems and their attendant nomenclatures. Commingled in America, these local customs could lead to chaos and inhibit American borrowing.

In the woolen industry, variations in mensuration started with wool sorting. Luccock listed ten grades of wool recognized in the East and North of England in 1805, adding that some staplers distinguished as many as seventeen. Nine grades appear in an 1812 price list made up by Thomas Colfox and Son of Bridport, Dorset. Robert Bakewell, a West Riding wool stapler, classified ten grades in 1818 and noted wide variations between sorters. Although the Luccock, Colfox, and Bakewell classifications each contained a similar number of grades, they were neither congruent nor identical in nomenclatures. For example *super head* ranked fourth in Luccock's system, second in Colfox's, and third in Bakewell's, while *downright* was sixth, ninth, and fifth in the respective classifications. Similarly, terminology varied. *Picked lock* was a term or grade peculiar to Luccock's experience, and *run-*

ning fine, fine skin, and *lamb* were evidently confined to Colfox's West of England trades.[1]

In card clothing the British and American cotton industries evolved different systems of counting cards in the nineteenth century. Before 1830 British cotton machine cards were rated by the number of wires in the breadth of the card sheet, a distance of three and a half inches for cylinder cards and two inches for top cards.[2] Diversity of mensuration systems in textile processing was most apparent in reeling. By the late eighteenth century, there were four count or numbering systems applied to wool, four to worsted, two to cotton, two to linen, and one to silk in Britain. This variety derived in part from ancient regional differences in weights and measures.[3]

A final example of the potential for technical confusion arising from the transmission of British textile technology to the United States was diversity in methods of counting the setts of British handloom reeds. In 1808 a Bolton cotton manufacturer testified that "what would be a sixty reed at Bolton would be a one hundred at Stockport; in Preston they are still worse, for almost every Manufacturer has a particular way of counting his own reed."[4] In Stockport and Manchester, the muslin reed was measured by the number of warp ends spaced in two inches; thus a forty reed had twenty dents or splits (spaces) to an inch. At Bolton and Manchester, gingham reeds were measured according to the number of dents in twenty-four and a half inches. At Rochdale, flannel reeds were counted by the number of beers in thirty-six inches; and here the beer was seventeen dents compared with the more common measurement of nineteen or twenty dents per beer.[5]

John Duncan, the Glasgow handweaver, guessed in 1807 that reed lengths matched the widths to which various cloths were traditionally woven, noting in Scotland three reeds: forty inches long for hollands, thirty-seven inches for linen, and thirty-four inches for cambrics. He added that the thirty-seven-inch reed, the length of the Scottish ell, was universally adopted in cotton manufacture. He seems to have been right, for by 1851 Glasgow cotton manufacturers measured their reeds by the number of hundred splits in thirty-seven inches.[6] Outside the cotton industry, there were yet other methods of measuring reed spacings, but little technical information on these has survived. Nomenclature was as multifarious as mensuration methods. For example, a group of twenty splits or dents, corresponding to twenty bobbins in the warp creel, was called a *porter* in Scotland and Ireland and a *beer* in England.[7] It is clear, then, that the migration of British workers to the United States and their admixture there could lead to argument and conflict as differing English technical traditions confronted each other. So too could artisan resistance to machinery.

British workers, it is well established, attacked machinery long before the Luddites first smashed stocking frames in 1811.[8] Sometimes the object was to put pressure on employers, perhaps for higher wages. Machine wrecking became a trade union technique or, in Hobsbawm's phrase, "collective bargaining by riot."[9] But it could be resistance to the machine itself when the workers saw it as a direct threat to their own employment or living standards. The problem of artisan resistance to technical change thus becomes a matter of identifying those sections of the textile industries where machine breaking was prompted by fear of technological unemployment.

In two sections of the cotton industry, mechanization inspired fears of technological obsolescence and organized resistance to the introduction of new machinery. Among calico printers, whose history of combinations dated before 1790, new printing machinery met sporadic resistance until 1815 at least. Cylinder machines, patented by Thomas Bell in 1783–1784, took some proportion of calico printing into the factory between the 1780s and 1810. The development was heralded by the installation of steam engines in several Lancashire printing firms in the 1790s and by the emergence of specialist calico-printing-machine makers in Manchester before 1809.[10]

But the cylinder machine was limited in the type of work it could do, and much handprinting existed alongside factory production, thereby eroding resistance to the new machines. As late as 1808, it was authoritatively reported, cylinder printing with more than one color was commercially not feasible because the fine adjustments among several color rolls were difficult to achieve. And the Bell machine printed a limited number of repeat designs, albeit with delicate lines, occupying no more than the surface area of its single cylinder. Hand block printing, or slightly faster flat-press copper plate printing, was therefore employed for designs with several colors or large-scale repeats before about 1813. The block printing trade was also perpetuated by a major technical improvement at the turn of the century. Instead of carving designs into a block's surface, copper or brass strips were set into the block to form the pattern in relief. The innovation was capital saving on the manufacturing side in that block designs could be used almost indefinitely.[11]

Remaining so laborious, hand block printing spurred attempts to develop mechanical substitutes. With a cylinder machine, a man and a boy could print two hundred pieces a day in 1808 compared with eight pieces a day with blocks or twelve pieces with a flat press.[12] Eventually modifications to the cylinder machine produced prints comparable to hand-blocked work. It was the introduction of these improvements, in the form of the surface machine, some time before 1813 that aroused the opposition of the block printers. That year the published rules of the Northern calico printers' union bound members not "to have any thing to do with those kind of patterns, or work, which has, or is intended to have, any thing to do with the surface machine, . . . [nor to] cover, with any block, any kind of machine work, done by any machine, only single coloured cylinder work."[13] At least one Lancashire firm was forced to abandon its surface printing machinery. But the printers' cause was weakened by the willingness of some block makers to manufacture wooden surface rollers for the new machine.[14] Then the master calico printers formed their own combination. Recalcitrant block printers were discharged, with sixteen hundred of eighteen hundred losing their jobs in Lancashire in 1815.[15] Thereafter little is heard of hostility to the surface machine, though the printers' combinations certainly did not disappear.

Resistance to new machinery in cotton weaving is much better known. Hand-loom weavers attacked power looms in 1792, 1812, 1826, and 1829.[16] The most widespread assaults, in 1812 and 1826, coincided with severe trade depressions, so the violence may not have been blind machine wrecking. Yet both the hand-loom weavers and the block calico printers were preindustrial workers with a hatred of machinery that was not wholly shared by factory operatives.[17] Mule spinners already operated semiautomatic machines and were concerned with safeguarding their wages when they sporadically went on strike in a number of Lancashire mill towns in the 1820s.[18]

Well known too is the resistance to new machinery in the woolen industry during this period. Led by croppers, or shearmen, against gig mills and shearing frames, it erupted violently in the West of England from 1802 through 1809 and in Yorkshire from 1802 through 1813. In Yorkshire it merged into the Luddism associated with the trade depression toward the end of the Napoleonic wars and disappeared after 1813.[19]

These outbursts of machine breaking had several implications for the transfer of textile technology to the United States. There is the possibility that emigration to America from the trades involved would have delayed the adoption of new machinery in the United States. With the croppers, this was unlikely. Even before the new machines spread widely (around 1806), the shearmen were few in number compared with the hand-loom weavers at their height in the 1820s: 4,000 croppers in Yorkshire and three or four times this number in the whole country, compared with 250,000 hand-loom weavers.[20] And while shearmen fought the new frames in England, in America more efficient shearing machines were already developed for use in conjunction with household cloth manufacturing.

Whether the immigration of handweavers and block printers would introduce active opposition to power looms and advanced calico printing machinery in America depended on the workers' original motivation for machine breaking and then on the number of those workers emigrating to the United States. If they possessed a rage against machinery, then their immigration would become an obstacle to innovation in the American cotton industry. But if their hostility to machines rested on fear of unemployment or lowered living standards, then higher real wages for less-skilled machine minding would melt their antagonism. In dealing with the Luddite period, Thomis argues that only the croppers nursed a blind opposition to machinery, but explicit evidence on motivation is hard to find in English sources.[21]

High numbers of emigrant weavers might discourage the adoption of the power loom in the United States by lowering the cost of weaving labor in relation to weaving capital. There was also the possibility that American employers would hesitate to hire British workers from trades or areas linked with machine breaking. Americans might even see the British experience as good reason for pursuing capital-intensive techniques in those trades more aggressively. The evidence on these possibilities is considered in parts II, III, and IV.

Preindustrial experience, in localized technical traditions or hostility toward machinery, was not the only British source of obstacles to transfer closely associated with the nature of the technology. Industrialization itself promised difficulties for international diffusion if the pace of technical change in Britain moved beyond American capacity to copy or if British innovations became inappropriate to America's factor endowment (that is, its relative proportions, and hence relative prices or costs, of land, capital, and labor).

The best available measure of technical change is patent registration, for, as Schmookler showed, inventive activity closely follows expectations of profit.[22] A sharp climb in the rate of patenting thus suggests a swelling entrepreneurial interest in modifying the existing technology; the extent to which it led to innovations can be gauged from contemporary industrial surveys. From table 3.1 it is clear that the number of patents issued for the cotton and woolen industries rose almost exponentially over the three periods corresponding to the spread of machine spinning (1790–1812), power-loom weaving (1813–1824), and the consolidation of mechanical spinning, weaving, and finishing (1825–1830).[23] Imitators faced not only the difficulty of locating and reading these proliferating patents but also of understanding their specifications, which assumed a level of technical knowledge that rose in complexity and grew in extent with the mushrooming of industrial innovations. American ability to keep abreast of this rate of technical change, in terms of awareness and understanding at least, depended largely on the channels of diffusion, in particular the numbers and trades of immigrant British artisans and the published forms of technical knowledge.

Evidence of changes in the direction of innovation also comes in part from patent registration. In the cotton industry, inventive activity was most highly sustained in spinning technology, with the rate of patenting more than doubling in the late 1820s, largely in response to bottlenecks in the fine yarn supply. Weaving ranked after spinning and occupied inventors most intensely during the mid-

dle period when the power loom appeared. In the woolen industry, finishing technology engaged inventive attention almost exclusively through the whole period, and especially after 1812.

Patentees' trade backgrounds add definition to the picture of technical change (table 3.2). Over the whole period, the proportion of inventors from the manufacturing sector mounted in relation to that from the capital goods side of the textile industry, hinting perhaps that inventive activity was approximating more closely the commercial realities current in Britain. Those on the edge of production, like merchants or propertied gentlemen, maintained their inventive interest, possibly because of involvement as suppliers of capital. Patentees from unrelated trades—an Oxford brewer, a Sheffield cleric, a Halifax surgeon, a Yeovil schoolmaster—disappeared altogether in the 1820s. If, as seems likely, they were left behind by the increasing complexity and costs of mechanical technology, it implied that American imitators would need rising capital levels and renewed reliance on skilled artisans as technology carriers.

Patentees' regional backgrounds reveal the preponderant interest of Lancashire and West of England men in inventive activity (table 3.3). These two districts provided nearly a quarter each of the 303 patents in the survey. London men, like those from Yorkshire, accounted for nearly 15 percent, and just over 9 percent originated from foreigners. But while the cotton regions increased their rate of patent registration by more than 40 percent over each of the three subperiods, the woolen regions quadrupled their rate over the first two periods and nearly doubled it again in the late 1820s. West of England men, working on improvements in raising

3.1. Changes in Inventive Activity in Britain's Cotton and Woolen Manufacturing Technologies, 1790–1830

	1790–1812 No. of Patents		1813–1824 No. of Patents		1825–1830 No. of Patents		
	Actual	Prorated to 24-year term	Actual	Prorated to 24-year term	Actual	Prorated to 24-year term	Total (Actual)
Cotton manufacturing							
Preparatory	10	10.4	4	8	0	0	14
Spinning	28	29.2	19	38	24	96	71
Weaving	22	22.9	19	38	7	28	48
Finishing	17	17.7	11	22	7	28	35
Total	77	80.3	53	106	38	152	168
Woolen manufacturing							
Preparatory	1	1.04	0	0	8	32	9
Spinning	2	2.09	3	6	1	4	6
Weaving	0	0	2	4	6	24	8
Finishing	16	16.7	38	76	31	124	85
Total	19	19.8	43	86	46	184	108
Cotton and woolen machine making	10	10.4	10	20	7	28	27
Aggregate figures for total cotton, wool, and machine making[a]	106	110.6	106	212	91	364	303

a. In sixteen cases (six for the years 1790–1830, five for 1813–1824, and five for 1825–1830), patents covered more than one processing level. In these instances, I assigned them to the level I deemed most important.

3.2. Backgrounds of Patentees, by Trade

	1790–1812 No. of Patentees		1813–1824 No. of Patentees		1825–1830 No. of Patentees		
	Actual	Prorated to 24-year term	Actual	Prorated to 24-year term	Actual	Prorated to 24-year term	Total
Machine makers	33	34.4	41	82	28	112	102
Manufacturers and artisans	44	45.9	40	80	47	188	131
Merchants and gentlemen	14	14.6	18	36	15	60	47
Unrelated trades	10	10.4	5	10	0	0	15
Unstated trades	5	5.2	2	4	1	4	8
Total	106	110.6	106	212	91	364	303

Note: In the cases of patents awarded to several patentees, the patent has been credited to the most technical trade represented, following the sequence given above.

machinery, especially played a major part in this performance, mostly in the middle years; by the late 1820s, Yorkshire inventive interest began to catch up. Thus, among other features, an aggregate analysis of the patents strongly indicates that some technical innovation in Britain aimed at higher-quality cotton yarns and goods and woolen cloths. If this were the case, Americans would find the new directions in these British innovations inappropriate to their circumstances.

The up-market movement in British textile manufacturing and technology is confirmed by other evidence. While Britain's water-frame spindleage, confined to low-count yarns, remained about 1.25 million spindles in the period 1795–1817, mule spindleage rose from under 4 million to 6.6 million.[24] Customs statistics show that British exports of printed and checked calicoes rose from 90 million to 189.7 million yards per year between 1816 and 1830.[25] Since anything coarser than a number 50 yarn for print cloths was unacceptable to British manufacturers in the late 1820s and since no continuous spinning machines could then produce yarns above this fineness, mule spindles must account for the doubled quantity of exported print cloths.[26] At the same time the average export value per yard of the print cloths fell by nearly 42 percent.[27] Doubled export yardage and halved unit price were made possible not only by falling wages (which slid on average by 35 percent) and plummeting raw cotton prices (which dropped by around 60 percent), but also by the application of capital to create a technology suited for fine yarn and goods production.[28]

3.3. Backgrounds of Patentees, by Region

	1790–1812 No. of Patentees		1813–1824 No. of Patentees		1825–1830 No. of Patentees		
	Actual	Prorated to 24-year term	Actual	Prorated to 24-year term	Actual	Prorated to 24-year term	Total (Actual)
Cotton regions							
Lancashire and Cheshire	31	32.3	29	58	14	56	74
Scotland	11	11.5	4	8	4	16	19
Midlands	8	8.3	5	10	9	36	22
Total	50	52.2	38	76	27	108	115
Woolen regions							
West	10	10.4	35	70	25	100	70
Yorkshire	11	11.5	11	22	19	76	41
East Anglia	1	1.04					1
Total	22	22.9	46	92	44	176	112
London	24	25.04	14	28	7	28	45
Wales	1	1.04					1
Ireland	1	1.04	1	2			2
Foreign[a]	8	8.3	7	14	13	52	28
Total	106	110.6	106	212	91	364	303

a. As stated in the patent specifications, except for G.B. Pats. 5316 (Houldsworth's patent for Arnold's differential) and 5822 (Hutchinson's patent for Danforth's cap spindle).

Note: Each patent has been credited with one patentee for the purposes of this table. When several persons shared a patent, the geographic location of the patentee with the most technical occupation has been preferred.

More than this, the technology was distinguished by versatility. Much of Britain's spindle capacity, whether mules or throstles, served lower-quality markets, as demonstrated by the quadrupling of plain calico exports from 1816 through 1830 to 219 million yards, for which the average price per yard fell by 55 percent.[29] But the higher unit profits in the easily overstocked fine yarn and goods markets drove manufacturers toward a technology capable of spinning or weaving a wide range of counts or a variety of goods qualities.[30]

This trend toward higher quality and versatility emerges again in an inspection of individual inventions for cotton spinning and weaving and woolen finishing, the areas already defined as capturing most inventive effort within the technologies under review. One example was the well-known pursuit of a fully mechanized mule, the most versatile spinning machine.[31] The goal of mechanical versatility also lay before the efforts to perfect a roving frame capable of producing a wide range of rovings in readiness for spinning. The central problem was to slow down the speed of the bobbins as their circumferences uniformly increased with winding and then to change the rate of decrease in speed to suit different counts; at the same time and rate, the up-and-down movement of the traverse rail had to be slowed. One partial solution came with the cone pulley arrangement for the bobbin shaft patented in 1797 by Aaron Garlick, a Cheshire manufacturer. This machine, nicknamed double-speeder, operated in a Macclesfield factory as early as 1798; by 1811, when it appeared in Rees's *Cyclopaedia*, it formed part of Britain's representative cotton manufacturing technology. Other patented solutions were those of Raynor (1813), who em-

ployed cones to retard the traverse and a series of graduated gears to slow the bobbins' rotation, and Green (1823), who ingeniously used the traverse rail to slow the bobbins by means of a grooved spiral escapement in tubes supporting the bobbins.[32]

An alternative method of improving yarn quality was introduced in Bodmer's patent of 1824. Of paramount importance was his sliver lapper, which allowed the production of more even rovings, eventually suitable for the cotton comb introduced in the 1840s. Bodmer's inventions were plagiarized widely while he was in Switzerland. But if widespread infringements and a patentee's tenacious struggle were any indicators of commercial success, then Bodmer's 1824 patent was one of the more profitable. In 1840 he gained a seven-year extension, thereby indirectly affirming the British quest for higher-quality production.[33]

A similar story could be told of cotton weaving and cloth finishing. For power looms, Horrocks's variable batten speed and the six-harness cam looms of Bowman and Roberts improved product quality and added technical versatility.[34] In woolen finishing, the same concern for product quality exhibited itself. At least three of the eight shearing patents awarded over the 1794–1801 period covered mechanized versions of the old hand shears because they cut the cloth more evenly than did the early rotary shears. And in gigging, where wire teeth replaced teasels, immense ingenuity was expended in attempting to replicate the raising action of teasel heads.[35]

The primary object of most, if not all, inventions was to save costs. This could be done by reducing either the expenses entailed in setting up a mill

3.1 Double cone pulleys (*q* and *t*) for retarding winding in the roving frame, patented in Britain in 1797 by Aaron Garlick, manufacturer of Duckenfield, Cheshire. As the belt (*r*), guided by a traversing screw, moved across the two conical pulleys, it reduced bobbin speed and so minimized roving breakages during winding. (G.B. patent 2166; courtesy Boston Public Library.)

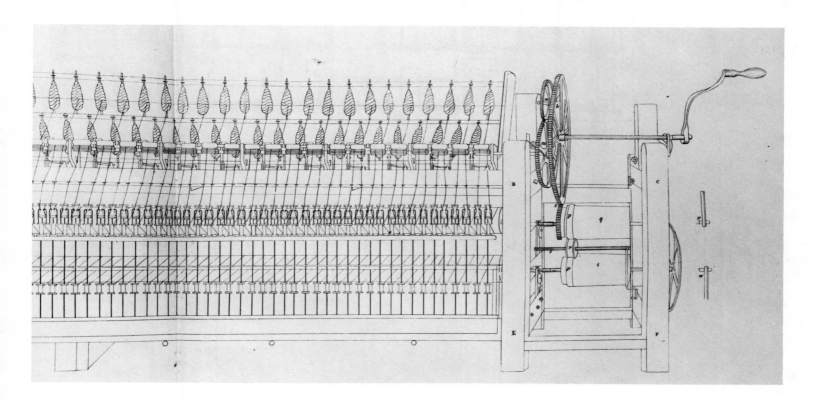

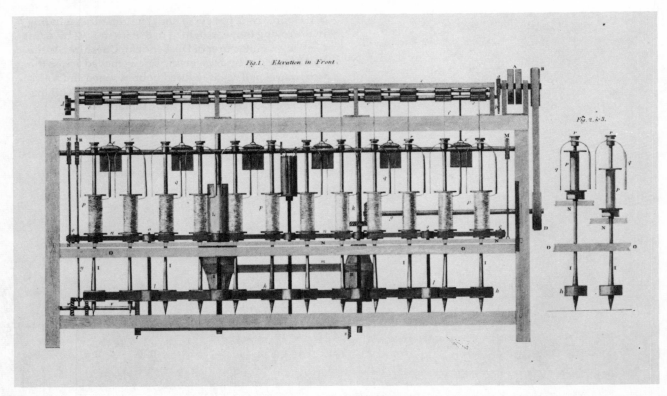

3.2 English double speeder for roving cotton; drawing of 1812. The double cone pulleys (*K,H*), patented by Garlick, now appear on vertical axles. Chains, alternately wound and unwound by gearing, lifted and lowered the bobbin rail (*N*) to perform the builder motion. (Source: Rees, *Cyclopaedia*. Courtesy Merrimack Valley Textile Museum.)

3.3 Cylinder of graduated brass gears and sliding pinion for retarding winding in the roving frame, patented in 1813 by Joseph Raynor, cotton spinner of Sheffield, Yorkshire. As the sliding pinion moved over the twenty-two gears of the cylinder, it slowed the speed of the bobbins. In Raynor's roving frame, the double cone pulleys were used to slow down the builder motion of the spindle rail. Retarding both bobbin rotation and spindle rail traverse minimized the physical forces acting on the roving and so allowed the production of very fine and even rovings, and hence yarn. (G.B. patent 3633; courtesy Boston Public Library.)

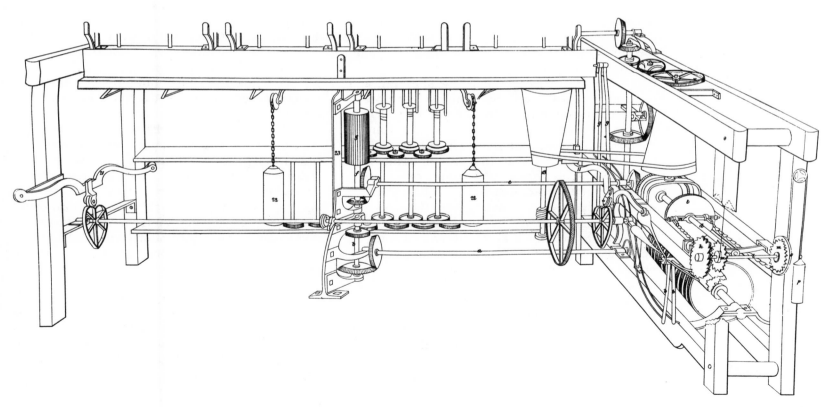

(fixed costs), primarily the capital costs of factory and machinery construction, or those entailed in running a mill (variable or overhead costs), chiefly the capital costs of power, machinery wear and tear, depreciation, insurance and interest, and the labor costs of operating, maintaining, and managing. The direction of efforts to lower these costs depended on whether and to what extent capital was cheap in relation to labor, or vice-versa. British inventions, of course, reflected British factor circumstances, and to some degree these can be surmised from the patent specifications.

The patented inventions included many forms of incremental wage-reducing devices. More mechanical work came through increased machine capacities—for example, in longer or pairs of mules and larger packages in roving, spinning, and weaving.[36] Equally, more mechanical work arose from running machines at higher speeds, explaining the interest in friction-reducing spindles and the disc substitute for the flyer.[37] Numbers of operatives might be cut by the continuous movement of materials through two or more processing stages either by combining different operations or machines into a single machine, as in the dresser or piecing machine, or by conveyor belts between machines, as in Bodmer's systems; and the integration of different processing machines might allow omission of a whole processing stage, as Bodmer wanted to dispose of carding.[38] Faster doffing and piecing enabled operatives to tend more processing units and so saved on wages; with this in view, several inventors inverted the throstle frame flyer.[39] Finally, the wage bill could be cut through feedback or self-regulating mechanisms, such as stop motions in roving and twisting frames and power

looms, and dobby and jacquard heads in the loom.[40]

Improvements for saving materials, a capital saving mostly, occurred at the levels of machine building and textile processing. Iron instead of wooden machine frames or brass instead of teasels exemplified the former. In processing machines, mechanical movements were more closely tailored, by ingenious combinations of known elements of mechanics, to fit a given action and thereby to lower breakages. Marsland's, Horrocks's, and Buchanan's variable batten speed motions were examples of this. And automatic fault detectors, the various stop motions, also curbed materials wastage.[41] Lowering power costs, a capital saving, emerges as another commercial target in the patents. Friction-reducing bobbins or flyers and power transmission to a line of looms by shafting instead of belting reveal this concern.[42]

Cutting overhead costs also required running machines constantly and organizing an efficient production line. Durable metallic parts were increasingly incorporated in equipment, for which versatility and variety of product helped to cushion against early obsolescence.[43] In the organization of production, cost savings derived from large-scale operations, vertical integration (bringing all processing operations under one management), and long production runs of standardized goods.[44]

The distinction between labor- and capital-saving biases cannot be carried too far. As Salter showed, many cost-reducing innovations affect both factors simultaneously.[45] One of the best examples of this is Horrocks's variable batten speed motion, a major incremental invention in power-loom technology and the device that characterized the Scotch loom.

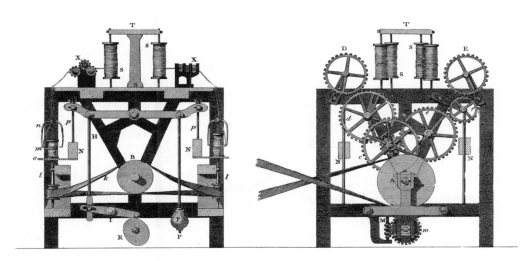

3.4 English throstle frame, as depicted in 1812. The tin roller from which the machine derived its name (see glossary) is marked *B* in figs. 1 and 3. (Source: Rees, *Cyclopaedia*. Courtesy Merrimack Valley Textile Museum.)

Fig. 3.

Elevation in Front.

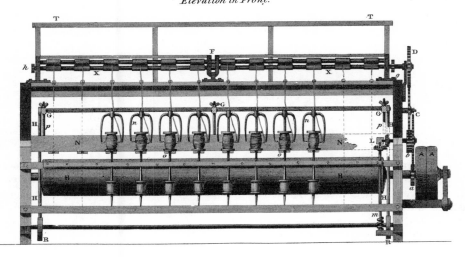

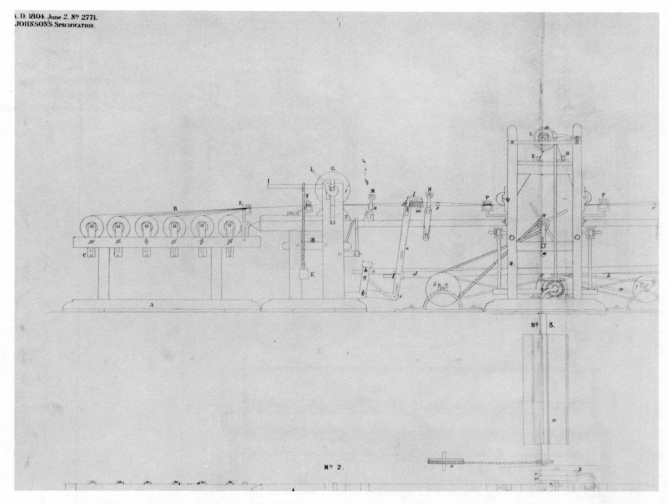

3.5 The machine that paved the way for power loom weaving: the English cotton warp dresser. This patent drawing of 1804 shows a slight improvement over the dresser patented the previous year under the name of Thomas Johnson, a weaver of Stockport, Cheshire. Johnson's employer, William Radcliffe, the Stockport manufacturer, shared in the invention of the dresser of 1803 and in the model of 1804 with its addition of crank-driven brushes for spreading the size evenly on the warp. (G.B. patent 2771; courtesy Boston Public Library.)

By means of an extra lever in its driving crank, the batten made a swift stroke, beat up the weft, and then returned quickly to a stationary position, at which point there was more time for the harnesses to open and form a larger shed and more time for the shuttle to cross the loom. Numerous economic benefits followed. A larger shed permitted larger shuttles, which could take bobbins straight from the mule spindles, eliminating rewinding onto small bobbins; larger cops, in turn, meant fewer weft piecing operations. Also, claimed Horrocks, "Waste from bottoms in the cops will be less."[46] The short, sharp stroke of the batten reduced the tension required for the warp, which thereby diminished warp breakages (and hence knotting costs) and power for the shedding mechanism; it also packed in more filling per inch, making the goods "stronger, more uniform, and better wrought in all respects."[47] Through savings in a mixture of capital and labor costs, the plain power loom reached commercial viability. In 1813, there were only twenty-four hundred power looms; in 1820, there were fourteen thousand.[48]

By moving toward higher-quality products and by making some of the cost savings noted, the two British textile technologies became inappropriate for American circumstances in several respects. Above all, quality products hardly ranked as a priority for U.S. infant industries. The American product market and the technology involved directed American manufacturers toward coarser textile fabrics made with a simpler rather than a more complex technology. Also American manufacturers were close to the cotton crop and felt less need to save raw material during processing; in fact they used better qualities of cotton for coarse goods than

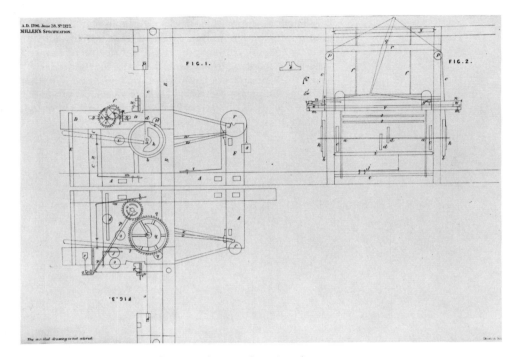

3.6 One of the early British power looms: the wiper loom in which shedding, picking, and beating-up motions were activated by cams; patented in 1796 by Robert Miller, calico printer of Dunbarton, Scotland. (G.B. patent 2122; courtesy Boston Public Library.)

3.7 The forerunner of Horrocks's variable batten speed motion: the batten motion patented in Britain in 1806 by Peter Marsland, cotton spinner of Heaton Norris, Lancashire. According to William Radcliffe, Marsland derived the motion from James Watt, Jr. Equipped with this motion, the power loom could weave yarns finer than coarse counts. (G.B. patent 2955; courtesy Boston Public Library.)

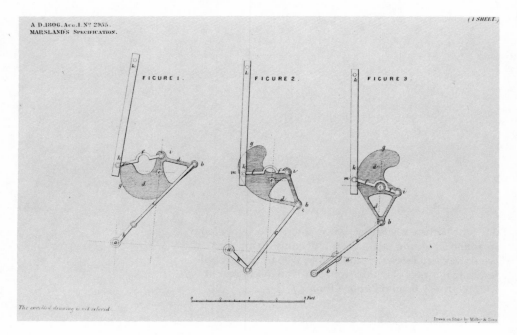

did their British rivals. Since cheap iron and nonferrous metals were not available to Americans before the mid-1830s, capital saving in the United States meant less use of iron and brass and more of wood in machine making, the reverse of the British trend. For Americans this led to shorter machine lives with more opportunities for making technical changes suited to their own factor situation. And the developing British technologies continued to require skilled labor, which Americans wanted to diminish because of its relatively high price. Before and after the introduction of Roberts's self-actor winding motion, the mule needed skilled operatives. Finally, the incremental nature of these improvements—the fact that they were numerous and individually of marginal value but in sum radical—compounded the problems of foreign imitators trying to sift the appropriate from the inappropriate technology. On the other hand, the tendency of cost-reducing innovations to save both labor and capital simultaneously might well have encouraged Americans to pursue technological solutions more keenly than if only one factor were saved with each innovation.

Industrialization also brought differing technical traditions in machine making. This was especially evident in cotton, the most capital-intensive branch of textiles. Hume's Select Committee of 1824 found, among manufacturers of Lancashire and of Scotland, an unshakable belief in the superiority of the Lancashire machine makers. The alleged superiority of the Lancashire men were seen as the result of an unparalleled division of labor. Peter Ewart, a veteran cotton spinner, attributed the superiority of British machinery to "the high state of the subdivision of labour, to which it is carried, and that is owing to the extent in which it is carried on."[49] As John Leigh Bradbury, a Manchester calico printer, pointed out, a spinning frame was built by at least six skilled tradesmen: "the spindle-maker, the roller-maker, the turner or bobbin-maker, the founder, the tinman, flyer-maker and others."[50] There were some exceptions to this general characteristic of the Lancashire textile machine-making trades, such as Richard Roberts's firm of Sharp, Roberts & Co.,[51] which used some of the new London machine tools.

In contrast the London engineers dispensed with the handicrafts of turning and filing and their intensive subdivision. Instead they relied on a new range of machine tools, handled by unskilled operatives, to accomplish the same jobs. The 1825 select committee reported to the Commons, "Messrs Martineau, Bramah, Maudslay and Galloway, all of them eminent engineers, affirm that men and boys . . . may be readily instructed in the making of Machines."[52] The capital-intensive, high-productivity machine tools in question were those developed by Maudslay and his pupils: industrial screw cutters, planers, and lathes distinguished by their capacity for precision work.[53]

The position gradually changed in the 1830s. In 1835, Ure credited Sharp, Roberts & Co. of Manchester with introducing a degree of interchangeability by employing planing and key-groove cutting machines to machine identical components.[54] Even so, Thomas Ashton, one of the older generation of cotton spinners, in 1841 judged half a dozen trades still necessary to make a spinning frame: "a man could not make a spindle and a roller."[55] William Jenkinson, a Salford machine maker, also asserted in 1841 that spindle manufacture was almost

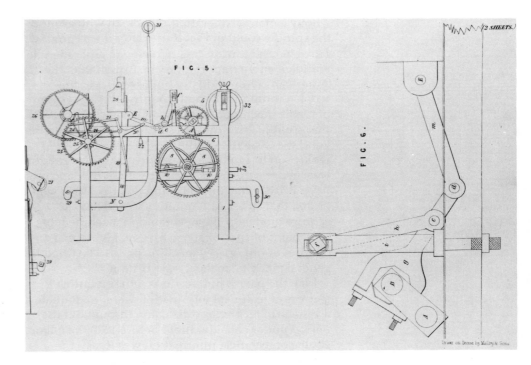

3.8 Variable batten speed motion patented in Britain in 1813 by William Horrocks, cotton manufacturer of Stockport, Cheshire. In figs. 5 and 6, *E* is the "lay sword" supporting the batten or lathe. In fig. 6, *A* is the center of the crank shaft and *A,B* the crank arm. (G.B. patent 3728; courtesy Boston Public Library.)

wholly a manual operation: "the man who ground the spindle could not set it true after it was ground." However, Jenkinson was referring to tapered mule spindles, more important to the British than the American cotton industry but essential in the woolen industries of both countries. His experience of textile machinery making and especially the fact that spindle grinding and setting still remained beyond the capacities of the new machine tools led Jenkinson to agree that machine making in London was "more a theoretical science" than in Manchester.[56] British machine building practices thus suggested that American imitators would have to acquire models of the new London machine tools, hire teams of specialist machine makers from Lancashire, or import key machine parts to be fitted by general machine makers from Britain.

Like the patent publications, other inanimate sources of technical information were inadequate as a vehicle of technology diffusion throughout the period under consideration. Before 1820 the major source of technical information was Rees's *Cyclopaedia*, published in forty-five volumes between 1802 and 1820.[57] All articles relating to the textile industries were written by men with firsthand experience of manufacturing. James Thomson, a relative of the Peels and partner in the family firm, who set up his own calico printing business at Primrose near Clitheroe, wrote the two articles "Cotton Manufacture" and "Manufacture of Cotton." Published in May 1808 and November 1812, respectively, these two contributions and several more on calico printing appearing in 1807–1808, also written by Thomson, summarized the most recent advances in cotton manufacturing technology. Although concealing exact operating

techniques, the plates and verbal descriptions in the articles gave readers a clear view of machine designs and basic operating principles. For a waterframe topography, the articles provided gear ratios and weights for rollers and calculated the draft ratio, but they omitted essential technical information on spindle speeds. These data did accompany the description of the water frame, but by 1812 it was obsolete in design, though not in principle. For its successor, the throstle, Thomson gave no such technical specifications.

The completion of Rees's *Cyclopaedia* in 1820 made available a large amount of information on machine topographies, processing principles, and general manufacturing techniques on a wide range of textile subjects. Two articles especially pertinent to cotton and woolen cloth manufacturing technology were published after 1812. That on "Weaving" (May 1818), by John Duncan, the Scottish weaving expert, briefly described the Johnson-Radcliffe dresser and Horrocks's power-loom patents, mentioning the latter's crucial variable batten motion. That on "Woollen Manufacture, Process of" (July 1818), by Robert Bakewell, a Yorkshire wool stapler and author, was also well informed and accompanied by five plates. It described the operations of the picker, carding engine, billy, jenny, woolen mule, warping mill, fulling stocks, gig mill, Harmar's shearing frame, Lewis's helical shears, and the frizing machine. Both articles covered the areas of greatest inventive activity in the period 1813–1824 but appeared too early to register the major steps forward or to appreciate fully the steps already taken, judging by the short paragraph given to Horrocks's loom. In any case, the *Cyclopaedia* never pretended to supply processing rules to manufacturers. As its editor observed on its completion, the work was a "Scientific Dictionary [giving] under each distinct head of science, an historical account of its rise, progress, and present state."[58] References were given to other publications—frequently the *Repertory of Arts*—and contributors consulted "the artisans and manufacturers themselves."[59]

After Rees, the next significant publication was Montgomery's *Carding and Spinning Master's Assistant* (1832). This offered an explication of much representative cotton spinning technology and included processing rules, but it did not describe the latest British improvements like Bodmer's preparatory innovations or Roberts's self-acting mule. Indeed all the most recent technology considered was of American origin: Arnold's fly frame differential of 1823, George Danforth's tube roving frame of 1824, Charles Danforth's cap throstle of 1828, and the dead spindle.

Montgomery published a second volume, *The Cotton Spinner's Manual*, in 1835, a year after a similar technical handbook, *The Cotton Spinner's Companion* by George Galbraith, was published (also in Glasgow). Omitting machine descriptions, Montgomery gave rules for calculating speeds and drafts in carding and spinning equipment and for calculating other necessary information such as sliver or yarn size and the costs of various yarn numbers. A work whose *"chief* object is utility," this small book was designed for the pocket of carding and spinning masters, as well as ambitious operatives.[60] After he went from Scotland to New England in 1836, Montgomery continued his note-making habit and in 1840 published his well-known *Practical Detail of the Cotton Manufacture of the U.S.A.* In this he compared the spinning

technologies of Britain and the United States at both the technical level of machine design and the economic level of mill profitability.

Cotton power-loom weaving and cotton finishing awaited a similar systematic, practical handbook treatment until after 1840. No textbooks on woolen manufacturing appeared in Britain until after 1850. The works of Ure, like Baines's on cotton, were essentially descriptive and did not provide the operating rules needed by managers or overlookers. Americans, of course, had Partridge, but his book was outdated in the 1830s by the advent of the woolen mule and power loom.

The publication of technical handbooks in America was facilitated by the federal copyright law of 1790, which legalized the American printing of foreign works without regard to their authors' literary property rights.[61] An American edition of Rees's *Cyclopaedia* was published at Philadelphia between 1810 and 1824 (though the bibliographers seem a little uncertain of these dates).[62] As far as can be judged from the addresses given by the 1,851 American subscribers to Rees, the greatest interest in publishing Rees existed in Massachusetts, Pennsylvania, and Maryland, where the proportions of subscribers substantially exceeded the respective states' shares of the nation's population (see table 3.4). Regionally, northern New England exhibited the highest interest in the "Scientific Dictionary," probably reflecting the region's strong educational traditions and the sort of cultural ethos that encouraged both the theoretical and practical approaches to new knowledge for which New England's universities and mechanics became equally famous.

A few other British technical publications were reissued in America. The second edition of John

3.4 Subscribers to the American Edition of Rees's *Cyclopaedia*

	American Subscribers		U.S. Population Proportion in 1820	
	No.	%	No.	%
Northern				
New England				
Maine	52	2.8	298,335	3.09
Vermont	6	0.3	244,161	2.53
New Hampshire	49	2.6	235,981	2.45
Massachusetts	323	17.4	523,287	5.43
Total	430	23.2	1,301,764	13.50
Southern				
New England				
Rhode Island	11	0.59	83,059	0.86
Connecticut	16	0.86	275,248	2.86
Total	27	1.46	358,307	3.72
Middle states				
New York	233	12.6	1,372,812	14.2
New Jersey	33	1.78	277,575	2.88
Total	266	14.4	1,650,387	17.12
Pennsylvania	355	19.2	1,049,458	10.9
Delaware	11	0.59	72,749	0.75
Total	366	19.8	1,122,207	11.6
Maryland	222	12	407,350	4.23
Other				
South and West	531	28.7	4,798,438	49.8
Location unstated	1			
Abroad	8			
Totals	1,851	100	9,638,453	99.97

Source: Rees, *Cyclopaedia*, vol. 46, "Ancient and Modern Atlas" (1824?) (cited in Shaw and Shoemaker, *American Bibliography*, pp. 65–66).

Nicholson's *The Operative Mechanic, and British Machinist; Being a Practical Display of the Manufactories and Mechanical Arts of the United Kingdom* (1825) was reprinted at Philadelphia "with additions" in 1826. Adhering to machine topographies and general principles of processing, it was in the style of Rees and added little to it except greater compactness and cheapness.[63] In cotton technology one such publication was Joseph Stopford and Nehemiah Gerrard's *The Cotton Manufacturer's Useful Assistant* (1832), a collection of yarn and warp tables first published in Manchester.

Americans became increasingly aware of the available sources for information. Thus among the business receipts of Jacob Wendell, a Portsmouth, New Hampshire, entrepreneur, there is one for five dollars (dated April 16, 1816) for the second number of Gregory's *Dictionary of Arts and Sciences. The Pantologia: A Dictionary of Arts and Sciences,* edited by Olinthus G. Gregory, was completed in twelve volumes in 1813.[64] Particularly novel was the partially deceitful citation of an English publication by New England mechanics in their attempt to evade Moody's dressing machine patent of 1818. In "special matter," lodged September 15, 1820, the defendants, Jonathan Fisk et al., claimed that Moody's dresser "had been used in England . . . & had been described in the public works following (that is to say in the Edinburgh Encyclopaedia & in Rees Cyclopaedia and in the Repertory of Arts.)"[65] But Johnson's dresser patents were never published in the *Repertory of Arts*. William Radcliffe, the inventor of this dresser, not only used Johnson's name to conceal his own interest in the invention but also prevailed upon the editor of the *Repertory* to forgo publication of these dresser patents.

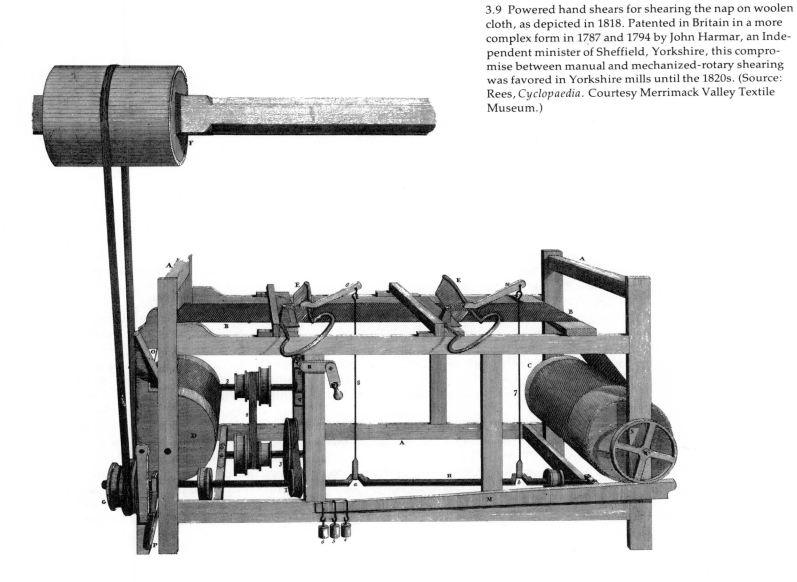

3.9 Powered hand shears for shearing the nap on woolen cloth, as depicted in 1818. Patented in Britain in a more complex form in 1787 and 1794 by John Harmar, an Independent minister of Sheffield, Yorkshire, this compromise between manual and mechanized-rotary shearing was favored in Yorkshire mills until the 1820s. (Source: Rees, *Cyclopaedia*. Courtesy Merrimack Valley Textile Museum.)

By the 1820s, Americans had a selection of their own manufacturing and engineering textbooks. Samuel Ogden, the immigrant Rhode Island machine maker and cotton manufacturer, proposed publishing his manuscript essays on cotton spinning in 1814. He envisaged a four-hundred-page volume selling at $2.50 a copy, in which "much valuable information would be laid open." [66] Evidently he found few sponsors, for the work never appeared. The following year, however, Ogden did publish a thirty-nine-page pamphlet in which he emphasized the importance of delegating more executive power to the mill manager and of managers exercising much closer control over their production systems through weighing, reeling, timing, personally inspecting, clearly instructing, and using piecework. [67]

In the 1820s, the *Franklin Journal* of Philadelphia pioneered in publishing British patents. Its first volume (1826) included a few English patent specifications. In its first years the journal also printed lists of British patents recently enrolled. To give practical help to its readers, articles by Peter A. Browne (a prominent member of the Philadelphia bar) on British and American patent law under the title "Mechanical Jurisprudence" appeared in the early volumes. [68]

Oliver Evans's *The Young Mill-Wright and Miller's Guide* (1795) was revised for its fifth edition in 1826 by British-born Thomas P. Jones, editor of the Franklin Institute *Journal*. [69] Three years later, two more introductory texts on mechanics appeared. Jacob Bigelow, a Harvard professor, published his Rumford lectures as *Elements of Technology* (1829). A close inspection of the sections on machinery and textile manufacturing confirms Ferguson's verdict on the book as "a pretty thin potion." [70] Its manufacturing chapter reads like a précis of Rees or Nicholson, and the section on mechanisms misses Arnold's differential. Also in 1829, Zachariah Allen, the Rhode Island manufacturer, published his *Science of Mechanics*, a work more surely rooted in practical experience but still more concerned with mechanical principles than with the rules and details of manufacturing methods.

British cotton and woolen manufacturing technologies in themselves therefore posed several problems for American borrowers: they were undergoing increasingly rapid modification; these modifications were made in directions frequently inappropriate to American circumstances, especially concerning product market situation and factor endowment; and the technologies were not completely described, in objective terms and to the extent of processing rules, in either British or American publications until after 1830. Americans therefore were compelled to select from and then modify the British technologies to suit their own circumstances. The limits of published technical information forced Americans to rely more on immigrant artisans than on any other method of transferring the technologies. But British artisans also presented drawbacks to American entrepreneurs: Commingled in America, British workers would bring diverse regional technological traditions into conflict, causing confusion; immigrant workers from the preindustrial trades of hand-loom weaving and calico printing were likely to impart resistance to technological change; and in machine making they might not be as cheap as the new machine tools or key parts imported from Britain.

II
Diffusion of the Technologies:
The First Three Stages

Introduction

An examination of the constraints on international diffusion points to the emigrant British artisan as the best available carrier of the new textile technologies in this early industrial period. Ascertaining whether and how he fulfilled this potential occupies parts II and III of this book. Part II explores the transfer of four different textile technologies to the United States: machine spinning, power-loom weaving and mechanized calico printing in cotton manufacturing, and newly mechanized techniques in woolen manufacturing. They represent the most radical technical developments in British textile manufacturing to reach America during this period. Cotton rather than wool predominated because the structure and properties of its fiber, above all others except the far more expensive silk fiber, yielded most readily to the application of mechanical processing techniques. From these case studies, the roles of individual immigrants and of other agents of technical diffusion come into sharper focus. So too do the conditions in which the initial stages of diffusion occurred.

In attempting to discover how institutional and technical obstacles to transfer in Britain and the problems of small markets and high capital and labor costs in America were surmounted, part II employs a stage analysis of each of the four technologies. When the stages are defined carefully, valid comparisons can be made and conclusions drawn about the nature of technology diffusion in a developing economy. As the introduction to this book indicates, the first stage consisted in the creation of the potential for international transfer. The buildup in America of the various carriers or vehicles—skilled immigrants, machine models, plans, verbal descriptions—just prior to the launching of pilot production lines marked the completion of this stage. Pilot production, usually in mills or factories but for some wool machines also in workshops, constituted the achievement of the second stage. This stage therefore deals with the first known and commercially successful manufacturing operations in America to use the technology in question. Demonstrable profitability and local acceptance led other manufacturers to copy the technology of the pilot firms, and so the third stage, internal spread or diffusion, unfolded. For various reasons the internal spread of a new technology might be aborted or delayed. Calico printing technology diffusion experienced such a delay. In this case, therefore, those firms that next followed the pilot firm, but, unlike it, activated the internal spread of a new technology through a region or industry, have also been regarded as pilot firms. As such they have been used to provide evidence of the first stage of diffusion. When Americans became familiar with the imported technologies, and this might happen simultaneously with or subsequently to the second or third stages, they made modifications to the imported technologies to suit their own economic, social, or technical circumstances. This fourth stage of modification invariably left behind the immigrant artisan, who generally adhered to the technical patterns with which he had grown up in Britain. Consequently the fourth diffusion stage is treated separately in part IV of the book.

Part III looks at the role of immigrant workers in technology diffusion from a fresh angle. Immigrants as individuals might have an impact very different from, even the reverse of, immigrants collectively. Certainly it is easy to imagine (quite wrongly) that Samuel Slater typified all British textile workers arriving in America in the early industrial period. Making the commonplace distinction between the individual and the group, therefore, part III establishes the size and composition of aggregate immigration into the United States from Britain's textile industries, and assesses its contribution to the growth of early nineteenth-century American cotton and woolen manufacturing.

4
The Diffusion of New Cotton Spinning Technology, 1790–1812

The crucial importance of the manager and machine builder in the Arkwright system and of the operative in the Crompton system, heightened before 1812 by the absence of published accounts of the two spinning technologies, could not be surmounted by importing machines without men. At Philadelphia a disassembled spinning mule confounded interested parties for four years and was eventually shipped back to Britain in 1787, leaving Philadelphians none the wiser but angrier. In New England three Arkwright machines, apparently imperfect models, for carding, roving, and spinning were built by two Scotsmen in 1787 for public exhibition but defied commercial exploitation until Slater arrived three years later.[1]

The late 1780s and early 1790s saw numerous efforts by Americans both to obtain workable models of the new machines and to recruit machine makers and managers. From Philadelphia, the federal capital from 1790 to 1800, emanated the most vigorous endeavors, several of which were linked with Hamilton and his assistant at the Treasury, Tench Coxe. In 1787 Coxe sent an agent, Andrew Mitchell (an expatriate Englishman), to England to procure models and patterns of Arkwright's machinery, but Mitchell was caught and his haul confiscated. Undaunted, Coxe recruited George Parkinson, an English mechanic and mill foreman, and encouraged him to patent a flax spinning machine, probably derived from the water-frame design.[2] Another member of the Pennsylvania Society, the Philadelphia merchant William Bingham, explored the possibility of importing Arkwright machinery from France. Through Jefferson he obtained a price quotation from John Milne, a Lancashire cotton manu-

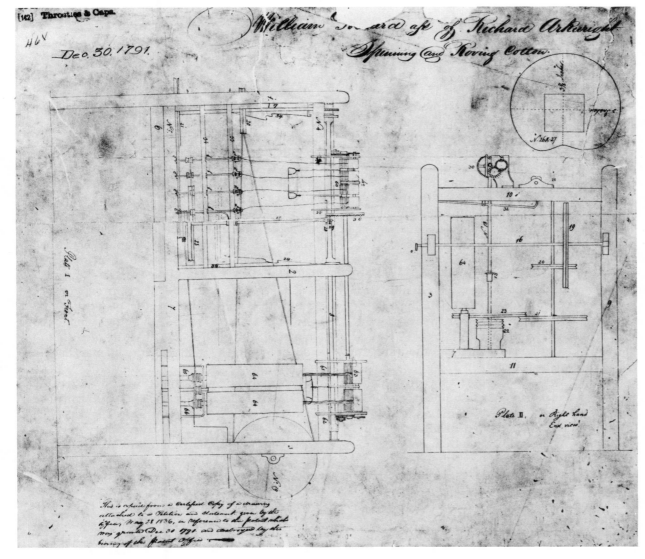

4.1 The first American version of Arkwright's can roving frame and water frame: William Pollard's patent of 1791. In plate I the roving can frame appears on the right and the spinning frame on the left when the drawing is turned sideways. The drawing shows the incorporation of two English improvements subsequent to Arkwright's patents of 1769 and 1775: a bobbin lifter rail patented in 1772 by Coniah Wood and its actuating heart-shaped cam, seen above plate II. (Courtesy U.S. National Archives.)

facturer and machine maker established at La Muette just south of Paris. But a few months later this source disappeared when France was overwhelmed by political violence. Possibly through his trade connections, another Philadelphia merchant, William Pollard, secured a water-frame model around 1790, which he proceeded to improve, patent, and promote as a commercial feasibility.[3]

Far more important than machine models, however, were skilled managers and mechanics, as Hamilton was informed by one of them in September 1791: "After you have made choice, Sir, of a person for the Directorship the first Necessary Consideration may be that of obtaining Mechanics from England, (if they cannot be got here) for the purpose of making Machinery and assisting in the Conducting of the Manufactory."[4]

At least five recruiters of skilled labor working for American projects were active in England in the late 1780s and early 1790s. Most effective was Thomas Digges, the disgraced son of a prominent Maryland family. As an American merchant in London during the Revolution, his tangled pursuits brought charges of espionage and treason from both sides. After the war and until 1799, he sublimated his kleptomania through industrial espionage. Initially he searched for tenants for his estates in Virginia and Maryland, but he widened his net later. Over a twelve-month period in 1791–1792, he claimed to have sent to America eighteen or twenty artists and machine makers. At Belfast in spring 1791 on one of his many forays into British manufacturing districts, he recruited William Pearce, a brash and intemperate mechanic who played an important part in diffusing British technology in the United States. Pearce allegedly worked for Arkwright and then

Cartwright before moving to Belfast and from there went to New York, Philadelphia, and New Jersey. Another recruit was James Hall, who became superintendent of calico printing in the Society for Establishing Useful Manufactures (SUM). Mechanics with special talents were given letters of introduction by Digges to influential Americans. So Pearce, in June 1791, carried his letters to Washington, Jefferson, an ex-governor of Delaware, the cashier of the Bank of New York, and two Philadelphia merchants.[5] Another activist, Abel Buell, the Connecticut silversmith and inventor, in 1789 raised £150 by a mortgage on his shop and went to England ostensibly to secure copper for the state mint. Whatever the original purpose of his visit, Buell returned fired with enthusiasm for cotton manufacturing, having allegedly recruited William McIntosh, an Essex worsted manufacturer, who arrived at New Haven in summer 1791.[6]

The American recruiting effort in Britain during the 1790s became related also to manufacturing enterprises located between the Susquehanna and Hudson valleys, particularly two embryonic mill towns, Alexander Hamilton's brainchild the SUM at the Great Falls on the Passaic in New Jersey and John Nicholson's at the Falls of the Schuylkill in Pennsylvania. Imaging that he was serving the cause of the national government, Digges fed some of his workers back to Hamilton and the SUM enterprise at Paterson, New Jersey. Hamilton even agreed with one of these, Joseph Mort, a calico printer, "that if desired by the Society, he is to go to Europe, to bring over Workmen, at his own Expence in the first instance; but with the assurance of reimbursement and indemnification"—this within months of Mort's arrival in America in 1791.[7] A year later Hamilton sent John Campbell, a stocking weaver, back to Scotland to recruit eight knitters, three framesmiths, and a setter; Campbell evidently fulfilled his assignment and also brought back his wife.[8] Nicholson, an inveterate speculator, envisioning a diversified mill town on the Schuylkill, employed a number of Englishmen and asked his friend and European agent Dr. Enoch Edwards to recruit others.[9]

Some immigrant workers initially contacted Americans in Britain to sound out their prospects and make settlement arrangements before they set out for America. Thomas Marshall, a manager at the SUM, astutely opened his American prospects by contacting, through a friend, two prominent New Yorkers while they were in London: Henry Cruger and William S. Smith, John Adams's son-in-law. When Marshall arrived at New York in July 1791, Cruger met him, arranged interviews with other merchants and potential patrons, and directed Marshall to Hamilton. Marshall was as important as Pearce, for he claimed managerial experience in working for Arkwright in Derbyshire.[10]

The first wave of immigrant artisans provided contacts with workers left behind in Britain. Apart from his familiarity with the latest improvements in Arkwright technology, another of Marshall's assets was his connection with other skilled workers in England. In particular he offered to bring over the foreman weaver from John Callaway's cotton mill at Canterbury; the sum of the skills of these two specialists would permit a vertically integrated manufactory at Paterson.[11]

American initiatives likewise attracted British textile workers to the United States through printed publications. Slater, who worked in Jedediah Strutt's cotton mill at Milford, Derbyshire, was attracted to America by the report in an English newspaper of a premium awarded in 1788 by the Pennsylvania Society to John Hague for building a successful carding machine.[12] But newspaper advertisements were rarely used by American promoters. They might not be permitted in English newspapers, for they encouraged violations of British laws. William Pollard decided against advertising his technical achievements and plans in American newspapers because he was "afraid too, that if one particular Branch sh'd be mentioned (as Stockings) that either the Governm't of England or the Manufacturers in that Article might form a Combination against it."[13] On the other hand Digges advised Hamilton to advertise premiums in American newspapers, which could then be distributed in Belfast, Liverpool, and Manchester, "for in these Countrys they hardly ever will publish any favourable acco' of America or insert a Paragraph wch. may lead the people to Emigration."[14] In Britain Digges advertised American opportunities by making personal contacts on his numerous tours and, in 1792, by distributing a thousand copies of Hamilton's *Report on Manufactures,* which he had printed in Dublin and sold on both sides of the Irish Sea at a shilling apiece.[15]

Incentives comparable to those current in Britain attracted skilled English managers and machine builders to work for the infant American mills. The premium that attracted Slater's notice was £100 for building a machine (a card) with which he had become familiar during a mill apprenticeship of seven years. At the age of twenty-one, he was lured to Rhode Island with an offer of all of the profits in a six-month trial period. After a trial period of four

4.2 Flat-top finisher cotton card used by Samuel Slater in Rhode Island and dated to the 1790s. The similarities between Slater's design and English models are evident when one compares this with illustration 1.2. (Courtesy National Museum of History and Technology, Smithsonian Institution, Washington, D.C.)

4.3 Water frame used by Samuel Slater in Rhode Island, a forty-eight-spindle model dated to the 1790s. This can be compared with the English antecedents shown in illustrations 1.5 and 1.6. (Courtesy National Museum of History and Technology, Smithsonian Institution, Washington, D.C.)

months, he was given joint ownership and half the profits for supervising a one-hundred-spindle Arkwright mill; his annual net profits for the 1790s were just over $800.[16] At Paterson, the SUM paid the following salaries to its key immigrant workers: Thomas Marshall, superintendent of the cotton spinning mill, £200 per year; William Hall, superintendent of fabric printing, £300 per year; Joseph Mort, an assistant to Hall, £100 per year. William Pearce, who with Marshall and Hall directed the construction of machinery—the full complement of which was to be four cards, four billies, four slubbing machines, twenty-five jennies, and sixty looms—received nearly $900 for machinery construction contracts, which he fulfilled in his spare time, having branched out alone.[17]

Besides salaries as managers or machine makers, the introducers of new technology could also expect returns from patent fees if they chose to patent. Patenting became easier after 1793 when the law was changed to dispense with governmental examination of the inventor's claim to novelty. The onus for this was placed on competitors who contested claims of invention in the courts, a position that remained unchanged until 1836. Low patent fees further encouraged the publicizing of inventions: $30 in the United States, as compared with £70–£100 in Britain. The sort of returns theoretically possible are disclosed in the case of Pollard and his Arkwright patents. For five hundred water-frame spindles and five hundred mule spindles, Pollard estimated his license fees to be £1,495.[18]

These financial returns, especially for those whose chances of reaching the top of Britain's mill management or of inventing a successful machine were slim, countered the risks of emigration and lured numbers of skilled textile workers to the United States. In addition some inventors, including Pearce and Douglas, saw Americans as sponsors for ideas that they were unable to promote in Britain. Douglas claimed to have invented three machines while in the United States in the 1790s, including a power loom and a powered cloth shear; Pearce went to the United States with plans of a water loom and a 180-spindle mule in his head or on paper.[19]

The barriers to Arkwright-technology transfer from Britain to America were largely overcome by the activities of recruiting agents, the readiness of workers to ignore the law in pursuit of better prospects in America, and the fact that the new technologies were embodied in the artisan. Failure to smuggle machines was of less importance to the introduction of new spinning technology.

Between 1790 and 1793, Americans at over half a dozen places between Philadelphia and Boston gained the small-scale capacity to build and manage both Arkwright and mule spinning technology. The contributions of Samuel Slater and William Pollard have been documented in detail, but they were certainly not alone in introducing a constructional knowledge of the new technology. Other immigrants, many with forgotten names, also took this knowledge across the Atlantic. One was John Daniel, who arrived at New York late in 1793 and advertised himself as "Machinerist for Cotton Works" who was willing to build machines of all sorts including billies, jennies, mules, drawing and roving frames, and water frames.[20]

While the transatlantic movement of technological knowledge was gathering momentum in the late 1780s, various conditions in the United States helped to establish an enduring industrial matrix in which imported knowledge and immigrant experience could be translated into mill production. The first important influence was the prevailing state of the American and Atlantic economies. The early 1790s, a period of fresh business activity, saw the launching of a number of textile manufacturing firms, as well as banking, turnpike, and canal companies. Conversely, economic depression after 1783, rather than the absence of skilled immigrants, explains why the few efforts made in the immediate post-Revolutionary years to set up cotton spinning mills failed.[21]

Few of the mills set up in the early 1790s survived the influx of British cotton goods imports when European war diverted British exports to the United States. Between 1793 and 1807, the North American share of British cotton exports was rarely below 30 percent and the total value of these exports steadily climbed from £1.6 million to £10.2 million.[22] Nicholson's mill town at the Falls of Schuylkill, the Livingston's cotton factories at Hell Gate, New York, and New Haven, and Jacob Broome's at Wilmington disappeared before 1800. The Beverly cotton mill closed in 1807, and the SUM at Paterson suffered decay.

Only strong capital resources and prudent management would preserve American cotton spinning mills until after 1807, when the embargo and continuing westward expansion provided protected and expanding markets for domestic manufacturers. The immediate effects of the embargo were disastrous, and manufacturing prosperity was an illusion projected by a combination of falling cotton and fixed factory yarn prices; but illusion turned to reality as wholesale commission houses in Philadelphia opened up the western trade.[23] The contrasting experiences of William Pollard and Samuel Slater illustrate the importance of sound capital backing and cautious expansion in alliance with household weaving in the years before 1807.

Pollard's proposal for a one-thousand-spindle (water frame and mule) cotton factory with eighty stocking looms and fifteen hand looms was taken up by the Pennsylvanian speculator John Nicholson. Early in 1793 he signed a personal agreement to advance the relatively large capital of £18,000 that Pollard needed, over a third of it for machinery. Over the next three years, Pollard engaged in constructing the machinery aided by native and immigrant craftsmen. But the project was hindered by its promoter's speculative fancies and fortunes. A mill was never constructed, although Nicholson considered placing it in his mill town at the Falls of Schuylkill, so Pollard operated his machines in rented rooms. By 1793 Nicholson's credit was falling. That year he was impeached for misappropriation of funds while controller-general of Pennsylvania. Consequently Pollard was unable to cash his notes and checks. As his cash flow dried up, Pollard found his workers deserting him. By autumn 1795, Nicholson's speculative empire was crumbling fast. With it went Pollard, who suffered eviction, the sale of his household furniture, and the disposal of the cotton machinery, which he had almost finished building under the 1793 contract.[24]

In contrast, Slater had the comparatively meager financial backing of William Almy and Smith Brown, son-in-law and cousin of Moses Brown, the Quaker merchant of Providence, Rhode Island, and the advantage of a going, though infant, concern. Slater first used an old fulling mill at the Pawtucket bridge, north of Providence. In 1793 the partners invested in a new mill at the confluence of the Pawtucket and Blackstone rivers in north Providence. Their original investment, unknown now, was probably small and came from Almy and Brown's retail store in the form of building supplies and notes on the store to pay wages. They met additional capital needs by ploughing back profits. A running account for the period 1793–1803 shows that the mill's expenditures for the decade amounted to under $100,000, two-thirds of it for the purchase of raw cotton. In 1793 expenditures amounted to only $1,565, of which half was for cotton and the rest for building supplies, wages, and other costs. When the account was balanced in 1803, a net profit of $18,000 remained, half due to Slater. And this was from a one-hundred-spindle mill initially. The capital investment, though small in relation to Pollard's grandiose scheme, flowed from a steady source and was well able to support the small scale of production with which it prudently commenced.[25]

Other factors played a part in Slater's success, for he was neither the first to build Arkwright's water frame in the United States nor the only British mechanic with the backing of American merchant capital in the early 1790s. He was the first in America, however, to achieve commercial profitability with Arkwright technology. A comparison with his contemporaries provides some explanation for his successful economic adaptation. He had competence, but not the driving ambition and personal instability that marked many of his rivals. Technically he was a close copier of the well-tried British techniques and machine designs with which he had grown up. Molded by the horizons of a rural Derbyshire factory and remaining a self-contained personality, he tempered his activities with caution. Slater patented nothing. In the 1790s he was not interested in the vertically integrated, large-scale enterprises tried at Paterson or the Falls of the Schuylkill. Diversification, as in Nicholson's company town, did not attract him either. Instead Slater concentrated initially on supplying yarns to domestic weavers and hosiers. For his first ten years in America, he remained manager and part owner of the Pawtucket mill. Only when he felt secure in his business did he start moving into partnerships in other relatively small mills. Slater's credibility, which served him in his role as a vehicle of technology transfer at all three stages (introduction, establishment, and diffusion), was greatly enhanced by his marriage within a year of reaching Providence to Hannah Wilkinson. She was the daughter of one of his machine builders, the Pawtucket blacksmith and Quaker Orziel Wilkinson.[26]

At the technical level, it is abundantly clear that America's traditional craftsmen were capable, under the direction of immigrant managers, of building Arkwright machinery. Spindles and flyers for one of the first Philadelphia water frames came from a local whitesmith, wrought-iron parts from a local blacksmith, and cast-iron weights from a New York iron founder. The machine was built under the supervision of William Pollard, a Philadelphia merchant who originated from Yorkshire where he learned worsted weaving. Pollard was assisted by a Manchester brass filer and cotton carder.[27]

Slater started with Moses Brown's two water frames, which he pronounced useless but cannibalized for parts. With the assistance of a local millwright, Sylvanus Brown, he started building a

4.4 Pawtucket Falls and Bridge, a water color sketch of 1812 signed D.B. The cupola of the mill built for Almy, Brown and Slater in 1793, and given an addition in 1801, is just visible above the bridge spanning the Blackstone River. To the right is the three-story Yellow Mill built in 1805 by a group of Rehoboth proprietors for cotton spinning. (Courtesy Rhode Island Historical Society.)

new frame. Spindles, rollers, and ironwork were made by Orziel Wilkinson. Slater's role in machine building lay in designing and in ensuring that components were assembled in their correct dynamic and static interrelationships. Brown and Wilkinson made the parts and put them together with the dimensions, settings, gear ratios, and surfaces directed by Slater. The greatest technical problem came in the construction of the carding machine, or rather the construction of the card clothing. Slater eventually perceived that the card clothing leather was poor in quality; that the hand-pricked holes were too large for the teeth, which consequently were loose; and that the bend in the teeth was not sufficiently acute to work the cotton. He returned the clothing to the maker, Pliny Earle of Leicester, Massachusetts, who crudely rectified the defects, and the problem was surmounted. It took Slater twelve months to get a production line in operation.[28]

Labor supply, so crucial to the successful introduction, establishment, and diffusion of the new technology, was an early problem. As Arrow has observed, "The channel has greater capacity if the receiver regards it as more reliable."[29] But one of the problems hindering the successful establishment of Arkwright and mule spinning lay in the unreliability of some of the immigrant workers.

Some immigrants were simply incompetent. McIntosh, the British manager recruited by Buell to manage a New Haven cotton and woolen factory, failed to install north lights or to ensure that the carding machines were properly acclimatized to the factory atmosphere, and this atmosphere became excessively dry.[30] Others indulged in experiments that neglected profitability. Such foibles, and pos-

sibly William Pearce, Hamilton had in mind when he cautioned the SUM board in August 1792 against a propensity to extravagance in machine building.[31]

Worse yet, other immigrants were rogues. Whatever the levels of professional competence of individuals, British machine makers and managers gained a reputation for sharp practices, if not knavery, in the 1790s. Wansey in 1794 encountered well-founded stories of the roguery of different managers at Paterson. Pearce in October 1793 had an altercation with the SUM and, abetted by Hall, decamped with some of the company's machinery to the Brandywine and the employment of Delaware senator Jacob Broome. The SUM eventually recovered the machinery from Wilmington.[32] On the Schuylkill, machine maker John Bowler surreptitiously built machines for outside customers while he was supposedly working for John Nicholson. Then in 1795 he absconded to Ireland with Nicholson's advance of $10,000, a sum never recovered.[33] The fears of one of Hamilton's correspondents early in the 1790s were evidently realized: "I repeat it, Sir, unless God should send us saints for workmen and angels to conduct them, there is the greatest reason to fear for the success of the plan [the SUM]."[34] The skills of immigrants were not enough. Without probity, they were suspect. Writing to Nicholson in 1793, William Pollard, the water-frame patentee and promoter of Arkwright spinning, explained the futile pursuit of fine-goods manufacturing in New York in terms of moral irresponsibility: "The Conductors of these Works were not interested in the event [outcome], or even had any Characters to loose, being English or Irish men, perhaps little further known than what themselves declared of their Abilities or integrity."[35] Pollard commended his Manchester cotton carder to Nicholson not only for his technical skill but also because he was "a very sober frugal Man."[36] Possibly the prohibitory laws were filtering only the less competent and less scrupulous artisans across the Atlantic at this time.

As in England, the first Arkwright spinning mills in the United States were operated by child and family labor. Marshall and Pearce at Paterson recommended the apprenticeship system to effect a saving in the wage bill. The state refused to give legislative permission for this step, but the company nevertheless utilized an unknown number of children. In Rhode Island, Slater also tried to introduce the apprenticeship system but discovered that the children would not tolerate English-type conditions. His child workers, consisting of seven boys and two girls in December 1790, were subsequently supplemented by the offspring of families who were persuaded to migrate to Pawtucket from the countryside. To tap this source of labor, mills in the early 1790s were established farther away from the coast in the country, nearer the farm population. Eventually whole families were brought and housed in the vicinity of the Rhode Island mills.[37]

Internal diffusion at a technological level meant not only replicating a manufacturing system but also passing the crucial new skills from transatlantic carriers to the indigenous population. While individual mechanics, like Pearce, who moved between New Jersey and Delaware, carried the new technology out from the first mills, the most influential agent of transfer at this stage was Samuel Slater. Partly through his wife's family and Quaker connection, partly through his own longevity in the industry, partly through immigrant contacts, Slater acted as the node from which Arkwright technology spread through New England and northern New York and even to Philadelphia.

Despite some secretiveness on Slater's part, his marriage into an American blacksmith's family, on whom he was dependent for machine-making skills, must have obligated him to share knowledge with his brothers-in-law. The new cotton mill management skills were also communicated, whether Slater liked it or not, to those American employees who worked in his mills and were interested and intelligent enough to learn by observation and imitation.

Formal diffusion from the Almy, Brown, and Slater mill in the Blackstone valley began in 1799. That year Almy and Brown bought an interest in the spinning mill set up at Warwick on the Pawtuxet River in 1794. The Warwick mill in 1799 had four spinning frames of sixty spindles each, along with the supportive cards, drawing frames, and roving frames. Whether the promoters of the Warwick spinning mill derived their original technology from Slater is unknown. The founding partners included James McKerries, a Scottish weaver who arrived in Providence around 1789 to work with Joseph Alexander, and John Allen, a Smithfield, Rhode Island, millwright. To be sure, when Almy and Brown entered the company in 1799, the Warwick mill mechanic, John Allen, was sent to the Almy, Brown, and Slater mill (built in 1793) to obtain details of Slater's equipment. Slater, who had no interest in the Warwick concern, was furious and reportedly threatened to throw Allen out of the window.[38]

But six months before his Arkwright technology was carried over to the Pawtuxet valley, Slater planned a mill of his own in the Blackstone valley.

In this, his father-in-law, Orziel Wilkinson, now a prominent ironmaster, and the husbands of his sisters-in-law, William Wilkinson (a blacksmith) and Timothy Green (a tanner), took ownership of half the land and capital. Under the firm name of Samuel Slater and Company, the White Mill at Rehoboth, Massachusetts (which became Pawtucket, Rhode Island, by a boundary change in 1861), remained in Slater's hands until 1810, when he sold his interest to his partners and his three brothers-in-law, Abraham, Isaac, and David Wilkinson.[39]

The kin network, reinforced by its Quaker connections, not only provided capital to spread Slater's cotton spinning technology but also supplied skilled machine makers and mill managers. Orziel Wilkinson, with his son David, had built a spindle-grinding machine in the late 1780s, and during the early 1790s Orziel expanded from blacksmith to ironmaster. In 1791 he set up a reverberatory furnace for wrought-iron production at Pawtucket and added a rolling and slitting mill three years later. Some of the earliest American wing gudgeons, for shafting in Slater's first factory, the old fulling mill, were cast in the Wilkinson furnace. David Wilkinson in 1794 perfected his screw cutting lathe for large industrial screws, such as were utilized in cloth presses in the cotton and woolen industries, and six years later he founded his own machine-making company to serve the southern New England textile industry.[40]

Apart from his wife's two brothers-in-law, several other relatives of Slater became capitalist managers in the textile industry. One was his own brother John, who came over from Derbyshire in 1803, bringing with him a knowledge of the latest cotton manufacturing technology, including mule spinning. Two years later, with the assistance of William Almy and Obadiah Brown, Samuel's original partners, the Slater brothers established a cotton spinning mill in northern Rhode Island on the southern branch of the Blackstone River. Known as Slatersville, the concern was managed by John Slater until the late 1820s at least. Another relative who distinguished himself in mill management was Smith Wilkinson, the youngest of Slater's brothers-in-law. At the age of ten in 1791, he went to work for Samuel, tending the breaker card. Fifteen years later, he joined his father and brothers in setting up the Pomfret Manufacturing Company at Pomfret, Connecticut, and was appointed its agent. The company's survival of the business crises of 1816, 1829, and 1837 was one measure of its successful management under Smith Wilkinson.[41]

In the first decade of the nineteenth century, Slater's employees and contractors also took his technology farther afield. Benjamin Stuart Walcott, a housewright by trade who had built the first custom-made Almy, Brown and Slater mill at Pawtucket in 1792 and 1793, carried the new spinning technology first to his native Cumberland, Rhode Island, about seven miles up the Blackstone from Pawtucket, in 1801; then to adjacent Smithfield, Rhode Island, in 1804; and finally to Whitestown, near Utica in upstate New York, in 1808. Charles Robbins, a Slater employee, went to New Ipswich, New Hampshire, in 1804 and then to Fitchburg, Massachusetts, in 1806, in each place superintending the construction and early operation of a cotton factory. William Mowry, a farmer's son and another Slater employee, seized the chance of industrial independence at Greenwich, thirty miles north of Troy on the Hudson in upstate New

York, where in 1807 he set up a Slater-type spinning mill.[42] Probably the most successful Slater employee was Alfred Jenks, who in 1810 moved from Pawtucket, Rhode Island, to Holmesburg, close to Philadelphia, taking with him "drawings of every variety of Cotton Machinery, as far as it had then advanced in the line of improvement."[43] Within a decade, Jenks moved to Bridesburg, part of the city of Philadelphia, and thus commenced the Bridesburg Machine Works, which regionally dominated the manufacture of cotton and woolen capital equipment.

Other agents of technology diffusion apart from kin and employees are less easily traceable. Doubtless there were a number of capitalist merchant promoters in the Providence region whose early interest in manufacturing or Quaker connections gave them access to Slater's work through his partners, Almy and Brown. One was Daniel Anthony, whose interest in Arkwright spinning, dating back to 1788, flowered in 1805 when his two sons promoted the Coventry Manufacturing Company on the Pawtuxet River.[44]

Independent machine builders also took Slater's technology through the New England region, if not farther afield. One was Perez Peck, the son of a farmer. After four years' apprenticeship to his brother Cromwell, a Providence, Rhode Island, blacksmith, he was hired along with his brother by Richard and William Anthony to build the machinery for the Coventry Manufacturing Company. Hitherto the Peck brothers had constructed no new cotton spinning machinery. Perez therefore importuned the superintendent of the Warwick Manufacturing Company, which had been controlled by Almy and Brown since 1801, to admit him to inspect the mill. Years later Peck recalled that he was closely watched on his hour-long tour but returned home "with what I had seen indelibly imprinted on my memory."[45] Apparently the Pecks then proceeded to build cards, drawing and roving frames, and two water frames of seventy-two spindles each. Either Perez Peck had a photographic memory, or else his story improved with age. Others who trained in the Providence mills were Seth Nason and Jesse Holton, who helped Samuel Batchelder build his second cotton mill at New Ipswich, New Hampshire, in 1808.[46]

Finally, Slater's operations diffused British cotton technology in America by channeling immigrants into the industry. Although these immigrants were few in number and did not always meet Slater's encouragement, at least two of them significantly contributed to the development of American textile manufacturing. Abraham Marland, the first, arrived with some capital in Boston in 1801 and immediately went to Pawtucket. Slater counseled him to invest his savings in land rather than manufacturing, but Marland, who came from the Yorkshire woolen industry, clung to his purpose. After two years as an employee at the Beverly cotton factory, he set up on his own at Lynnfield, Massachusetts, but moved to nearby Andover in 1807. Before ill health forced him to switch back to woolen manufacturing in 1811, Marland had established on the Shawsheen River at Andover a mill out of which grew the Marland Manufacturing Company (1834–1879).[47]

Another immigrant associated with Slater was John Blackburn, a machine maker recruited by John Slater in 1803 to go to Pawtucket and build a slide lathe for turning rollers. According to David Wilkinson, Blackburn's machine was a failure. But Blackburn proved of value in building a mule for the Pawtucket mill in 1803–1804 and in 1805 two more for the Slatersville mill managed by John Slater. That same year Blackburn joined Luther Metcalf and five others in forming the Medway Cotton Manufactory, which went into operation in 1807 with 820 spindles.[48]

Slater's business operations formed the single most fruitful node of technology diffusion in American cotton manufacturing before 1812. But there were other channels of diffusion besides Slater and his kin, employees, associates, and contacts. After 1807 the Slater network of mills was outnumbered by new and independent companies. Even before 1807, other immigrants, unrelated to Slater, arrived in America during the lean decade prior to the embargo and acted as pulse points of technological transfusion. Two of the more notable of these were James Beaumont and Samuel Ogden.[49]

Diffusion by immigrants was not always a smooth, rational process. A cotton factory established in 1806 within thirty miles of Providence informed the federal authorities: "This establishment suffered much in the outset, in being put to much expense by English workmen, who pretended to much more knowledge in the business than they really possessed. At present only two are employed, and Americans, as apprentices, &c. are getting the art very fast."[50]

The diffusion of the basic Arkwright and mule spinning technology, together with improved manufacturing conditions after 1807, encouraged the spread of innovations developed in Britain after the early 1790s. A whipper, or variant of the batting frame, was built by the American machinists Hines and Cady in Rhode Island around 1807 or 1808. By 1812 the picker, derived from the woolen picker, was in use.[51] Double-speed roving frames were built by Daniel Large, the immigrant steam engine maker from Philadelphia, in 1807.[52] Ogden built throstles, variants of the water frame, in Rhode Island in winter 1807. Kelly's semiautomatic mule was imported by Beaumont in 1803–1804, when he set up a small spinning mill at Canton, Massachusetts. Its machinery was built by a Scottish immigrant, John Clark.[53] Beaumont allowed farmers' wives and children into his factory to see his 144-spindle mule at work and later recalled that his visitors "having espeyed the Mule would scream, 'do marm come here and look at this great big high wheel, that has everso many spindles drawing out at once and nobody to them.'"[54]

Mule spinning seems to have made much less headway than water frame spinning before 1810. The federal census estimated sixty-seven mills running at the beginning of that year, with a total of thirty-one thousand spindles in operation. Since the census also estimated that every eight hundred spindles gave employment to five men and thirty-five women and children, the predominance of Arkwright technology is inferred. Nevertheless mules gained ground during the War of 1812, especially in Rhode Island and states to the south.[55]

The two alternative techniques posed a difficult choice for American manufacturers. Mules had the great advantages over throstles of spinning a much wider range of finenesses (from the extremely coarse to the very fine) and of producing a much more uniform yarn (suitable later for cotton prints). Before the 1820s mules may have produced as much as 20 percent more yarn per day, spindle for spindle, than throstles. Soon after the War of 1812 American throstles reached higher speeds (and therefore higher productivity levels) than English throstles, and they rivaled mules for productivity.[56] An unchanging advantage for the semipowered mule was that it needed much less power than the throstle; one horsepower reportedly drove seven hundred to a thousand mule spindles, compared to only a hundred throstle spindles.[57] On the other hand, mules entailed much greater operative skill and therefore higher labor costs than throstles. Unfortunately, almost nothing is known about the bases on which American manufacturers in aggregate reached their decisions about mules and throstles. Certainly abundant water power and a scarcity of mule spinners would have predisposed them toward throstles; reversed circumstances would have pointed to mules.

5
The Diffusion of Cotton Power-Loom Weaving Technology, 1810–1820s

By 1812 the United States had a sturdy cotton manufacturing industry centered on Rhode Island and based on factory spinning and household hand loom weaving. Its capacity in late 1810 was variously estimated at eighty-seven thousand and one hundred thousand spindles in 87 or 330 mills, respectively. In both aggregate and unit size it was puny compared with Britain's cotton industry, which in 1811 had 1.25 million water frame and 4.6 million mule spindles, while several of the largest Manchester cotton mills possessed spindleages equivalent to America's aggregate capacity.[1] Nevertheless, in absolute terms America's cotton industry was growing rapidly.

The output of New England cloth is computed to have increased fifty-fold between 1805 and 1815, rising from 46,000 to 2,358,000 yards per annum.[2] Furthermore there was room for the industry to grow more than 60 percent simply by displacing household spindles. By 1815 America's cotton consumption was reported to be 27 million pounds (Britain's was 81 million pounds). Assuming that the U.S. factory spindleage in 1815 was close to its 1820 level (stagnation arising from the intervening depression) of 330,000 to 340,000 spindles, then some 10 million pounds was spun on household wheels or jennies.[3] And 10 million pounds of raw cotton might be spun on 200,000 factory spindles if the rate of 50 pounds per spindle per annum, true for one of Slater's mills in 1820, were typical of both spinning systems and all mills.[4] Such a calculation, of course, ignores the effects of demand changes made by a growing population and the supply changes of technical innovation, renewed foreign competition, and improving transportation networks after 1815.

Confronting the expansion of the American cotton spinning industry was the bottleneck in production posed by hand loom weaving. In 1818 a Massachusetts merchant operating the putting-out system had

his weaving . . . done by families of Farmers some of whom live 100 miles off. He pays 6cts. pr. yard for plaids &c. & 5 cents for shirtings for yarn 13 or 14. Payment made in yarn at the same rate mentioned above. He sends out a man with his yarn, who travels through Massachusetts & into New Hampshire, delivers a given quantity of yarn with each of the Customers & on his return takes up in his waggon the goods ready wove.[5]

The solution to the bottleneck was a complete and viable power loom weaving technology. In 1812 the British were on the way to solving many of the technical problems of power weaving. Warping and dressing were already mechanized and powered. Although work on the power loom had begun in the late 1780s with Cartwright's experiments, it was still being developed in 1813. Even so, according to later estimates, about twenty-four hundred were then in operation in Britain.[6] The primitive state of their technology was discovered by a German visitor to Manchester. In one weaving shed where a fourteen-horsepower steam engine drove 240 looms, he found one operative to every two looms. Besides this, half the looms were idle because of mechanical troubles, "yet the foreman . . . told us that the owners [Chadwick & Co.] were satisfied with their looms and had hopes of increasing their efficiency."[7]

Such was the situation when Americans tried by at least three routes to acquire knowledge of power loom technology. First, they tried to build their own power looms, independently of British efforts. The water loom of Philo Clinton Curtis of Oneida County in upstate New York was one of the earliest, patented in November 1810. Another American power loom inventor was Thomas Robinson Williams of Newport, Rhode Island, a watchmaker, who in his late teens constructed a tape loom in his father's garret (1809–1810). While apprenticed to the machine makers Cromwell Peck & Company, of Coventry, Rhode Island, Job Manchester designed a powered picking movement (1812). Four years later he invented a twill power loom.[8]

Second, they sought the assistance of immigrants who arrived, at least as early as 1811, bringing experience with water looms. John Murray, a Scot, set up water looms near Bloomfield, New Jersey, while John McDonald built others in Baltimore.[9] Thomas Siddall, a member of the Lancashire immigrant family, set up a power loom in Germantown near Philadelphia in 1813–1814, which was reported "like the looms in use in England & in Boston."[10] Knowledge of the improved English power loom, incorporating Horrocks's variable batten speed motion, was brought by William Gilmour, who emigrated in 1815 from Dunbarton, Scotland, to Boston and then to Rhode Island.[11] An alternative, less successful, power loom came with an Englishman named Fales who arrived in Rhode Island in 1818.[12] This influx shows the ineffectiveness of the British prohibitory laws in checking the outflow of technology because it was embodied in the artisan. Little is known about the returns these carriers received, though Gilmour was rewarded by the Providence manufacturers with a purse of $1,500.[13] Expectations of remuneration must have been raised by the growing American interest in power

loom weaving. This was registered partly in the rising number of patents for loom improvements: in 1808, one patent; in 1809, two patents; in 1810, three; in 1811, four; in 1812, eight; in 1813, eight; and in 1814, twelve.[14]

Third, American merchants visiting Britain on the eve of the War of 1812 observed the efforts to achieve power loom weaving and perceived the potential of this possibility for their own country. One was Nathan Appleton, who was in Britain early in 1812.[15] Another was Richard Ward of Paterson, who reportedly returned from Europe with plans for power loom weaving.[16] But the mainspring behind the most far-reaching developments in American power loom weaving was Francis Cabot Lowell, a Boston merchant and Harvard mathematics graduate who went to England in 1810, ostensibly for health reasons.[17]

There is no evidence that industrial espionage was in Lowell's mind when he left Boston. In only two of the extant letters written during his stay in Britain from June 1810 through June 1812 did Lowell mention factories. However, European war and deteriorating Anglo-American relations were damaging his trade investments and prospects, and his visit to Britain opened him to the possibilities of textile manufacturing. In spring 1811, he was shown the Glasgow factories by a William Smith. The following February he told Appleton of his in-

5.1 One of the first American power looms: the loom patented in 1810 by Philo Clinton Curtis of Oneida County, New York. Notable features include its adherence to the hand-loom design and the crudeness of its mechanical motions, as in the levers and ropes. (U.S. patent drawing 1391, courtesy National Archives.)

tention to visit Manchester to obtain "all possible information" on cotton manufacturing "with a view to the introduction of the improved manufacture in the United States."[18]

As Gibb suggests, Lowell's status as a Boston merchant visiting England for health reasons would open doors normally closed to rival mechanics and manufacturers.[19] By close observation, Lowell learned enough to satisfy his trained mathematical and commercial mind that factory spinning and power loom weaving could be a profitable investment in America. Lowell certainly did not memorize everything he saw. He obtained a drawing of Radcliffe and Johnson's patent dressing frame.[20] And probably, like Bodmer, made careful notes in private.[21]

At this early stage in the development of the power loom, Lowell thus offered almost optimum reliability as a vehicle of technology transfer. His family background, financial involvement, and his keen mind honed on Harvard mathematics and mercantile risk taking, combined with a personal inspection of the state of the art in British factories, made his judgment almost impeccable. He lacked only engineering experience and practical knowledge of weaving, and these were to be supplied through a protégé of Jacob Perkins. Against this combination of New England capital and talent, isolated American mechanics or immigrant operatives offered much smaller chances of successfully introducing and completing the technological potential for power loom weaving.

In the establishment of prototype power weaving production lines, the general economic situation played an important part. The economic warfare between Britain and France, by which each penalized neutrals for trading with its enemy, led to the American Embargo (1807) and Non-Importation (1809) Acts, severely hurting American trade. Consequently New England merchants searched for alternative investment opportunities, a search intensified by the deterioration of Anglo-American relations in 1811 and war the following year.[22] Elsewhere too the break with Britain led to new investments in cotton manufacturing—for example, in northern Delaware and around Baltimore. Wartime protection from foreign competition encouraged experiments in power loom weaving. Firms like Lowell's Boston Manufacturing Company that managed to establish mill production of power loom woven goods before the war ended stood a better chance of surviving the subsequent revival of foreign competition than those who did not. Siddall's power loom failed in the experience of one manufacturing firm partly because its technical problems remained when the inundation of British goods began in 1815.[23]

Gilmour's "Scotch" loom was successfully promoted in 1816–1817 when the 1816 tariff promised protection to American cotton manufacturers. After 1816 imported cotton goods were subject to a 25 percent ad valorem duty and goods under twenty-five cents a yard were revalued to twenty-five cents and charged six and one-quarter cents a yard. Since coarse cottons were then twenty-five to thirty cents a yard, the tariff gave American manufacturers the chance to capture the country's coarse cotton goods markets by using the power loom to lower labor costs.[24]

The situation changed again after 1819. The financial crisis of that year hit the agricultural sector more severely than it did manufacturing, for credit was more overextended in the former. Raw cotton prices fell by over 35 percent.[25] Thereafter cloth prices fell too, removing the least efficient manufacturers and pressuring the more efficient ones to save costs and, at the same time, maintain profits by increasing the production of goods for which demand was greatest. This pointed to the use of capital-intensive equipment and organization and large-scale production, which were applied to power loom weaving in the two following decades.[26]

As with Arkwright spinning technology, the combination of a secure capital base and technical skill were essential prerequisites in establishing power loom weaving. But now the capital requirements were higher because the establishment of power loom technology was no longer a matter of simple duplication. Since British power looms were still being developed, American entrepreneurs moving into factory power loom weaving effectively shared the burden of development costs.

The firm of Duplanty, McCall & Company, cotton spinners on the Brandywine who attempted to use Siddall's power loom, made the mistake of both undercapitalizing their weaving operation and underestimating development expenses. They started in June 1814 with one of Siddall's looms, which cost $101 (including $25 for patent rights), intending to use it to complete their government contract to supply the army with fifty thousand yards of twilled cotton drilling. By early December 1814 their efforts with and without Siddall's help to improve its performance and to build others were dashed by the influx of imports.[27] In contrast, Lowell's Boston Manufacturing Company (BMC) was capitalized at $100,000 with the object of developing an integrated and powered spinning and weaving production line. By 1820 the BMC's capital investment was increased to $400,000. In aggregate, the switch to power loom weaving meant, on average, a tripling of the investment levels in cotton spinning and a 34 percent increase over investment levels in spinning-and-handloom-weaving concerns (table 5.1).

Apart from a knowledge of preexisting power loom technology, the most crucial labor input of the early research and development stage of power loom weaving was a combination of the skills of machine builder and textile worker. This combination was present in both major breakthroughs in power loom weaving: those of the Boston Manufacturing Company at Waltham, Massachusetts, and Gilmour at Providence. Paul Moody, superintendent of the Waltham mill's machine shop, had the unique experience of training as a weaver with the Scholfield brothers, immigrant woolen manufacturers from Yorkshire, in the 1790s, and later learning his trade as machine maker with Jacob Perkins, the most famous of Yankee mechanics.[28] The reliance on immigrants for knowledge of the power loom was very slight at Waltham.[29]

At the Lyman Mills in Providence in 1816–1817, Gilmour needed the help of a practical weaver. Gilmour, a Scotsman, designed the parts of a warper, dresser, and loom on an empty mill room floor. Wooden patterns were then made, the parts were cast in iron and brass, and finally Gilmour assembled them. But he was

a thorough machinist, entirely unacquainted with the practical operation. . . . The warper worked badly, the dresser worse, and the loom would not run at all. In this dilemma an intelligent though intemperate Englishman, by trade a hand weaver,

5.2 The fountainhead of the Waltham system of manufacturing: the Boston Manufacturing Company mills on the Charles River at Waltham, Massachusetts, ca. 1830. (Courtesy Lowell Historical Society.)

The first mill (to the left in the painting), built 1813–1814, measured 90 feet long by 45 feet wide and was four stories high above the basement. Its bell cupola and clerestory monitor echoed features in earlier Rhode Island mills though its brick construction marked a new departure.

The same height as the first mill, the second one, built 1816–1818, measured 150 feet long by 40 feet wide and incorporated the novel addition of an external but adjoining stair tower. The 3,584 spindles in this mill became the unit of measurement of one mill power in northern New England.

The separate machine shop constructed in 1816–1818 is the building on the river's edge in front of the first mill.

5.1. Average Capital of Three Types of American Cotton Manufacturing Firms, 1820

State	Cotton Spinning Firms		Cotton Spinning-and Weaving Firms		Cotton Spinning-and Power Loom Weaving Firms	
	Average Capital	No.	Average Capital	No.	Average Capital	No.
Maine	$37,000	4	$ 22,000	1	$ 22,000	1
Vermont	3,375	4	13,886	7	19,000	3
New Hampshire	16,471	17	22,886	14	37,040	5
Massachusetts	18,650	14	29,128	44	63,873	11
Rhode Island	13,555	24	29,788	45	48,133	15
Connecticut	18,647	17	39,015	24	48,279	11
New York						
Northern district	25,000	2	42,498	28	59,374	11
Southern district	45,750	4	86,571	7	105,750	4
New Jersey	42,000	5	33,000	9	42,500	4
Pennsylvania						
Western district	15,667	3	6,000	1		
Eastern district	10,669	13	44,064	11	38,333	3
Delaware	37,833	6	67,667	3	44,000	1
Maryland	32,600	6	228,000	5	260,000	4
South	19,000	2				
West	6,523	32	43,068	5	76,669	2
Overall	17,552	153	39,902	204	63,657.6	75

Source: USNA, Census of Manufactures, 1820, MS returns.

came to see the machinery. After observing the miserable operation, he said the fault was not in the machinery, and thought he could make it work; he was employed. Discouragement ceased; it was an experiment no longer. Manufacturers came from all directions to see the wonder. To this day [1861] the same loom, with trivial alterations, is in use in all our mills.[30]

The other labor elements demanded by power loom weaving were managers and operatives. The former were supplied by the preexisting Arkwright cotton mills in Rhode Island. In northern New England the Boston merchant aristocracy, skilled in the management of finances and familiar with trading patterns, at first supplied its own managers (agents): Patrick Tracy Jackson (Lowell's brother-in-law) for the Boston Manufacturing Company and Kirk Boott for the Merrimack Manufacturing Company and the Locks and Canals. Samuel Batchelder for the Hamilton Manufacturing Company was one of a few outsiders before 1830; he began with his own small mill in New Hampshire.[31]

The supply of unskilled operatives was solved in two different ways. In northern New England, Lowell and his successors employed teenage girls, housed in boardinghouses under the eye of respectable matrons. Farther south, the traditional Rhode Island pattern of family labor, concentrated in mill villages, prevailed.[32]

In developing the power loom, success at the technical level came to those choosing the lowest targets for the technology. Overambition delayed and even extinguished efforts to bring a power loom into production. Thus, Duplanty, McCall & Company, being fine spinners, expected Siddall's power loom to weave their fine yarns. But despite the inventor's claims, Siddall's loom would not weave number 30 warps. Siddall therefore lowered his sights and asked for a warp of "No. 26 in the bundle" to try on his prototype looms at Frankford near Philadelphia. But eight months later, in March 1816, the Brandywine company's Philadelphia agent learned that it could not weave this either.[33]

In contrast, Lowell and Moody planned their integrated production line to spin coarse yarns (number 14) for weaving into heavy sheeting, thirty-seven inches wide, forty-four picks per inch, and weighing about three yards to the pound. Coarse cloth like this was easier to make and to sell. Although the latter consideration was a paramount one, the former was also important because of the type of loom that the BMC adopted.[34]

Lowell designed a simple, crude power loom, utilizing cams for the major movements. Appleton and Batchelder recalled a cam drive to the batten that beat up the weft into the fell of the cloth. More important, Job Manchester remembered the reason for the BMC's preference for the cam and its constant batten speed: the early Waltham loom had no warp protector stop motion. Consequently the batten was not positively driven, and this allowed slippage on belts when the shuttle jammed in the warp shed. Instead of being smashed out of the warp, the jammed shuttle halted the batten.[35] Such a crude loom, clearly derived from Miller's British patent "wiper loom" of 1796, was easier to develop than one with a variable batten speed (which Horrocks introduced in Britain during the War of 1812, while Lowell was developing his machine). It did, however, demand the use of strong warps to withstand the stresses and strains imposed by the loom's movements and especially the force of the batten when the shuttle jammed in the warp.

In order to operate economically, the power loom needed several other pieces of manufacturing equipment, in particular a winding machine to prepare weft bobbins for the shuttle and warping and dressing machines to prepare the warp. Without these, the power loom's speed would become dependent on bobbin winding by a variant of the hand spindle wheel and on manual sizing of the warp, for which the loom had to be stopped.

Americans seem to have developed their own bobbin winders, which would not have been difficult since no more than variants of a water frame were needed. The Waltham company first employed a device invented by Stowell of Worcester, Massachusetts. John Thorp and Silas Sheppard of Taunton, Massachusetts, also patented a winder (1816).[36]

Warping and dressing frames were more vital accompaniments for the power loom. Thomas Siddall patented a warping machine, presumably to serve his power loom, on March 27, 1815, though neither specification nor drawing has survived. Gilmour built all three machines together at the Lyman Mills in Providence in 1816–1817. The BMC's warp preparation technology also derived from British models. In his dressing frame patent of January 17, 1818, Paul Moody stated that his machine "resembles in many particulars one intended for the same object, said to have been invented in England by Thomas Johnson and for which he obtained a patent . . . [dated] 2 June 1804." This was Radcliffe and Johnson's improved dresser in which size-spreading brushes were crank driven. Moody further admitted in his 1818 patent, "I have never seen a machine made according to the said Johnson's patent, but I have seen the specification

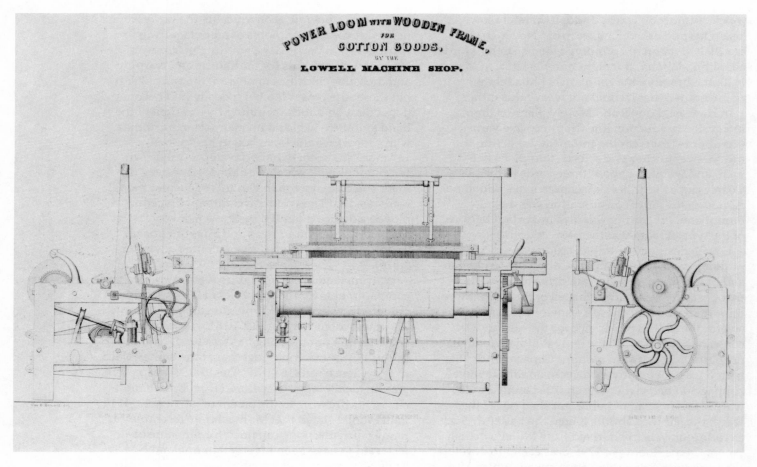

5.3 One of the early wooden-framed power looms built
in Lowell, an illustration first published in 1848 by the
Lowell Machine Shop. While this machine may have
conformed in some respects to the design of the original
Waltham loom developed between 1813 and 1815, it does
not seem to incorporate the distinctive belt-driven batten
of the first Lowell and Moody power loom. (Courtesy
Baker Library, Harvard University.)

and drawings accompanying his patent, from which I constructed a machine resembling [Johnson's]."[37]

By 1820 the pilot-firm stage in the establishment of power loom weaving was almost over. Even in 1815, it was later claimed, the BMC's vertically integrated factory completely processed a piece of cotton fabric at half the price of merely weaving it on a hand loom.[38] The profitability of Lowell's experiment became public knowledge on October 7, 1817, a month after the founder's death, when the BMC declared a dividend of 17 percent. This level of return was maintained. Between 1817 and 1826, the company's annual dividends averaged a handsome 18.75 percent, a level rarely reached again in the nineteenth century by other New England firms.[39]

This establishment stage in the transfer of technology clearly shows that Americans now began to choose among British technological alternatives. Although many, especially in Rhode Island, tended to follow British patterns, as Slater's lead directed, the Boston capitalists demonstrated a surprising degree of rationality in declining to follow the British up-market. Despite the strong temptation to opt for mule spinning and the power loom weaving of fine goods, especially during the War of 1812, Lowell and his associates perceived its dangers.

Their selection of simpler alternatives from the range of available British technology was not confined to the power loom. Moody solved the mechanical problems vexing the roving frame by opting for uniformity of product rather than the versatility sought by British mule spinners. Patent litigation showed that Moody independently arrived at his four-cone regulatory system after being informed by a Lancashire immigrant of the British double-cone mechanism.[40]

The crucial feature of Moody's speeder, as Batchelder observed years later, was that it "was attended with great cost for every alteration of the size of roving for finer or coarser work."[41] For this reason the Waltham system mills "were generally adapted to the manufacture of a single article of sheeting or drilling, of the same number of yarn, without any means of changing from one fabric to another, according to the wants of the market."[42] Moody's speeder therefore predisposed Waltham-type mills toward manufacturing uniform yarn in long production runs.

In short, the simpler technology Lowell and Moody selected for the BMC was designed for the manufacture of long runs of standardized coarse goods. Aiming to break into the bottom of the market and displace hand loom weavers, this technology would secure a strong industrial base for American manufactures.

Despite the setback of industrial depression in 1815–1816, power loom technology spread quickly. By 1820 at least 1,665 power looms had been built, most in New York and New England. (See appendix D.) In the lower Hudson valley in spring 1819, the agent of the Matteawan Company at Fishkill, Dutchess County, told a Brandywine mill manager, "We have also the Scotch looms in good condition." He rated their efficiency highly and recommended them to the Delaware company: "They require very little repair. I have no doubt they will be found indispensable to every Cotton Factory in this Country."[43] The allusion to "little repair" was possibly related to mechanical aspects of the Gilmour power looms. More likely it meant that the Gilmour loom,

5.4 Detail of view of Lowell, drawn by E. A. Farrar in
1834. To the left, along the Merrimack Canal (above and
south of the Merrimack River) stands a line of boardinghouses built for the girls employed by the Merrimack
Manufacturing Company. To the right of the boardinghouses are the five mills of the Merrimack Company,
distinguished by their clerestory lights. (Courtesy Merrimack Valley Textile Museum.)

with its Horrocks variable batten speed, caused
fewer warp breakages than did the Waltham loom.
The following year the Matteawan factory ran 44
power looms that produced forty-three hundred
yards of shirting and sheeting per week.[44] Lower
down and on the opposite, west, side of the Hudson, the Ramapo Works, originally a nail and
metalworking establishment, at Hempstead in
Rockland County, had 73 power looms in 1820
making shirtings and sheetings and also stripes and
plaids. The timebooks of the Ramapo Works show
that it ventured into power loom weaving when it
took up cotton manufacturing. Loom turnings and
castings were being made and looms assembled as
early as the period July 1817–May 1818, and parts
for a dressing frame, like rollers and size rollers,
were being built over the same months. Fifteen
looms and two dressing frames were ready by June
1818.[45] Clearly the preexisting industrial structure,
especially networks of manufacturers, merchants,
and mechanics, facilitated this rapid diffusion.[46]

Gilmour's iron-frame loom prevailed in states
outside northern New England for several reasons.
First, its incorporation of Horrock's variable batten
speed imparted a technical capability for weaving
finer warps than its rivals could handle.[47] Second,
its cost was low. In 1817 it cost $70, compared with
the Waltham loom, which then cost $90 and sold at
$125. By the mid-1820s, Gilmour's loom was being
built by independent machinists for $16.50 per
loom, on top of which there must have been the
cost of materials.[48] Third, no protective devices hindered the spread of Gilmour's loom. It was not
patented, either by Gilmour or his employer, Judge
Daniel Lyman, and its existence and advantages
were not shielded by secrecy. Quite the reverse:

Lyman allowed Gilmour to sell, for a trifling $10, the rights to his drawings to David Wilkinson, the Pawtucket machinist and brother-in-law of Samuel Slater (who, incidentally, resisted the use of power loom weaving until 1823).[49]

Wilkinson's involvement was a fourth factor in the successful diffusion of Gilmour's loom. Trained in the construction of textile and industrial machinery, Wilkinson pioneered the use of new industrial machine tools. His machine shop spread the new power loom technology in three ways. Wilkinson himself made valuable incremental improvements to the loom, including construction with heavier parts. Then, as he recalled later, before 1829 his shop built machinery for cotton factories all over New England and as far away as Trenton, Baltimore, Pittsburgh, and even Georgia and Louisiana. An unknown proportion was power loom equipment. Third, his shop-trained mechanics, like Jonathan Thayer Lincoln, set up their own shops in the 1820s and 1830s.[50] As the loom spread, so new incremental improvements were made.

The new technology of the Waltham power loom was channeled into northern New England by a rather different method. Here the Boston merchants in the 1820s launched a number of land and power promotional corporations that modeled their vertically integrated factories on the Waltham company. Such corporations offered the reliability of their directors' mercantile reputations and of their agents' well-tried manufacturing and engineering experience, thus becoming the new vehicles of technology diffusion. At Lowell the ten great textile manufacturing corporations launched in the 1820s and 1830s shared a common source of technological information, skill, and machine-making resources after

1825. The Proprietors of the Locks and Canals on Merrimack River, another corporation, then took over machine-building operations and became one of the earliest capital goods firms in the United States to diffuse technology among industries as well as firms in one industry.[51] The increase in numbers of power looms during the 1820s testified to the effectiveness of diffusion: from under 2,000 known power looms in 1820, the American total rose to a high proportion of the 33,506 cotton looms reported in 1831.[52]

6
The Diffusion of New Calico Printing Technology, 1809–1820s

Calico printing with cylinder machines was extraordinarily labor-saving. The cylinder machine employing one man and one boy printed and dried one piece of cloth in two minutes, equivalent to the production rate of one hundred hand block printers and one hundred tear-boys.[1] Employed in Britain since the 1780s, this technology reached America in 1809. Joseph Siddall, evidently an earlier immigrant, went to England that year to recruit workers and obtain equipment. The brothers Issachar and James Thorp arrived in Philadelphia in October 1809 and became established calico printers in Bristol township six miles north of the city. By 1820 the Thorps and their partners the Siddalls had fixed capital valued at $8,000. Their capital equipment consisted of a copper cylinder printing machine, a number of engraved copper rollers, a surface printing machine, two hot calenders, a dash wheel, and squeezers. But in 1820 this capacity was underemployed, with six men and a wage bill of $2,000 per year. If demand picked up, the partners reckoned they could print twice the quantity of 50,000 pieces or 1.4 million yards of calico that they printed in 1812. Why the company failed to expand is not wholly clear. Most likely the postwar depression and the return of foreign competition were prime factors. Its reliance on a Brandywine cotton mill and power loom factory, in which the Thorps were partners with the Siddalls and Hodgsons (also immigrant families) must have been another factor. Just as the spinning and weaving operations ran into difficulties during the depression, print profit margins were probably also cut.[2]

After 1817 the Philadelphia calico printing and cotton finishing industry received a fresh fillip with the arrival of Jeremiah Horrocks from Manchester. Other firms followed, and by 1832 there were a dozen cotton finishing establishments in the northeast part of Philadelphia county.[3] Since the Philadelphia transfer remained a diffusion cul-de-sac until the 1820s, this investigation turns to the New England firms, which exploited the new technology with much greater vigor and commercial success.

The growing demand for textiles, fueled by population growth and western settlement, assured profitability for power loom weaving. The enormous profits made by the Boston Manufacturing Company spurred its Boston promoters to switch more of their capital from commerce to industry. Using the spinning and weaving technology developed at Waltham, they started by developing the land and water-power site at Lowell on the Merrimack River in Massachusetts with a new company, the Merrimack Manufacturing Company (chartered in 1822 with a capitalization of $600,000). In making this move and forming a model for succeeding promotions, the Boston capitalists changed the Waltham pattern in several ways. Besides increasing the scale of manufacturing operations and placing the machine-making function in a separate company (the Locks and Canals), they made the important decision to add bleaching and calico printing to their manufacturing operations.[4]

The reasons for this step have not been fully explored. It is plain, however, that calico printing offered several advantages to both parent and offshoot companies. A new product eliminated competition between the two firms, an important consideration for interlocking directorates; it increased the value added by manufacturing; and, as Ware observes, it extended the range of textile

products without necessitating complex technical changes at the spinning or weaving levels of production. In this way it reduced the need for a roving frame capable of weaving a range of high warp numbers (though the Merrimack Company made finer yarns than did its parent). Finally it promised some relief from the effects of the falling price of unfinished ("grey" or "brown") cotton goods, which dropped by nearly 50 percent between 1816 and 1826.[5]

Of the three early stages in the transfer of calico printing technology, the first, the introduction of the necessary technological skill and equipment, seems to have caused most difficulty. For the Merrimack Manufacturing Company at Lowell, it lasted from 1822 until 1826. For the Dover Manufacturing Company at Dover, New Hampshire, it spanned three years—from spring 1825 when calico printing was chosen as a manufacturing operation until summer 1827 when the last batch of English equipment and recruits reached Dover. In the experience of these companies and of the Taunton Manufacturing Company of Taunton, Massachusetts, the skilled-labor shortage and the unreliability of available methods of transmission compelled them to send members of their managements to England.

But there was another reason why the first transfer stage was the most difficult to organize and mobilize: the new nature of the technology involved. Cotton finishing necessitated some acquaintance with the unfamiliar subjects of industrial chemistry and commercial art and design, as much as with mechanical technology and economic information about that technology. This complexity compelled the recruitment of teams of specialists and highly experienced managers to coordinate them. And because dyeing had not yet made the transition from an art to a science, the coloring aspects of calico printing were practiced by traditional craftsmen whose knowledge of the partially understood chemical processes came from the accumulated empirical discoveries of previous generations.

The Taunton and Merrimack companies led the way in the race to acquire the new technology. In 1823–1824 they used the resources of skill nearest at hand. The latter relied on its parent company to supply their first printery superintendent. But Allan Pollock, who trained with Ebenezer Hobbs, the founder of the BMC's bleachery in 1819–1820, was insufficiently familiar with calico printing, although he visited England in 1821–1822 and there bought a loom.[6] Within his first year at Lowell, Pollock's replacement was considered. At this stage, either in 1823 or 1824, the Merrimack Company hired an engraver, Aaron Peasley, who was believed to be familiar with the latest improvements and was most likely an immigrant.[7]

The Taunton Company meanwhile secured an immigrant calico printer, John Thorp, as their first print superintendent. Whether he was related to the Thorps from Lancashire who, with the Siddalls, established the Germantown calico printworks near Philadelphia in 1809 is unknown. Thorp and his son claimed the ability to construct printing machines, engraving machines, and engraved rollers. However, they fell out with the Taunton directors and by summer 1825 left the company.[8] Technical incompetence may have contributed to their parting, or, more likely, their English attitudes of secrecy and superiority caused alienation. A manager in the Dover company reported the Thorps'

opinion of themselves "that they are the only men in this country or in England who can put them [the Dover printwork plans] right," and described the elder Thorp as "wily and close."[9]

Though a year behind its rivals, the Dover company also turned to the means of transferring the technology closest at hand.[10] In spring 1825 they sent an agent to the rival American finishing establishments. The agent, Arthur Livermore Porter, professor of chemistry and pharmacy at the University of Vermont and a graduate of Dartmouth College and Edinburgh University, aimed to educate himself and possibly to recruit workers.[11] He interviewed the Thorps and in November 1825 pirated Aaron Peasley from Lowell. Besides Lowell, Taunton, and Boston (for the Charlestown Bleachery), Porter went to Philadelphia and Baltimore.[12] At Philadelphia he contacted a blockmaker and printer, John Schaffer, whose ability to achieve cylinder calico printing was in doubt, but the Dover company envisaged beginning with block printing. Negotiations with Schaffer recorded the company's plan to send him to Germantown to "see cylinder printing which may be very instructive."[13] However, Schaffer wanted an annual salary of $800 for his services.[14]

By summer 1825 the labor supply problem was seriously jeopardizing the three companies' efforts to introduce the new technology. The short supply of competent calico printers familiar with the management and operation of the latest techniques pushed up the price of available workers, encouraged the less competent to offer their services, and led to intense labor piracy. The agent of the Dover company reported the situation to his company treasurer in Boston in a very revealing letter:

6.1 View of the Lower Bridge and Factories of the Dover
Manufacturing Company, on the Cocheco River, Dover,
New Hampshire; lithograph based on original by
Thomas Edward, 1830. (Courtesy Boston Athenaeum.)

6.1 Waltham-System Cotton Goods Prices

Year	Price per Yard of Waltham Sheeting[a]	Average Price per Yard of Merrimack Prints[b]
1816	$0.30	
1819	.21	
1820		
1825		$0.2307
1826	.13	
1829	.085	
1830		0.1636
1835		0.1604
1840		0.1209
1843	.065	

Source: Appleton, Introduction, pp. 16, 34.
a. Boston Manufacturing Company, Waltham, Massachusetts.
b. Merrimack Manufacturing Company, Lowell, Massachusetts

I hope you will get the refusal of Mr. Thorp's services if he is a printer. I apprehend however you will not be able to get him even if you wish to ever so much as I have recently understood Taunton were much in want of Printers.

I have written Mr. Schaffer of Philadelphia again. He will come as a printer & block cutter for $800 to 1000 pr annum. We should begin to print by the 1st Sept'r had we work people & blocks. I have no doubt we shall use cylinders soon, must however, begin with blocks.

Are we not as much at liberty to engage Mr. Peasley as his time has expired with the Chelmsford company . . . [incoherently complains of Lowell company's conduct]. Mr. Fiske before he went from here [said] that he had the offer of a room if he would return to Chelmsford, had been discharged from them by Mr. Worthen their general overseer. Not a month passes but we are broken in upon in this way. New Market, Brooks & others, Great Falls & others, Nashua &c. Mr. Chase I fear will be bid out of our hands; why should we stand with our arms folded & let all other establish'ts lead us? It is quite certain that we shall not for many years find a man of first rate genius to lead in this or any other important business suing for a place in our Establish't for want of other good employ.

If we obtain a man of Mr. Peasley's ingenuity we must purchase him at a dear rate, but what is that to getting two or three years' start in this business? He is probably well acquainted with every improvement in the art of cylinder printing & arrangement of works &c. & if his engagements are at an end, why should we not compete for his future service? So far as I am acquainted this is the order of the day in regard to others toward us.[15]

The labor supply problem compelled the three companies to change their strategy. Instead of relying on skill available in America, they actively sought calico printing managers, machine makers (builders of printing and engraving machines and engravers), designers, and printers in Britain. In this they were assisted by the removal of restraints on free emigration (1824); now most respected and well-known men could leave Lancashire openly and in the knowledge that they were free to return. Furthermore the companies supplemented established recruiting networks, like the Merrimack's merchant connection and the Dover company's English recruiters, by sending members of the three companies' managements to England to select key workers and obtain machine models.

First to go was Charles Richmond, director of the Taunton company, in 1825. He took with him specifications, drawings, and possibly models of two machines used in his mills: Asa Arnold's differential and George Danforth's tube-roving frame. The latter he patented through an American contact in Manchester, Joseph Chesseborough Dyer. The differential he secretly communicated to Henry Houldsworth, Jr., a Manchester cotton spinner who had visited the United States in 1822–1823. Through these dealings Richmond returned with information, parts of a cylinder printing machine, and a Mr. Yates to superintend the Taunton printworks. Possibly this was T. K. Yates, a thirty-two-year-old calico printer who arrived at New York from Liverpool on October 17, 1825.[16]

Meanwhile the Merrimack Company opened negotiations with John Dynley Prince, who was spotted earlier in 1822 by Timothy Wiggin, a company stockholder and Boston merchant, then visiting England. According to his naturalization records, Prince was born in Leeds in 1779.[17] He was apprenticed to Thomas Hoyle, a prominent pioneer of chlorine bleaching who "created one of the

6.2 John Dynley Prince.

biggest dyeing and calico printing works in Manchester, at Mayfield, Ardwick" in the 1790s. Later Prince went into partnership with two elder brothers who removed to the Garrison Print Works in Derbyshire. The business collapsed at some unknown point because debtors defaulted; afterward Prince returned to Manchester where, as manager of Green & Neville's printworks, Wiggin met him sometime between 1822 and 1825.[18] Reportedly when Prince named a thousand pounds for his annual salary at Lowell, Wiggin gasped, "Why that is more than our Governor [of Massachusetts] receives." To which Prince, with unassailable logic, replied, "But can your Governor print?"[19]

Prince's terms seemed too high until 1825 when Pollock's difficulties were compounded by the loss of Peasley, who went on to build a printing machine at Dover, and by the Taunton company's English strategy, if the Merrimack company were aware of it. After hasty consultations with his directors, Kirk Boott, the Merrimack agent, sent to England for Prince in November 1825. Prince arrived at Lowell in June 1826. His successful recruitment prompted the Dover company to reach across the Atlantic and later led Boott himself to visit Britain for other workers and equipment.[20]

The Dover company was less fortunately placed than were its rivals. It had no inventions to trade in England, and its transatlantic recruitment network failed to secure competent and reliable operatives. One recruiter, Swan, sent out inappropriately skilled workers—including a mule spinner and hand loom weavers—in the mid-1820s.[21] Chadburne, another recruiter, apparently doublecrossed the company, but his "villany was fortunately discovered before we sustained through him

an irreparable injury."[22] Part of the villainy entailed sending out incompetent men like Gilbey, a foreman block cutter who "could not complete a block except in the most ordinary style, if he could do it at all" and was worth only five shillings a day.[23]

Eventually, like its rivals, the Dover management sent one of their own number to Britain; this appeared the most efficient and reliable way of securing the necessary technology.[24] Company letterbooks record the decision to send Porter to England (March 1826), the instructions given him for his mission (April 1826–July 1827), and intermittent notes on the workers and equipment he obtained. His prime target was to secure a printery superintendent

who has been thoroughly taught, has had great experience, and at the present moment stands exceeding high & is well known to the trade, and who has a perfect knowledge of every piece of machinery, its capacity and duty, as well as thorough knowledge of the highest classes of work produced at this present day in this or any other country and how to obtain the same for us in the most economical manner—and nothing short of a superior to Prince or Yates.[25]

Williams, the Dover agent, specified in great detail what else Porter should aim to do.[26] First, he was to compare the Dover bleachery (which had a capacity of five to six tons a week) with English bleacheries, noting the construction, dimensions, layout, and lighting of buildings; the cloth wringing machinery; the number, size, construction, and position of vats and boilers; the number of workers employed and how they were paid (by the piece or the day); and actual costs and prices charged per pound or yard for the various descriptions of goods.

Second, Porter was to investigate calico printing, beginning with dye rooms, where he was to record quantities of cloth daily dyed in two series of seven vats each, the amount of indigo required for each piece of cloth, methods of indigo grinding and of dipping and clearing the vats, and whether indigo colors could be printed as well by cylinder as by block printing. Williams's interest in indigo, the source of most blue pigment, arose from the high cost of the raw material, a point that he emphasized to Porter, asking him to note arrangements and dimensions of vats and wash wheels, number of workers employed, wages, and production costs and charges in the blue dye room and the madder (pink-brown pigments) house. Also the Dover agent was enjoined to discover the advisability of employing a man to take charge of color making, one "who perfectly understands the compositions of the various dyes . . . and discriminating with taste in the selection of figures; a good judge of cylinder work, the adjustment of machines and capable of applying a remedy."[27]

On the mechanical side of calico printing Porter was to obtain the rolls for the White engraving machine Peasley was making; fifteen to twenty engraved cylinders "of delicate and handsome figures" and a similar number not engraved (thirty-five inches long and four and a half inches in diameter); Spencer's circular lathe for engraving cylinders; a transferring machine "to transfer from one dye to another at once"; seventy-two steel and seventy-two brass doctors; a set of block cutters' tools; wire plates for shaping wires for printing blocks; and a set of blocks for shawl prints.[28]

Finishing was the area on which Porter was to concentrate, but Williams asked him to gather any information about the "economical manufacture of cotton for printing during any part of the process from the picker house to the cloth room."[29] Finally, if time allowed, Porter was expected to travel to France to familiarize himself with the most fashionable designs. On his return he would be regarded, Williams hoped, as "one thoroughly versed in the most delicate style of work whether applied to bleaching, dyeing, printing, finishing or packing the goods."[30]

Soon after his arrival in England, Porter received additional instructions. Williams and his machine makers (Peasley, Fiske, and Chase) requested a three-color printing machine and accessory apparatus, parts of a padding machine, a hydraulic press, a friction calender, and a glazing calender. Clearly the Dover company expected to obtain from Britain the most important elements of the new technology: key workers, critical machine parts, economic information about management and operation, and market information.

Porter made two mistakes in his machinery purchases in England. Unwittingly he bought a slide lathe, an American invention of which the Dover company already had a dozen examples in its machine shop, and he failed to purchase a printing machine auctioned by Liverpool customs after its seizure on the way to France. Otherwise he made purchases as instructed. Not much equipment arrived before October 1826. The following February Williams informed Porter that some items had reached New York but that a three-color printing machine and possibly a two-color one were needed. At the end of March 1827, Williams registered Porter's procurement of a rolling machine (for the nail-works), a sinking machine, an engraving machine, a drying machine, a slide lathe, a three-color printing machine, a clam machine, a hydraulic press, brass bowls for a mangle, thirty cylinders, and blocks, doctors, mandrels, and blanketing. He thereupon extended Porter's list to include two more printing machines, a five-bowl friction calender, and six more pieces of blanketing. He expected the Dover machine shop to be able to build printing machines but thought it expedient to have as many as three, besides Peasley's, to carry on the work meantime. Through spring and summer 1827, Porter's purchases arrived in Dover. In April the first of the English printing machines was set up. Compared with Peasley's, it was a "mammoth." Fiske, the company's machinist, reckoned that in New England the patterns alone would have cost £400–£500. By the end of June 1827, a large quantity of equipment, invoiced from Manchester before April 12, had arrived at Dover or New York. Much had been smuggled through British customs; for the years 1826–1827, Board of Trade records show that only eight steel rollers and one hundred copper rollers were exported under license to the United States while two copper engraving machines intended for New York and Havre were denied export licenses.[31]

Porter also secured a printery superintendent who appeared to promise well. Peter Bogle from Chorley, Lancashire, reached Boston in March 1827. Williams greeted him with enthusiasm: "Mr. Bogle's twenty years of experience with Dr. Porter's sagacity overlooking ought to afford stability and confidence in the business," he told the company treasurer.[32] Bogle's annual salary was around £300.[33] In addition, Porter hired J. F. Street, an engraver; Joseph Lawton, a designer; Thomas Lons-

dale, a machine printer; James Bamford, a madder dyer; Gilbey and McCormack, block cutters; and Robinson and Hough, block printers.[34]

The methods by which Porter and, earlier, Richmond, smuggled machinery out of Britain were discovered by Kirk Boott when he accompanied Prince to England in 1826–1827. Their objects were to hire a "first rate Engraver," to collect information about calico printing and manufacturing, to obtain an engraving machine, and to collect Prince's family.[35] Boott first sought an engraving machine. After examining a number, he chose one costing £500 made by John Worswick. By the end of January 1827, he also had a plan to ship the machine out of the country illegally. He described it in a letter to one of the Merrimack directors. It was "to get the cases, containing the machine, secretly conveyed from the maker's to a friend's warehouse in Manchester, who will immediately send them to another friend's in Birmingham, where I will have them repacked and then forwarded to Liverpool as 'hardware.' "[36] In February he wrote, "I have heard of a Jew-broker in Liverpool, who will ship and guarantee for fifteen or twenty per cent, and I am not sure I shall not conclude to pay this insurance."[37] By mid-February the engraving machine, dismantled and cased, was in Birmingham. Presumably it reached Lowell as safely as Porter's machine purchases arrived at Dover. Once more the British machinery export licensing system had little effect in halting the flow of equipment to America.

Boott arrived back in Boston on April 30, 1827. Sailing cabin class with him on the *Emerald* from Liverpool were Prince, traveling as a "gentleman," and his family (a thirty-four-year-old wife, six sons,

6.3 Rotary calico printing machinery, 1836. This American print from White, *Slater,* clearly derived from a British one published a year earlier in Baines, *Cotton Manufacture.* In both, the mill interiors, equipment, sex and posture of operatives, and print cloth designs are identical. The only difference is that British operatives are shown dressed in knee-length breeches while American operatives are given trousers, a symbol perhaps of American democracy. (Courtesy Merrimack Valley Textile Museum.)

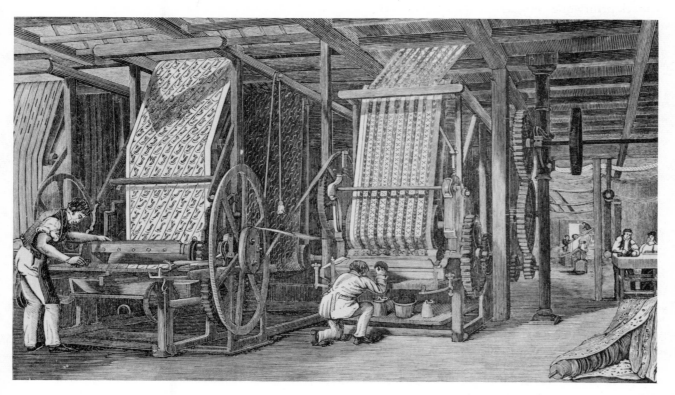

and two daughters); Richard Worswick (thirty), described as a turner but in fact an engraver, presumably from the Manchester engraving firm of this name, and his wife and son; Stephen Dickinson (forty), single, a block cutter; and Edward Payne (forty), a calico printer, and his wife. Among the steerage passengers were a pattern drawer (William Paul), three calico printers, and two engravers. In addition, Thomas Slater, who later entered the Merrimack printworks' engraving department, also traveled on the *Emerald* but went steerage and was described in the passenger list as a "letter press printer." [38]

While trusted agents became the major method by which the new calico printing technology was introduced to New England, a fresh influx of immigrant calico printers in 1826–1827 also supported the transfer. News of opportunities in New England coincided both with the freeing of emigration in 1824 and a 50 percent fall in Lancashire calico printers' wages between 1822 and 1825. [39]

At Dover one of the first of these unexpected arrivals in 1826 was Thomas Greenhalgh. Aged thirty-five, he arrived at Boston at the end of April with a forty-two-year-old wife and four children under the age of ten. [40] Evidently he went straight to Dover for he was there within five days of landing at Boston. Within a week the company agent sent him to southern Massachusetts to collect "the people who gave you the promise of their services before you left England." [41] Williams directed Greenhalgh to bring fellow workers from Fall River and Taunton before they were engaged and not to stop short of labor piracy and industrial espionage. A "good block cutter and a first rate designer" were to be secured, as was valuable economic and technical infor-

mation about rates of pay, capital equipment, and product specifications. [42] Why these workers were obliged to Greenhalgh was unclear; perhaps his Methodist connection linked them together, or perhaps they emigrated as a calico printing team. By December 1826 the Dover company had one designer, two cutters, six printers, one machine printer, and four other calico finishers, all from Thomson & Chippendale's Primrose Printworks near Clitheroe, Lancashire, from whence Greenhalgh presumably originated. Significantly, this factory was a leader in the Lancashire cotton industry; and one of its founders, James Thomson, had written articles for Rees's *Cyclopaedia*. [43]

Once a technological potential was acquired, prototype production lines were relatively easy to establish. The basic complementary industrial structures in factory spinning and weaving, with their resources of capital, mill management, labor organization, and machine shops, already existed. No research and development costs were entailed because the technology already existed in England. In financial and organizational terms, the addition of calico printing was no more than a grafting operation, at least for the Waltham system or larger mills in the favorable economic conditions of the 1820s and 1830s. At the levels of technology, marketing, and labor relations, however, a number of problems confronted managers engaged in adding calico printworks to their mills.

One fundamental technical problem related to the choice of yarn for print cloth. The British example in moving up-market with the production of finer yarns looked hard to resist since finer yarns, made on mules, permitted the manufacture of more delicate prints. When Kirk Boott was in England in

1827, he found mule spinning "universal" and heard "on all sides, that we shall never make cloth, with water twist, suitable for printing."[44] Yet the Merrimack and other New England mills succeeded in printing on cloth made from water- or throstle-spun yarn.

English manufacturers cannot have been doubtful about the capacity of throstles to spin middling yarns. Montgomery, the Scottish manufacturer, reported that throstles could spin as high as number 50, provided that they used better cotton.[45] So American success may have rested on their comparative advantage with regard to the supply of cotton. There is another possibility. Mule-spun yarns were more uniform than were throstle yarns, and uniformity of yarn was obviously important in printing; without it, exact and regular definition in repeat patterns, especially those with fine lines, was impossible. The agent of the Ware Manufacturing Company, of Ware, Massachusetts, told his Boston selling agents in 1829 that it was not in the interest of the company "to attempt the printing cloth. At present our machinery is not in order to spin No. 35. We are at this time to work improving the machinery and hope by and by to make even yarn. Then it may be for our interest to attempt the printing cloth."[46] Manufacturers at Lowell improved the uniformity of their throstle-spun yarns by using stretchers, presumably of the throstle type. Both the Merrimack and Hamilton companies had stretchers by 1830. Whether it was because of better cotton, or the use of stretchers, or a finishing technique prior to printing, New England cotton manufacturers succeeded at a technical level where their English rivals predicted failure.[47]

The Waltham-system printers' preference for throstles was largely dictated by the American labor situation, and strengthened by improvement in throstle spindle productivity. John Williams, agent of the Dover Company, noted in 1827 that English mule technology meant "dependence on a set of foreign mule spinners till we can instruct others."[48] And immigrant mule spinners were notoriously militant before the perfection and eventual introduction of the self-actor.[49]

Also, at the technical level, there arose some confusion, but not from the intermingling of immigrant workers, for all came from Lancashire (as far as can be ascertained) and the cotton industry was technically more homogeneous than was the woolen. Instead differences between British and American dyestuffs terminologies flummoxed and misled American managers. This problem surfaces occasionally in agents' letterbooks; for example, Greenhalgh, printery overseer, was asked whether there were other names for peachwood (he replied that camwood and peachwood were the same).[50] Differences in nomenclature are more evident in the workbooks of Samuel Dunster, one of the American apprentices hired in the late 1820s.[51]

The numerous machine-level technical problems, particularly relating to the chemical quality control aspects of calico printing, have been considered elsewhere; so too have the design problems Americans encountered.[52] Here the difficulty lay in originating and executing salable designs. Immigrant Englishmen were one source of new designs. The European visits of Richmond, Porter, and Boott must also have brought fresh ideas. And imported cylinder shells, already engraved, were another.

Once the New England pilot firms acquired the technical capacity previously possessed only by the

Philadelphia calico printing firm, internal diffusion soon followed. Compared with the Philadelphia pilot firm, the New England companies received greater injections of capital and enjoyed a longer period of business expansion. More immediate considerations made calico printing an attractive proposition to investors in textile manufacturing.

The profitability of calico printing with the new technology was the major incentive. The Merrimack Company's earnings averaged 14 percent of net worth per year between 1828 and 1835, the best for the period, and they remained among the highest in the antebellum period. Print prices did not fall as fast as did sheeting prices, as table 6.1 shows.[53]

In addition, the trade depression of 1829 gave a new fillip to colored goods since a number of companies switched to this line to bolster profit margins. At this point the Hamilton Company at Lowell added calico printing to its production line. So did other cotton mills in the vicinity of Boston and Providence.[54]

By 1831–1832, three of New England's major calico printing centers turned out nearly 20 million yards of prints: at Lowell, 90 percent of the Merrimack and Hamilton companies combined production of 10.1 million yards; at Fall River 4.4 million yards; and at Taunton 6.3 million yards. Patrick Tracy Jackson, the Boston Manufacturing Company's agent, estimated in 1831 that aggregate annual American output was 25 million yards. Under the protection of the 1833 tariff, this figure was quadrupled by 1843; at that date 100 million yards of prints amounted to more than a quarter of Britain's printed-goods exports.[55]

The actual process of diffusion among firms cannot have caused much difficulty at Lowell. The continuing influx of skilled immigrant printers, the training of print managers under Prince (who, unlike Yates and the Dover immigrant overseers, achieved distinction and longevity in his position), and the interlocking directorates and specialized machine shops at Lowell all facilitated the extension of the center's calico printing capacity.

The case of the Hamilton Manufacturing Company at Lowell illustrates some of these generalizations. The Hamilton started in 1826 with two mills, one spinning number 20 yarn for twilled stripes and stout bleaching and the other numbers 40 to 44 for dimity or fine jean.[56] By October 1827 falling cotton and cotton cloth prices led the company treasurer to inform the stockholders: "The fine goods which we made are found to be worth very little more than those of a coarser description and costing much more; it was thought expedient to change one half of Mill No. 1 in which we spun No. 40 yarn to the same coarseness as the other half, of No. 22. . . . From the experiments we have made we are satisfied, that to make the most of our establishment we must have a large proportion of our goods printed."[57] He went on to report that contracting the printing to other firms (which included the Dover Company) was fraught with "much trouble, and we are subject to pay others a large profit on the printing." The stockholders approved his decision to set up a printworks, a dye house (to dye yarns for stripes), and a bleachery (to bleach yarns and cloth).

Land and water power were purchased from the Proprietors of the Locks and Canals, whose machine shop built the equipment, estimated to cost $50,000. In January 1831 the company's print and bleachworks, with gearing and fixtures, was valued at $49,000; its machinery at nearly $13,000; its padding and dye house at $5,600; its beetling machine and house at $2,000; and its dye house at $1,500.[58] Certainly some of its equipment came from the Merrimack Company. A company waste book records the payment of $1,715.48 to its sister company for doctors, rollers, files, engraving rollers, turning rollers, and the use of calender patterns.[59] The Hamilton Company also made payments to England for printing and bleaching apparatus, the interest on which to April 9 was $39.72.[60] More than likely it imported engraved and plain shells for its printing machines, the former to supplement its range of print patterns and reduce designing and engraving costs. These could be especially heavy, but perhaps less so in America than in Europe where the dictates of fashion were more frequently changing. One Manchester calico printer spent £5,000 on the design of three thousand patterns and the engraving of five hundred in one year (1833) alone.[61]

The indispensable ingredient of diffusion, the training of indigenous workers, occasioned great difficulty at Dover. Williams, the agent, struggled against attitudes of secretiveness each time a key immigrant worker arrived. For the Dover management, it was essential to overcome immigrant secretiveness about technical matters. In England it might be tolerated because the most skilled were more permanently employed than in the States. But a higher social and geographical mobility in America demanded the institutionalization of knowledge, a crucial point of which Williams was shrewdly aware: "It is highly important that the knowledge & experience which every person brings to the institution should be retained by it that its influence may be seen & felt long after such persons have left it."[62] The migratory habits of immigrant English workers suggest that other American mill managers would have met the same problem.

7
The Diffusion of New Woolen Manufacturing Technology, 1790–1830

Americans made relatively few efforts to acquire the appropriate skills for establishing the new woolen technology. This was not altogether surprising, for the profits in cotton manufacturing appeared far greater and the first effort to establish the woolen technology at Hartford, Connecticut, encountered unexpected difficulties. American recruiting activities apparently extended no further than U.S. ports, which in 1787 Tench Coxe urged his fellow improvers to visit in order to meet disembarking immigrant artisans.[1]

The introduction of the new technology was dependent on the state of the woolen trade in Britain, or rather Yorkshire, where the most capital-intensive technology was established. But most Yorkshire clothiers enjoyed expansive conditions between 1775 and 1790. Their adoption of the new carding and spinning machinery gave them an increasing share of British woolen cloth exports, which doubled in value over this period.[2] Another factor handicapped the transfer of woolen technology: in contrast to Arkwright spinning technology, woolen technology in the 1790s hinged on the presence of a range of skilled workers.

The combination of these circumstances evidently delayed the introduction of woolen carding technology until 1793–1794. At Hartford, in 1788, Colonel Jeremiah Wadsworth succeeded in hiring only English army deserters and ex-prisoners of war with textile-trades experience. That experience was already outdated when Wadsworth organized his manufactory by the improvement of the carding engine and the appearance of the billy. An American manager at Hartford complained of his English workers' trying wild experiments in efforts to catch up with technological changes in Britain.[3] One immigrant woolen manufacturer, Samuel Mayall, reportedly arrived at Boston in 1788 or 1789 and set up a wool carding machine on Bunker Hill. For unknown reasons he departed in 1791–1792 to Gray, Maine, over 120 miles north of Boston. Most likely, he found an alliance with household manufactures easier to maintain in the rural outback than in Boston.[4]

Only a very close examination of conditions in Yorkshire will show the influences that drove some men and families to emigrate in the 1790s. Few left the West Riding, where trade continued to expand, with setbacks only in 1793–1794. But with far-reaching consequences one family, the Scholfields, left Yorkshire for the United States in spring 1793. Looking back in 1810, John Scholfield wrote, "At the time I left England I dreaded the consequence of so hasty a move, but I have had no reason to reflect on it since."[5] Expectation of diminishing opportunities seems to have played a part in their move. Their clothier father had seven sons, all of whom potentially competed for local openings. And, in the area where the Scholfields lived, on the border between the West Riding and Lancashire, the cotton industry was displacing some woolen manufacturers in the 1790s.[6] The outbreak of the European war in 1793 could well have tipped the balance in favor of emigration that spring. In August their father wrote to them in Boston, reporting that in Yorkshire "trade of all sorts are very bad and provishons of all sorts very dear. Things are strangely altered sins the [sic] left England."[7]

Another possible reason for the Scholfields' migration emerges from an examination of the cost and profit calculations recorded in a notebook by one of the brothers while still in Yorkshire. Calculations for their cheaper cloths, retailing at three to four shillings a yard and under, showed an expected profit of up to three pounds eight shillings per piece of forty yards. But for more expensive and better cloths, retailing at six shillings a yard, they incurred a discount charge of up to two shillings and sixpence a yard, imposed by the merchant who transferred the cloths to more distant markets, including America. Consequently the Scholfields' profits on more expensive cloths were little different, and sometimes less, than on their cheaper goods. Thus the move to Boston could well have been an attempt to bypass distribution costs and sell more directly to New England consumers of fine cloths.[8]

The Scholfields' migration in the 1790s was not an isolated one. With the clothier brothers, Arthur (aged thirty-six) and John Scholfield (thirty-five) —and John's wife and six children—went John Shaw, a spinner and weaver.[9] In the 1790s too John Mayall left Yorkshire to join his brother at Gray, Maine. At Newburyport, Massachusetts, in the 1790s, the Scholfields attracted other English immigrant workers to the factory, including three other brothers.[10]

There is other evidence that the transatlantic movement of the new woolen technology was left almost entirely to small Yorkshire manufacturers. Henry Wansey, a Salisbury gentlemen clothier and pioneer West of England user of the carding machine and jenny in the early 1790s, reflected the position of his fellow clothiers in 1796 when he published a very pessimistic view of manufacturing prospects in the United States. It was based on a tour of the country between Boston and Philadelphia in 1794. Twice he was invited to settle and start woolen manufacturing. In Philadelphia the Virginia congressman Colonel Josiah Parker asked him to set up woolen manufacturing at Norfolk, Virginia, where he might have *"the work of slaves* for almost nothing" (the italics clearly expressed the liberal Wansey's shock and disgust).[11] Earlier on his tour he met Wadsworth who made "very handsome offers" to induce Wansey to settle near Hartford. Wansey declined them: "Many objections occur to me: besides the giving up the society and friends I am used to, a concern of this kind would require twice the exertion and fatigue, and twice the capital; and certainly, were I resolved to leave this country . . . there are many other concerns to be engaged in, equally profitable, without half the capital, or a quarter of the trouble and exertion."[12] William Partridge, a Gloucestershire dyer who emigrated in 1808 and stayed in the American woolen industry until after 1850, claimed in 1823 that he knew of only four West of England woolen manufacturers in the United States.[13] There were more, but his firsthand knowledge of both trade and immigrants reliably indicates America's dependence on the Yorkshire industry for the potential to develop the new woolen technology.

Besides the influence of the Yorkshire industry on the timing of the transfer, the crucial labor elements (card machine builders, spinning operatives, and supplementary finishing workers) meant that isolated workers, such as Mayall, would find it very difficult to set up the technology in America. The optimum was a team of workers. This the Scholfields provided. The brothers may have perceived the need to emigrate as a team, or they may have been guided by unrelated motives. It is not impossible that they also knew Mayall at this point; in

1799–1800 John Mayall, Samuel's brother, was in Boston, and the Scholfields used him to purchase mill supplies.[14]

The Scholfields had other assets besides their combined trade skills. Their kin group and young children represented the greater reliability of a more effective channel of transfer. And this was not merely a matter of inspiring confidence in members of the host community. Dependents helped to brace the will to work and succeed. After a few months, when they moved north of Boston, a distinguished New England clergyman, the Reverend Dr. Jedediah Morse, gave them a certificate or testimonial, dated November 22, 1793, for them to take to Newburyport, which read in part, "since May last . . . they have conducted [their business] like *honest, ingenious, industrious* men."[15]

Far more mobile, and perhaps unreliable, was the solitary carrier of technology. James Douglas, a native of Dumfriesshire and a machinist, was one of these. He moved first to Ireland where he worked in a woolen mill and then, in about 1792 or 1793, emigrated to Philadelphia. Here he introduced the gig mill and invented or introduced designs for a cloth shearing machine, a water-powered loom, and a brick-making machine. To be fair, part of his failure to make a lasting adaptation to America arose from the patronage he secured in Philadelphia: the British consul, Phineas Bond, persuaded him to return to England in 1797.[16]

Given the unpromising start of low American interest in woolen technology the Scholfields astutely used their capital and labor resources to establish themselves as a pilot manufacturing firm in America. One outstanding feature of their early months in the United States was the cautious manner in which they established and consolidated their position. They set themselves up at Charlestown, Massachusetts, as a household carding, spinning, and weaving unit, demonstrating their skill and establishing a reputation. With very little capital (presumably only their combined savings) they were in business within four months. A running account started by the partners on June 20, 1793, shows expenditures amounting to £112 3s. 7½d. by December 1, 1793, the date they left Charlestown. Of this sum £71 3s. 6d. was spent on raw wool, purchased from nearby Watertown. The rest was expended on tools, materials for machine making, and manufacturing supplies like oil. Evidently they made no attempt to smuggle tools or equipment out of England. Files, planes, chisels, a saw, a vise, pliers, an auger, and a rule were purchased locally in Charlestown or Boston. So too were lumber, nails and screws, brass and iron wire, iron for the shuttle ends, cane for reeds, and cording for harnesses. With these traditional tools and these commonplace materials, the partners made a forty-spindle jenny, warping bars, and a loom, all completed on August 4 at a labor charge to the account of £12 3s. A few accessories came from local specialists: a reed, a reedstock (partly made by a smith), and a pair of hand cards. When the Scholfields closed their first account on January 27, 1794, £140 13s. 7½d. had been spent on tools and materials, and £86 18s. 6½d. was debited for spinning and weaving, work credited to stock. An income of £155 5s. came from sales of cloth and stock in hand, leaving £72 7s. 2d., which the partners regarded as a credit balance.[17]

The first cloth they made was a twenty-four-and-a-quarter-yard broadcloth, which they rated at 14s.

6d. a yard but sold for slightly less (perhaps for cash) at £16 6s. on October 28, 1793. In this manner they avoided the middleman's charges of over 30 percent of the wholesale price. Compared with imports, the selling price was moderate. Superfine 7/4 cloths were imported from Britain at 18s. to 19s. a yard and sold in Philadelphia for 40s. to 45s. a yard between 1791 and 1805.[18]

Impressed by their potential, Dr. Jedediah Morse, minister of the First Congregational Church at Charlestown, pesuaded his friend William Bartlett, a wealthy Newburyport merchant, to support an expansion of the Scholfields' operations from household to carding mill–workshop. This provided the greater capital demanded by an increase in scale and the application of power. In December 1793 the partners left for Newburyport, taking their tools and equipment with them. The following month, on the basis of enthusiasm generated by the construction of a prototype card in a local stable, the Newburyport Woolen Manufactory was incorporated by the state of Massachusetts with an authorized capital of £90,000. A factory building, three stories high and measuring one hundred feet long by forty feet wide, was built at Byfield (about five miles inland from Newburyport) on the Parker River to house the carding mill and spinning, weaving, and finishing workshops.[19]

The carding equipment at Newburyport was built by immigrant labor. After the original partners assembled the prototype single cylinder, hand-powered card and set up the jenny and loom they brought from Charlestown, two more carding engines were constructed for the mill. Presumably based on Yorkshire multiroller models, they were attributed to James Standring, Samuel Guppy (from a family of Bristol metal merchants), and John Warren Armstrong (a Wiltshire clothier), no doubt directed by the Scholfields.[20]

The Scholfields' card was different from that used by Wadsworth at Hartford. The design of the two Hartford cards, built by a local general builder in spring 1793, earned the derision of Wansey. He thought them "of the oldest fashion. Two large centre cylinders in each, with two doffers and only two working cylinders, of the breadth of bare sixteen inches, said to be invented by some person there."[21] The use of only one worker with the main cylinder or swift severely reduced the machine's efficiency; the carding action would occur at only one point on the swift's surface. The Scholfield card remedied this by employing a number of small cylinders (workers and clearers) over the main cylinder, as in England. In fact the Scholfield-Standring card design is well established from surviving models in the Smithsonian Institution's National Museum of History and Technology and the William Penn Memorial Museum at Harrisburg and from an English patent of 1804, which includes a "Plan of Old Engine." This drawing shows a frame construction and roller deployment (three workers over a swift) identical almost to those of the artifacts.[22] The Scholfields also introduced the billy at Byfield. In 1798 they purchased an axletree for a billy for 9s. Two undated plans of a billy in the Scholfield papers show a machine very similar to that used in Britain's cotton and woolen industries before 1812.[23]

Fellow immigrants assisted the Scholfields in the operation as well as the construction of the Newburyport manufactory. One was Thomas Kenworthy, who received a single payment of £3 12s. 6d.

(May 7, 1794). Most likely this was Thomas Kenworthy, the immigrant spinner who reached Boston in 1789 and for whom Moses Brown and William Almy, on hearing of his arrival from a Scottish immigrant weaver, sent a horse and chaise to bring him to Providence. Kenworthy's activities at Newburyport are unknown, but in 1801–1802 he renewed his acquaintance with the Scholfields after running away from a British man-o'-war in New Orleans.[24] Another was Abraham Taylor, who emigrated in 1797 at the age of nineteen. When the Scholfields left Newburyport for Connecticut in 1798, he went with them. By the War of 1812, however, Taylor was back at Byfield as a woolen manufacturer with five in his family.[25] One hint that other workers besides Taylor and two other Scholfield brothers (Isaac, who emigrated in 1794, and James, who came later) were at Newburyport is found in the Scholfield correspondence. After John and Arthur Scholfield left for Connecticut, John Taylor (perhaps related to Abraham) wrote:

Mr. Bartlett was up here this morning and informed us that the owners of this factory had agreed to carry it on another year with the present stock (viz.) 7000 Doll. so that the few that we left (viz.) Leach, Lees, James Hall, your brother [Isaac], Hill, Sanderson and I may have imploy. . . . [Brother James expects to meet John and Arthur Scholfield at Rygate]. . . . We have no news from old England but expect some every post.[26]

The Scholfields established themselves cautiously, relied upon American merchants for capital to set up a mill-workshop operation, and did not attempt manufacturing independence for a few years. In this adaptation they were remarkably like Samuel Slater. At the technological level too, the Scholfields' behavior was similar to Slater's. They did not stray from the technology with which they grew up in England. Fortunately for them, its basic shape remained unchanged throughout their lifetime. The caution paid off. Savings enabled Arthur (and perhaps John) to invest in land in Vermont. And although the late 1790s saw other textile manufacturing companies in difficulty (including the Newburyport one), Arthur and John had enough savings to set up on their own at Montville, Connecticut, in 1798. Furthermore another brother, Joseph, saved enough at Newburyport in the 1790s to return home in January 1799 and to pay off his creditors (from whom he may have fled), getting "the praise of all the Countrey Round a Bout."[27]

At the third stage of technology diffusion, woolen carding and spinning machinery spread in much the same way as did cotton spinning technology, through kin and trade networks. Hitherto the role of the Scholfield brothers has been hailed as preeminently important in diffusing industrial woolen technology through New England. However, their surviving papers suggest that they were not the only agents of technology diffusion. Arthur Scholfield, the only brother to move from manufacturing to machine making, started building carding machines soon after he went to Pittsfield, Massachusetts, in 1801. In 1805 he charged $150 per card, excluding clothing (which cost $40 locally and nearly $90 in Boston), and $40 per picker. In summer 1807, his orders amounted to twenty-two cards, besides pickers, valued at $1,500.[28]

Two rival carding-machine makers, Standring and Gookins, spread the new English woolen card through New England in 1802. Arthur Scholfield told his brother in November that year:

I have been at a stand what to do about it for they are setting up Machienes all round me but have finally concluded to have one. Standring & Gukins are spreading all over the Country. They have secured almost every stand that is good for any thing in the western part of New Hampshire, Vermont, Massachusetts & New York state as far as Whitestown: They reserve a part in almost all the Machienes they make.[29]

The Standring involved was possibly James Standring, the English immigrant who built the Byfield carding machine under the Scholfields' direction in 1794. Or it may have been Benjamin Standring (perhaps a relation), who obtained an American card patent in 1803. Another clue to the identity of the Standrings involved in carding technology comes from the artifact evidence. The Scholfield-type card at Harrisburg has two doffer comb blades with "E. Standring Newburgh N.Y." stamped on them. They were not the original blades, however; on removing one, I found old screw holes underneath. The "E" could be an ill-formed "B." Standring's partner was probably Richard Gookins, who took out a U.S. patent in 1806 for "machine for making bats or frames for wool hats" and, like an itinerant mechanic, gave his address as "New Hampshire, Massachusetts."[30]

The Scholfields' rivals perceived profits in the sale of carding machines to local mill operators. This perception was based on the high productivity of the Scholfields' card, though the higher efficiency of the machine card was not immediately reflected in a fall in the cost of carding one pound of wool. In 1793 the Scholfields at Charlestown rated their hand carding at between fourpence and fivepence (equivalent to eight to ten cents) a pound; in 1801

124

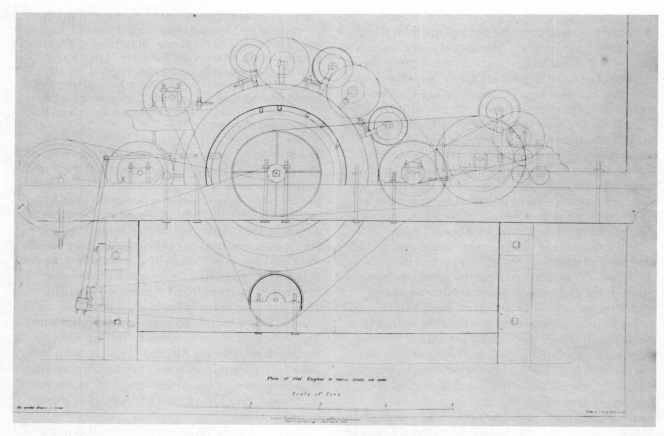

Plan of Old Engine to same Scale as new

Scale of Feet

7.1 Single-main-roller woolen carding machine, included in the drawings for a British patent of 1804 and then described as "Plan of Old Engine." It closely resembles the typical English woolen card of 1818 (see illustration 1.8). Here the feeding cloth is at the right-hand end of the machine. (G.B. patent 2766; courtesy Boston Public Library.)

7.2 Single-main-roller woolen carding machine with a thirty-inch-wide feed, built by the Scholfields ca. 1790s–1820. The feeding cloth is at the left-hand end of the machine. Comparison with illustration 7.1 confirms that the Scholfield card was a very close copy of the English carding machines of the 1790s. (At National Museum of History and Technology, Smithsonian Institution, Washington, D.C.; photo by author.)

Arthur Scholfield charged twelve and a half cents a pound for machine carding at Pittsfield. Competition from other cards did, however, bring down the price of carding a pound of wool to under ten cents by 1802. The immediate advantage of the Scholfield card was the 50 percent increase it permitted in the handspinner's productivity. If she carded her own wool, a woman at home could spin four skeins (each of seven knots, 560 yards long) a day; if the wool were machine carded, she could spin six skeins a day.[31]

Small operators of rival grist or fulling mills saw a chance to increase their income with the addition of a wool card to serve local housewives, and they rapidly adopted the machine. Writing in March 1803, the Leicester, Massachusetts, card clothing manufacturer Winthrop Earle informed John Scholfield that he had orders "to furnish Cards for fifty Machines this Spring."[32] By 1810 the country had 1,776 carding machines in operation: 40 percent in the six New England states, 23 percent in New York, and 19 percent in Pennsylvania.[33]

The Scholfields were effective carriers of carding technology because the various brothers and their families moved from Boston and Newburyport to different parts of New England: Arthur and John to Montville, Connecticut, in 1798; Arthur to Pittsfield in western Massachusetts in 1801; James to North Andover, Massachusetts, in 1802. John's children had spread through Connecticut by 1810. John, Jr., with four young sons, at Preston, eight miles northeast of Norwich, Connecticut, had a woolen mill with two carding machines, a billy, two jennies, and four looms. James, with two children, was at Canterbury, fourteen miles north of Norwich, and was starting a carding mill. Joseph was left at

Montville where he was dressing cloth and likely to take over from his father, John, Sr., who was running another mill at Stonington. When John, Sr., wrote his will in 1819, he had three manufacturing establishments, all in Connecticut, at Montville, Waterford, and Stonington, and interests in a grist mill and sawmill at Stonington.[34] Besides carding machinery, the Scholfields later set up other improved woolen equipment. The first woolen jack in the United States was built by James Scholfield after he moved to North Andover in 1802.[35]

Much slower was the diffusion of the woolen manufactory form (carding and spinning, weaving, and finishing workshops) established at Newburyport. In 1810 two dozen such mills were reported in the census: fifteen in Connecticut and all but three in the other New England states. As with cotton, the War of 1812 gave a great impetus to manufacturing; by 1820, 253 woolen firms with integrated carding, spinning, and weaving were reported. Of these 115, or 45 percent, were in New England and 53, or 21 percent, were in New York.[36]

Woolen manufacturing technology was not diffused by the act of machinery construction, except in the case of the carding machine. Much of the technology was a combination of equipment and operator skills. Consequently one factor was the time taken to learn new skills. About this little is known. Partly it depended on the ability of learners, partly on the number of immigrants available to teach, and partly on the willingness of immigrants to impart their skills.

Arthur Scholfield retained some of the attitudes of secrecy and suspicion with which he had been brought up, but not consistently so. In 1802 he wrote from Pittsfield to his brother John in Mont-

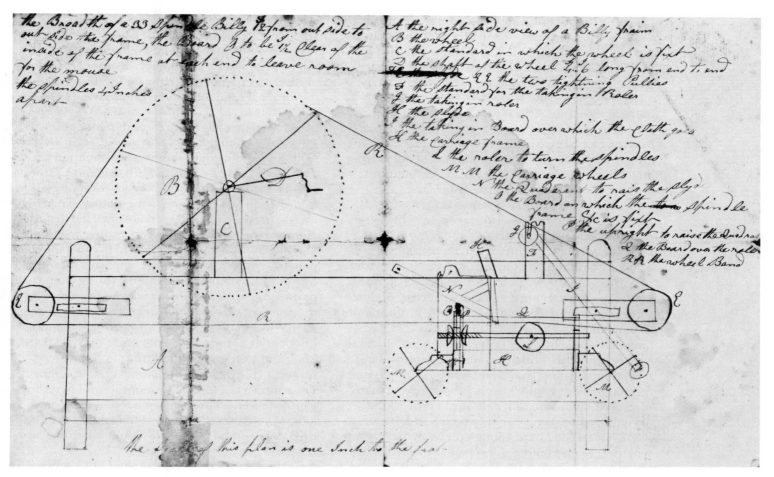

7.3 Undated ink plan of a slubbing billy, surviving in the
American papers of the Scholfield family. Similarities
with the English billy may be seen by comparing the plan
with illustration 1.9. (Courtesy Connecticut State
Library.)

ville, cautioning him about a worker named Smith heading for Montville: "I don't think it will answer your Purpose to imploy him. All he wants is to get into the manufacturing Business and Machionery. I have kept him as blind as I could but he is an inquisitive Yankey and knowes everything." Yet in the same letter, Arthur recommended the bearer, Silas Harmar, to his brother, requesting that John provide Harmar with "Instructions to Build a Spinning Machiene" since he, Arthur, could not help him.[37] Possibly Harmar was a fellow immigrant; certainly he was especially trusted.

John Scholfield received requests for technical information, but the absence of copies of his letters prohibits any firm conclusions about the extent of his secretiveness. In 1810, when merino sheep breeding was at its height, an enterprising breeder in North Carolina wrote to him for details of "the nature of your machinery, the price it cost, & your process in manufacturing, & such other information as you will be so good as to furnish me."[38] One long-standing customer expected John Scholfield to supply him with limited amounts of technical information. General Sylvester Dering, a Long Island merino breeder whose fleeces John regularly manufactured, once sent him a skein of yarn, asking, "What Reed or slay it ought to be put in & if you think it is twisted hard enough for warp," and whether, all else failing, John would weave it.[39] Other Long Islanders wrote to him for a carder and cloth dresser.[40]

The problem of persuading immigrant workers to abandon their secretive attitudes and willingly teach their skills to Americans was certainly well understood by the management of one early woolen manufactory whose records have been closely studied. It was organized by French immigrants who must have been aware of British secretiveness before they went to America. Éleuthère Irénée du Pont, his brother Victor, and two other Frenchmen needed an experienced manager for the woolen mill they built at Louviers on the Brandywine in 1808–1809. They hired William Clifford, a West of England woolen manufacturer, and in his contract they made arrangements to ensure that Clifford's skills and knowledge were rapidly imparted to their own manager and workers.[41] To this end Clifford was obliged to give his employers "every possible information" on manufacturing cloth and cassimeres; to take charge personally of all the departments in the woolen mill except fulling and dyeing, of which he had no experience; to instruct the company's workers; to "make as compleat a master as he himself is of the art of manufacturing cloth such partner or partners" who so wished to engage in the practical side of the business; to communicate all the information he had "respecting the new discoveries or improvements made in Great Britain"; and to manufacture cloth "similar in every respect to the best cloth manufactured in England." The contract ran for twelve months, and in return Clifford received four dollars a week, free board, lodging and washing, and, at the end, a four hundred dollar gratuity if he "faithfully complied with his promises."[42]

Clifford's marriage to Victor du Pont's reportedly unattractive daughter, Amelie, in September 1812 might also be expected to surmount secretiveness on Clifford's part, as Slater's example suggested. But for a complex mixture of reasons, Clifford refused to be wholly open, although he was rehired after a year for an annual salary of six hundred dollars. Only one of the four French partners, Raphael Duplanty, wanted to learn mill management and manufacturing techniques, but Clifford evidently did all he could to deter him. And Clifford prevailed. E. I. du Pont was not blind to his niece's husband's secretive ruses. If Duplanty "becomes thoroughly competent we will be out of leading strings and no longer dependent on him [Clifford]," he told another partner shortly before Duplanty left to try his hand at cotton spinning.[43]

Clifford may have been exceptional. Another English manager on the Brandywine, Isaac Bannister, showed little regard for trade secrecy. When his partner, William Young, defended himself against charges of labor piracy, Young explained that the shortage of skilled workers made him rely on Bannister for teaching his employees their trades. Three months after the Youngs and Bannister formed their partnership, Young reported that "those that have applied to us with the exception of J. Draper have all been taught the branches in which they were employed by our J. (= I.) Bannister."[44]

On the other hand, in 1832, one immigrant woolen manufacturer not only voiced reasons for his attitudes of secretiveness but also claimed that such closeness was prevalent throughout the American woolen trade. John Bancroft (1774–1852), who emigrated from Salford, Lancashire, to Delaware in 1822 (and in doing so switched from chairmaking to flannel manufacturing), declined to answer all the questions posed by the federal census of manufactures in 1832 because "it would make the foreign manufacturer and his government acquainted with the weak points of our situation; and without having any unkind feeling towards my native country, or any other country, still I do

not wish to furnish them with that knowledge which might lead to bad consequences to myself and family."[45] He then went on to argue that innovation led to secrecy among all American woolen manufacturers:

This trade has undergone a surprising revolution within a very short period, every one in it acquiring further knowledge, and no one inclined to communicate his own discoveries and improvements to the others; and some of the processes now in use in the United States are superior to those practised in foreign countries; and it is not desirable to call the attention of foreigners to them, because these foreigners, in the matter of wages, have decidedly an advantage over us, and the improved processes would, if brought into their service, increase the advantages.[46]

How true Bancroft's claims about secretiveness were is hard to tell. Certainly a number of manufacturers shared his suspicion of a census, but the majority seem to have responded to the federal questionnaires, both in 1820 and 1832. As an immigrant with relatives in the textile trades in Britain, Bancroft may have been more keenly aware of Britain's relative technological position than were native American manufacturers.

Diffusion of woolen technology in America received a notable fillip in 1823 when William Partridge, a Gloucestershire dyer, published his *Practical Treatise on Dying*. Over a third of it was devoted to practical hints about woolen processes and machinery employed in the West of England with advice to Americans for improving their own cloth manufacture. His dye recipes supplemented the earlier American publications of Asa Ellis, Elijah Bemiss, the Bronsons, the Yorkshire dyer James Haigh, and the Swiss dyer John Rauch.

The diffusion of new technology might well be hindered by immigrants who fled from technological unemployment in Britain. In the early 1820s, Partridge believed that many English woolen workers arrived in the United States with "this prejudice against machinery" still "very strongly entertained by them."[47] Consequently, he claimed, the nap was raised by hand in many factories, the gig mill resisted or run slowly, and only small jennies (sixty spindles) installed.[48] One of the Hollingworths, a family of weavers and clothiers who emigrated from Yorkshire to New England in 1826, initially resented spinning his yarns from wool carded by one of the new American condenser cards but did not explain his hostility. Abel Stephenson, a Yorkshire clothier who emigrated in 1837, certainly harbored a deep and lasting resentment against power-driven machines in his own trade.[49]

Another barrier to the diffusion of skill was the existence of technological confusion resulting from the admixture of immigrant workers from the various woolen manufacturing districts. Although only traces of evidence survive, it was a problem in the early American woolen industry. The extent of the potential for confusion in America's technological melting pot was discovered by William Partridge, who emigrated to the United States in 1808. In one New Jersey factory as late as the early 1820s, he found "they calculated yarn by the bier."[50] In other words, they were so confused as to employ a reed measuring unit in a yarn numbering system, an absurdity that amazed Partridge. In the manufacture of such staples of the woolen trade as broadcloth and cassimere, he noted, "There will be various opinions given by European workmen on the mode of making these goods, some will be recommending

one system, and some another, which serves to show there is considerable latitude in the practice pursued in other countries, as well as in this." And his Gloucestershire system, he asserted, was "as good as any practised in [England]."[51]

Emphatically Partridge advised American manufacturers to employ workers from only one British region (the West, the North, or Ireland) in one factory:

I cannot conceive a more uncomfortable situation, than for a manager, who is not perfect in the business, to be surrounded by a mixture of Irish, Yorkshire, and west of England workmen. Whatever advice he might receive from the one party would be condemned by the others. If any description of machinery were recommended by the one, the other would be sure to suggest something different as being better; their opinions on spinning, weaving, braying, fulling, raising the nap, shearing and pressing, and packing, would be all at variance with each other, and it would be much if the manager did not make his election from the worse of the three. I know manufactories that are now labouring under this dilemma, and all of them have suffered more or less in the same way. A few years, however, will remove the injury arising from this cause. It is only to make and encourage American workmen, and the evil will be gradually receding.[52]

Although secretiveness and conflicting technical traditions in the immigrants' backgrounds threatened to delay the diffusion process, the presence of immigrants might attract other workers from Britain's woolen industry. Clifford was responsible for bringing several key workers to the Louviers mill and its vicinity, men who remained after he had gone.[53] Chief among these was William Partridge, who was lured from Connecticut, where he was dyer for the Middletown Manufacturing Company. In his first year on the Brandywine, Partridge's earnings equaled those of Clifford; on July 31, 1813, the Louviers mill paid him $1,680.97 "for his services at the factory & also for drugs furnished by him at different times."[54] Partridge does not appear in the company's ledgers after 1815, and by the early 1820s he was in New York, which became his home until his death in 1858. Despite the unpleasantness surrounding Clifford's association with the company, Partridge remained in touch with the du Ponts and when visiting England in 1820 wrote a letter of introduction to Charles du Pont on behalf of a Mr. Jones, "a woollen manufacturer & one of the most celebrated in the West of England."[55] This must have been John Jones of Staverton, between Bradford-on-Avon and Trowbridge. Partridge emphasized that Jones's motive in wanting to emigrate was "on acct. of the many very oppressive & disgraceful measures of the English government"; and that he was "a Gentleman of character of strict honour." In Partridge's estimate, "Any concern put into his hands cannot fail to answer if the situation of the country will admit of it."[56]

Again, however, the reliability of the carrier of the technology complicated purely economic or technological influences. Clifford's departure from Louviers in 1814 both left the woolen factory in highly inexperienced hands and deterred the French partners from employing another immigrant as manager. In March 1814 Clifford was exposed as a bigamist who had deserted his wife and child in England. His real name, which he reported to the U.S. marshal when registering as an enemy alien, was Nathaniel H. (Clifford) Perkins. After his departure, the Louviers mill was left in the hands of

Charles du Pont, then a seventeen year old, who had been learning the business from the secretive and high-handed Clifford as best he could. Within a year of Clifford's disgrace, the company switched to making kersey cloth for the army, moved downmarket, and succeeded despite the difficulties with the first manager.[57]

After the Clifford episode, the Louviers proprietors regarded immigrant manufacturers aspiring to factory management with considerable skepticism, judging by William H. Cox's case. In a phonetically and badly spelled letter, Cox, citing his experience in Britain sorting wool and in Rhode Island superintending a woolen factory, offered to fill the superintendency left vacant by Clifford. His postscript, "James Shaw saw my work here [Providence, R.I.]; thre's none Like it in the union," evoked E. I. du Pont's scathing endorsement, "Englishman: knows every thing better than every body."[58]

By the 1820s the best of the American woolen manufacturers had reached technical parity with, and even overtaken, their Yorkshire rivals. Nevertheless, Americans remained interested in developments in the English woolen industry, and a number went to England to gather or share technological and economic information. Through the eyes of Zachariah Allen, a Rhode Island woolen manufacturer who kept a diary of his visit to Britain in 1825, can be seen an American perspective of the state of British woolen technology and the methods by which it was transferred to the United States.[59]

After having visited a number of mills in Saddleworth and Huddersfield, Allen noted, "I am convinced that in making cloth in America we fail more from the hurry and want of care & attention with which the processes are completed than from want of skill & good machinery." At Bradford-on-Avon, one center of the West of England woolen industry, he summed up the manufacturing technology tersely: "I did not perceive anything new or interesting in this place."[60] The Rhode Islander was decidedly unimpressed by the persistence of jenny spinning, hand loom weaving, and hand cloth shearing, which he found even in Gott's mill in Leeds. To his surprise, no manufacturers were attempting to apply the cotton power loom to woolen weaving. This reluctance to keep up with the cotton industry he attributed chiefly to workers' resistance based on fear of technological unemployment, as in the case of the cloth workers when gigs had been introduced in Yorkshire twenty years earlier. In other instances, the rejection of a new technique sprang from danger of impaired quality. Benjamin Gott and his aged head fuller both told Allen that steam milling had been tried twenty years earlier and was considered unsatisfactory because cloth color was injured in the process. And at Kirkstall Abbey, hand shears were used in conjunction with a diagonal rotary shear because the latter was likely to damage the cloth when traversing it longitudinally.[61]

In Yorkshire, whose cloths he would be most interested in imitating, Allen's attention was centered on finishing equipment. He sketched in detail a steam dresser or brushing machine at Huddersfield and then visited the patentee, John Jones at Leeds, discovering that the price of one dresser was £118, the brushes alone costing about £35 because "he used wire twisted in with the bristles to give greater action for brushing flannel before burling."[62] Finding a new mill under construction at Kirkstall Abbey near Leeds, Allen measured one of the cast-

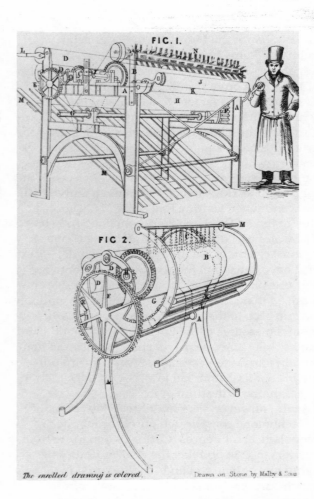

7.4 Cloth brusher, or dresser, patented in 1824 by John Jones, brush manufacturer of Leeds, England. The machine removed loose fibers, laid the nap in one direction, and fixed the nap by means of a steam apparatus (a well-known system, according to Jones's specification). (G.B. patent 4897; courtesy Boston Public Library.)

iron beams and learned that they were tested with twelve to fifteen ton weights. The item that attracted his attention most was the gig. A machine shop at Leeds had gigs and cards under construction, and Allen recorded, "From several visits made to these Machines we have been able to get a pretty accurate draft."[63] The drawing and details of construction filled seven pages of his diary. Also in Leeds he showed much interest in the layout, equipment, and processing techniques of the large mills belonging to Hirst and Gott.[64]

Apart from woolen mules, little mechanical technology in the West of England impressed Allen. The fine quality of the region's cloth, which he greatly admired, he attributed to nonmechanical skills and processing techniques. Consequently he spent no more than a month in the West of England before returning to Yorkshire in August 1825.[65]

Allen utilized various ploys to carry off this technology. In touring English manufacturing districts, he visited both manufacturers and machine makers, seeing the equipment at work in the mill and discovering the details of its construction in the machine shop. In diary jottings and sketches, he recorded technical information. And like the Dover, New Hampshire, management in their pursuit of calico-printing technology, Allen was careful to record the costs of the technology, noting English machinery prices for later comparison with American ones before deciding whether to build or import machines and parts. Apparently Allen conveyed little actual hardware back to America. After writing the gig specification in his diary, he added, "I have procured exact iron patterns of the frames, keys &c. to take to America."[66] Otherwise there are no hints that Allen was on a

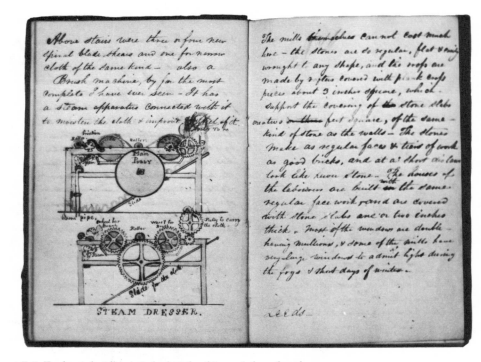

7.5 Zachariah Allen's ink sketch of Jones's brusher (see illustration 7.4), seen when the American visited a Huddersfield woolen mill in 1825. (Source: Allen's journal, Rhode Island Historical Society.)

134

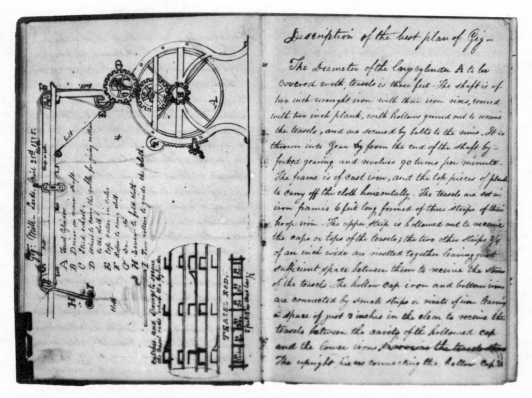

7.6 Ink sketch of a gig mill built by a Leeds machine shop
in 1825; drawn by Zachariah Allen, the Rhode Island
manufacturer, during his European tour that summer.
(Source: Allen's journal, Rhode Island Historical Society.)

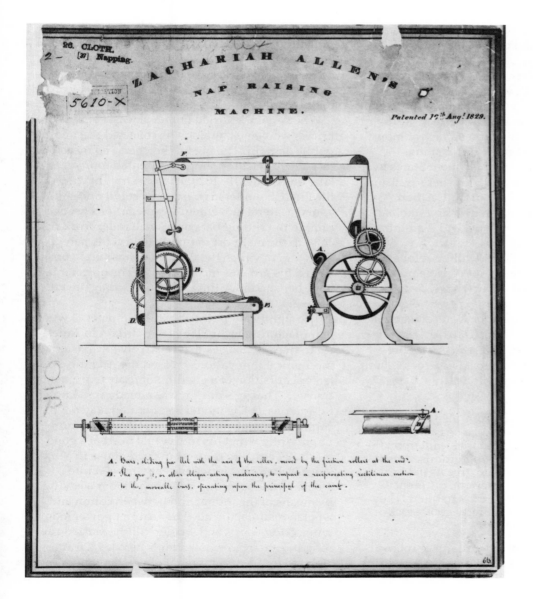

7.7 Zachariah Allen's American patent gig mill of 1829, based on the machine he saw in Yorkshire in 1825 (see illustration 7.6). The inset drawings show grooved cams intended to lend a lateral motion to the teasel bars of the gig drum. (U.S. patent drawing 5610.)

major shopping mission, which he, far less than the massive New England cotton corporations, could afford.

In getting into English mills, Allen received the cooperation of William Davy, the American consul at Leeds. Davy's contacts with local manufacturers, and indeed his move from Hull to Leeds in 1818, arose in connection with his difficult task of monitoring at source the arrangements for smuggling Yorkshire woolens into the United States by way of Canada. Davy took Allen to Hirst's mill and quite possibly introduced him to Gott. Albert Davy, his son, conducted Allen over an establishment for printing cassimere shawl and then into Marshalls' famous flax mills.[67]

Allen did not hesitate to use mildly dubious methods to gain information. At Leeds he procured yarn and warp tables used in Hirst's weave room through an independent reed maker who dealt with Hirst's head weaver.[68] Where the locus of technology was still embodied in the artisan, as were many cloth finishing techniques, and secrecy still prevailed, doubtful methods of persuasion seemed unavoidable. Thus one July entry in Allen's diary reads:

Stroud. Cains Cross inn. July 26th 1825. In the evening two clothiers called to see us from Lewis mill to whom we had spoken in their mill upon the subject of manufacturing. Having promised them a pound or about $5 if they would give us the best information in their power upon the different branches with which they were acquainted. They commenced, after drinking a pretty good stock of beer and brandy, as a preface.[69]

There followed five recipes: for milling or fulling cloth, for using the "washer machine," for scouring cloth of the grease, for raising the nap, and for the finishing process for blue cloth. Only a practical manufacturer could assess their value. Allen was sufficiently convinced of his informers' competence, integrity, and degree of sobriety to record their information.[70]

When Allen returned home, at least some of the British woolen technology he collected reached the public domain. The gig he sketched and measured in a Leeds machine shop he patented four years later on August 10, 1829. It was a straight copy of the English model in every respect but one. Allen added a lateral movement (by means of a groove cam) to the sliding bars on a small roller working close to the main drum (marked A on his patent drawing, in which the draftsman has surely positioned his grooves incorrectly).[71] The object must have been to move the cloth sideways against the main gig cylinder.

The significant point about the transfer of woolen manufacturing technology from Britain to America in the late 1820s is that relatively few basic mechanical innovations crossed the Atlantic. Allen's acquisition of a gig did not represent a major transfer. The gig's design was already well known in America.[72] The model he obtained may have incorporated some marginal improvements; more importantly it met his individual need of a good machine, equal to the best in current use in Yorkshire. Americans independently moved into woolen mule spinning and power loom weaving by means of the interaction between cotton and woolen manufacturing. Thus satinet power looms, with cotton warps and woolen filling, started running in American woolen mills before 1820 and assisted in the development of woolen power

looms.[73] Much more important to American manufacturers was the British command of manufacturing techniques, especially in finishing. The transfer
of technology in this form largely depended on developing a stable immigrant work force, which
American agriculture effectively prevented.

Transmission of the Technologies; Some Conclusions

Clearly Britain's institutional constraints against the leakage of technology were riddled with loopholes. British customs stopped some machinery, but a lot more seems to have been illegally exported by well-known smuggling routes. Few men seem to have been prevented from crossing the Atlantic when the prohibitory laws were in operation. Even those who were caught could find ways of circumventing English legal processes. One of the two McConnel and Kennedy spinners arrested by customs on board the ship *Union* at Liverpool on March 9, 1811, and committed on March 12 for trial at the next Lancaster assizes, arrived in America early the following June. Evidently Dennis Manion jumped bail and at the second attempt successfully evaded customs; during the War of 1812 he worked as a spinner at the Globe Mill in Philadelphia.[1] While emigrants' motives have not been extensively probed, it has been shown that high financial returns, as high as and sometimes higher than those for comparable work in Britain, awaited those who introduced desirable British improvements to the United States.

Americans recruited British artisans more energetically in the cotton than in the woolen industry, presumably because there were higher and quicker returns in the former. Larger capitalists, like the Boston Associates, adopted more aggressive and rational methods of recruiting British workers because they, far more than smaller entrepreneurs, could afford to dispatch managerial-level agents to Britain, where their established mercantile networks facilitated recruiting and shopping. And American visitors to Britain were careful to seek out best-practice technology. American publications, like the reissue of Rees's *Cyclopaedia*, also represented attempts to rationalize the international movement of technical information. But as long as British improvements were confined to the mill mechanic's back room, the factory floor, or a creaking patent system, the artisan remained the prime international carrier of the new technology, at least in the introductory stage of diffusion.

Skilled immigrants and well-rooted American entrepreneurs, however, always combined in launching prototype production lines. In the case of power loom weaving technology, immigrants assumed less importance than did American capitalists; woolen machine carding exemplified the opposite polar case. The establishment of pilot mills and a new technology were achieved most successfully when American market conditions favored business expansion: when English competition shrank under pressure of war or the American tariff, and when transportation networks and population growth widened product markets. On the other hand, even when English competition was unrestrained, as it was in the 1790s, manufacturers like Slater or the Scholfields who possessed the necessary technical skills could survive by complementing household manufactures and prudently ploughing back profits. To be sure, American capital complemented immigrant skill: Pollard and Nicholson (both British-born but well established in America at the time of transfer) or Almy & Brown provided capital for factory spinning; Francis Cabot Lowell for power loom weaving; the Boston Associates for calico printing; and William Bartlett and the Newburyport merchants for woolen machine carding.

Technology diffusion occurred on a moderately crowded stage. The notable carriers of technology,

like Slater and the Scholfields, competed with numbers of others possessing the latest industrial skills, though they were still in a small minority among the total number of textile immigrants. Besides Slater, there were William Pearce, Thomas Marshall, and John Daniel; besides William Gilmour, there were the Siddalls, John Murray, and John McDonald; besides the Thorps and John D. Prince, there were more Thorps, T. Yates, and Peter Bogle; and besides the Scholfields, there were the Mayalls and William Douglas—and perhaps dozens more.

These studies of technology diffusion confirm what is well known about rates of technology transfer (times between the setting up of pilot production lines in the originating and host economies). In the early nineteenth century, the new textile technologies crossed the Atlantic rapidly, within months or a few years, once American entrepreneurs were determined to secure them. This determination depended very much on economic conditions in America and on the state of technological change in Britain. Most of the technologies reviewed here were well developed in Britain before they were imported to the United States. But in the case of the power loom, Americans were forced to share in research and development, and in so doing they had an important opportunity to modify this pivotal textile technology to suit their economic circumstances.

Internal diffusion of the technologies rested primarily on kin and trade networks before 1807 or soon after. With increases in the scale of production, wrought by vertically integrated and powered spinning and weaving after 1812, capital goods firms, often the offshoots of the manufacturing companies of the 1820s, became more important agents of technology diffusion; so too did selling agents. Nevertheless a few key immigrants played a significant part in technology diffusion, especially in calico printing and woolen manufacturing where skills were not so swiftly mechanized.

Immigrants could be difficult employees for American masters. Individual cases show them earning the reputation of incompetence, inexperience, and dishonesty. But the confusions promised by conflicting technical traditions in England apparently were limited mostly to the American woolen industry, and there with short-term effects only. No cases of physical aggression toward new machinery occurred among the immigrants studied. Most serious for American entrepreneurs and managers were the immigrants' attitudes of secretiveness displayed in both the cotton and woolen industries. Americans found it difficult to pin down immigrants, either with contracts or high wages, and it was always hard to tell whether the immigrant's information or skill was worth having anyway. The extent to which some immigrants would go to guard their technical secrets is suggested by John Bancroft's habit of writing in shorthand in his mill notebooks.[2] On the other hand, by 1830 secretiveness seems to have been shared by native manufacturers.[3] Conversely, numbers of skilled immigrants were willing to train American-born workers.

These case histories of technology transfer have defined the role of individual immigrants from the British textile trades, but they provide little ground for assessing the impact of the total immigration of British textile workers. Indeed it might be con-cluded that despite their failings, immigrants were invariably a positive asset to, and largely responsible for, the early growth of the American cotton and woolen industries. This widely held impression is rigorously tested in part III.

III
Diffusion of the Technologies: The Impact of Aggregate Immigration

Introduction

Although it is true that a few key British workers, as individuals or small family and trade networks, conveyed nearly all the new textile technology across the Atlantic, it is quite a different matter, and an illogical conclusion, to assume that most or all British textile workers emigrating to the United States in the early nineteenth century carried with them skills indispensable to the growth of the U.S. textile industries. Yet this is the consensus opinion of most authorities. Heaton's work seems to be the basis of the notion. After studying lists of enemy aliens in the United States during the War of 1812, he pronounced, "There is a large company of men who have come, full of the new knowledge, from Lancashire, Lanarkshire, Yorkshire, and the West of England."[1] Berthoff lent strength to the view by observing that because the woolen industry continued to rely on traditional skills in weaving and finishing, "There were Englishmen, therefore, wherever woollen was woven in the United States during the 1820s and 1830s."[2] Thistlethwaite went further, asserting, "Following the original mechanical geniuses who built spinning frames, carding machines, looms, and printing machines, thousands of artisans came to provide an essential stiffening of skills for the labour force in textiles."[3] Most recently, Jones echoed the impression in a book based on a television series on American immigration: "Each of the basic American industries—textiles, mining, iron and steel—leaned heavily during their formative stage upon the technical know-how brought by British artisans, operatives and managers."[4]

In this part, these sweeping assertions are subjected to close scrutiny and tested. Quantitative evidence for the size and composition of immigration into America from the British textile trades is assembled in chapter 8. In chapter 9 it is utilized more exactly to estimate and explore the impact of aggregate British immigration at the industry, region, and firm levels. In this way I hope to define the characteristics and contributions of British textile immigrants as a group rather than as individuals.

8
Profiles of Immigration into America from Britain's Textile Trades, 1770s–1831

Evidence of the transatlantic movement of British textile workers has survived for three short periods between 1770 and 1830. This material gives three cross-sectional views of aggregate immigration. The first cross-section of emigration, recorded in the British Customs passenger lists from December 1773 through April 1776, gives the emigrant's name, age, social rank, occupation, family, former residence, port destination, vessel data, and reason for leaving Britain.[1] The lists have previously been studied by Mildred Campbell, who estimated that of the 12,000 emigrants recorded, half were sailing for the West Indies or North America.[2] Of this 6,000, 80 percent were men, 12 percent women, and 8 percent (450–500) children. The majority went as indentured servants, paying for their passage with three or four years' servitude in the New World. Of the 6,000 bound for North America, 5 percent went to the West Indies, and perhaps some 5,000 went to the American colonies that later formed the United States.

Three qualifications must be borne in mind in using the 1773–1775 data (no textile workers appear after September 1775). First, they were taken at a high point in eighteenth-century emigration, being compiled for this very reason. Emigration from Ireland and Scotland, as well as England, peaked in the early 1770s, as Dickson and Graham, respectively, have shown.[3] Consequently projections or extrapolations cannot be made; the same proportions of skilled workers would not necessarily leave Britain when the volume of emigration contracted. Second, those returns record emigrants, not immigrants. Some proportion very likely died on the voyage or arrived at a port different from the one stated in the returns. Thus, there is some discrepancy, presumably small, between actual immigration and the pattern adumbrated in the emigrant lists. But by and large, as Campbell showed, the data provide a reliable picture of emigration from Britain to North America on the eve of the American Revolution. Third, there is the possibility that the illegality of artisan emigration dissuaded some emigrants from disclosing their true occupations. The concern apparently did not affect a majority of emigrants, however. A very wide range of textile and machine-making trades was represented in the lists. Furthermore, Customs made no attempt, as far as is known, to halt emigrants. Although the number of manufacturers looks, at first glance, suspiciously small, it may well reflect the English economic situation rather than any dissimulation. At this time both cotton and woolen industries in England were experiencing the beginnings of a new expansion associated with the new technology.

The next cross-sectional view of British immigration comes in the War of 1812, but the extant sources are subject to greater distortions and incompleteness. After the United States declared war on Britain on June 18, 1812, the Department of State on July 7 ordered all immigrant males from Britain over the age of fourteen to register themselves with the marshal of the state in which they were living.[4] On the returns were listed names, ages, occupations, families, dates of arrival in the United States, American addresses, readiness to apply for U.S. citizenship, and, occasionally, physical attributes like color of eyes or hair, and even the marshal's assessment of the immigrant's political loyalty. Heaton estimated that about 10,000 men and youths

reported themselves. The returns, however, are incomplete; those for New Hampshire have been lost, and those for Connecticut, Pennsylvania, and Rhode Island appear to be partially lost. Heaton's estimates also did not take account of the presence of duplicates in the returns, chiefly in those for New York.[5]

The major difficulty in using the 1812 returns for a view of immigration is that they record a solitary census of existing immigrants. Consequently there is no way of estimating changes due to death, naturalization (for which five years' residence was required after 1801), change of occupation, or reemigration—what can be termed the attrition rate. Only the difficulty of naturalization can be eliminated by choosing immigrants who arrived within five years of registration. This has been done in comparing these immigrants with those of the 1770s and 1820s. One major advantage of the 1812 returns is that they give an accurate occupational profile. In the United States, British immigrants had no reason to conceal their skills; quite the opposite, for possession of the new industrial skills could win American approbation.

For the 1820s, a third profile of British textile workers arriving in the United States comes from the American ships' passenger lists (which are now on microfilm at the U.S. National Archives).[6] Instituted by the federal Passenger Act of March 2, 1819, and commencing in 1820, the passenger lists supply the name, sex, age, occupation, immediate family, date of arrival, vessel data, and sometimes the former British address of each passenger disembarking in the American port. But these lists too have shortcomings as census evidence. Most, but not all, captains seem to have been conscientious in performing the clerical chore of compiling the lists. As with their late-nineteenth-century successors, the minority of lazy or harassed captains cut corners by the generous use of ditto marks under occupational nouns as "lads," "laborers," or "farmers."[7] And in the early 1820s some ships' captains recorded "none," presumably meaning unemployed or illegal emigrants.

Several other limitations are inherent in these 1820s lists as a source of migration data. During 1824 and 1825 and before, it is quite likely that some emigrant artisans believed that they were breaking the law (the prohibition on artisan emigration remained in force until June 21, 1824) and therefore concealed their true occupations from the ship's captain. Certainly in 1824–1825 a lower number of textile workers emigrated than later in the decade. But it is impossible to unravel such legal effects from economic influences, like the improved trade conditions prevailing in Britain in 1824–1825. Also, passengers' nationalities were never consistently stated. Sometimes English, Welsh, Scottish, and Irish were distinguished; sometimes all were classified as British. These vague national origins make the possibility of tracing the workers' backgrounds extremely difficult. What, for example, can be done with weavers from Liverpool with Irish-sounding names? Did they come from Ireland or Lancashire, and did they represent experience from an industrial or a preindustrial context? Only in a small minority of cases were places of birth cited. Information on destinations was rarely more than the law required. However, in rare cases, towns of intended residence were noted.

Another drawback of the passenger lists is their incompleteness. Out of a possible total of 288 months' lists for the three ports, some 31 months' lists are incomplete or wholly missing: for New York, February 1828; April–June, September, November, December 1829; July–November 1830 (18 lists only extant for these months); January, April, May, August 1831 (one list extant for these months); and for Boston, January–March 1826, July–September 1827, July–December 1828, October–December 1831. Evidence of annual fluctuations cannot easily be found.

The data thus allow aggregate profiles of the occupational and age structures of immigrant workers. In that they record names of family members traveling together, the lists also offer evidence of networks of distribution. Obviously the margin of error in extracting kin relationships from the lists is high. The only links between adult inividuals are shared surnames and propinquity in the passenger lists. Only an uncommon surname would link a wife following her husband to America. Therefore none but emigrants traveling on the same ship, sharing the same surname, and recorded sequentially on the passenger lists have been classified as related. I have applied this rule to isolate the single or unaccompanied emigrants needed for comparison with the 1770s and War of 1812 groups.

A final methodological point must be made about the use of the names in the passenger lists: I have excluded individuals wherever the ditto marks in the occupation column suggest, by their abundance or palaeographic obscurity, that a captain was failing in his clerical duty.

Previously the passenger lists for the 1820s have been utilized only by Charlotte Erickson.[8] Taking the years 1827, 1831, and 1841, she has concentrated on the backgrounds of English workers. Her preliminary findings suggest that immigration from England in the late 1820s was characterized by preindustrial craftsmen and farmers traveling in families, a migration that achieved economic absorption in rural America comparatively easily. Equally important, Erickson's work, and mine below, shows that official American immigration statistics grossly understated the true situation. Figures for immigrant weavers illustrate the point. William Bromwell, who made an abstract in the 1850s using congressional reports, State Department abstracts, and customs house papers (and therefore has been criticized for overstating the volume of immigration), computed 2,588 male weavers and spinners of all nationalities entering the United States from 1824 through 1831.[9] But in

the passenger lists for New York, Philadelphia, and Boston I found 2,429 male weavers (not including specialist weavers) and 182 spinners from Britain alone. These figures must understate the real position because of the missing passenger lists.

In spite of the shortcomings of these sources, they can be used to reconstruct the volume and quality of immigrant textile workers. Changes in the range of immigrant skills afford one indicator of the quality of the supply of immigrant labor. Table 8.1 shows the result of an analysis of immigrant trades from the 1770s through 1831 (listed in appendix B). In this, as in all of the other tables, the various operative trades have been grouped in categories corresponding to the major stages of production—preparatory, spinning, weaving, and finishing—in

8.1. Trades Conveyed from Britain to the United States, 1770s–1831

	1773–1775		1809–1813		1824–1831	
	No.	%	No.	%	No.	%
Operatives						
Preparatory	4	8.89	6	10.53	9	8.26
Spinners	5	11.11	6	10.53	14	12.84
Weavers	16	35.56	12	21.05	18	16.51
Finishers	9	20.00	8	14.03	22	20.18
Total	34	75.56	32	56.14	63	57.80
Managers	2	4.45	8	14.03	11	10.09
Machine Makers						
General	3	6.67	7	12.28	20	18.35
Textile	6	13.33	10	17.54	15	13.76
Total	9	20.00	17	29.82	35	32.11
Total	45	100	57	100	109	100

order to facilitate the analysis. Workers from textile trades other than cotton and woolen have been included because their fiber-processing skills were in many respects similar, and they therefore represented a potential supply of skilled labor for America's cotton and woolen industries.

The doubling of the number of textile occupations over the period reflects the increasing specialization of labor accompanying industrialization in Britain. Exceptional is the relative diminution of numbers of weaving specialists. In fact the range of weaving trades was the only one that did not double in size between the 1770s and 1820s. This could reflect a decreasing number of weaving specialists or, a variation of the trend, the hand loom weavers' search for any and all sorts of work to stave off the unemployment that increased in the British cotton industry during the late 1820s with the spread of the power loom. The increase in the number of finishing trades was also notable. Among machine makers, the dramatic rise in the number of machine-making trades may be a function of the trades chosen. In selecting these trades, problems of comparability had to be resolved. Clockmakers, for example, were potential textile machine makers in the 1770s, but it is very doubtful whether they still supplied gear-cutting skills in Britain or America after 1810. Hence I have included them only in the 1770s lists. Furthermore I have, with some arbitrariness, omitted a number of general machine-making trades. Wrights, carpenters, joiners, smiths, blacksmiths, founders, ironfounders, steel workers, brass founders, copper founders, braziers, filers, and file cutters were all left out and for two reasons: colonial America possessed these trades already, so they added little to the existing stocks of

skill, and their involvement in textile machine building was very indirect, uncertain, or small.[10] Spinsters, whose title now proclaimed marital rather than occupational status, have also been omitted from the 1770s profile.

As important as the range of trades are the numbers of workers in them. Was there really "a large company of men . . . full of the new knowledge," as Heaton claimed? Throughout the period, weavers consistently formed the largest group of textile workers emigrating to the United States. In the 1770s and from 1809 through 1813, two-thirds of all textile workers were weavers; by the 1820s the proportion fell to a half. In absolute terms the weavers' average annual rates of immigration to the United States rose from over 100 in the 1770s and the period 1809–1813 to over 330 in the 1820s, as table 8.2 illustrates.

Circumstantial evidence indicates that in the 1820s most of these workers were still hand loom weavers. As Bythell showed, Britain's work force of hand loom weavers peaked in the mid-1820s at around 200,000 to 250,000.[11] Simultaneously power loom weaving was expanding. Baines estimated that English and Scottish power looms quadrupled from 14,150 in 1820 to 55,500 in 1829, with operatives running at least two looms each.[12] Wage differentials made power loom weaving far more attractive than hand loom weaving. In the latter, piece rates were falling; in the mid-1820s, they stood at a third of their 1815 level, which in turn was well under half of 1790s levels.[13] By 1825 a hand loom weaver's average net earnings in Manchester were between four shillings threepence and six shillings sixpence a week, compared with seven shillings sixpence to ten shillings sixpence a week

8.2. British Immigrants in Textile Trades Entering the United States, 1770s–1831

	1773–1775		1809–1813		1824–1831	
	No.	%	No.	%	No.	%
Operatives						
Preparatory	36	6.91	10	1.24	76	1.52
Spinners	22	4.22	70	8.68	427	8.54
Weavers	334	64.11	525	65.14	2,695	53.91
Finishers	30	5.76	30	3.72	434	8.68
Total	422	81.00	635	78.78	3,632	72.66
Managers	12	2.30	79	9.80	366	7.32
Machine Makers						
General	38	7.29	58	7.20	848	16.96
Textile	49	9.41	34	4.22	153	3.06
Total	87	16.70	92	11.42	1,001	20.02
Total	521	100	806	100	4,999	100

for Manchester power loom operatives (male or female between fourteen and twenty-two years) in 1826–1827.[14] In short, conditions and earnings in British weaving trades strongly suggest that hand loom weavers rather than power loom weavers would choose to emigrate in the 1820s.

Whether the immigrant hand loom workers were carriers of obsolete or new industrial skills is not easy to answer. One English hand loom weaver built the first spinning jenny in America, and another helped to perfect the Scotch power loom in Rhode Island. But relatively few weaving trades were taken over by the power loom, even by the 1830s. Baines in the mid-1830s reported that only in weaving calicoes, velvets, shirtings, and smallwares was the power loom dominant in Lan-

cashire. Printing calicoes, muslins, stripes, checks, ginghams, counterpanes, quiltings, cloths with fancy and figured weaves, shawls, and mixed cotton and linen or cotton and worsted fabrics were still woven on hand looms.[15] Suggesting a lack of such hand loom skills among the immigrant British weavers entering the United States is the fact that in both the 1809–1813 and 1824–1831 periods, over 93 percent of them described themselves plainly as "weaver," with no indication of specialization.

To the extent that the power loom had not taken over the weaving of very fine yarns or intricate drafts, hand loom weavers might not be conveying obsolete skills to the United States. But weavers whose hand looms were not immediately competing with power looms had much less motivation to emigrate than those who were being displaced. It is probable that the majority of emigrant weavers were carrying outmoded hand loom techniques with them, and a substantial proportion of immigrant weavers must have had no industrial experience whatsoever, since they came from rural backgrounds or, as Bythell observed, used weaving to supplement other urban trades when seasonal or cyclical unemployment came—though in this last case it is impossible to tell which occupation (weaving or the joint occupation) was the primary one.[16] Finally, not much hand loom weaving required special skill. All but fancy and figured hand loom weaving could be mastered in a matter of three to six weeks, so that immigrants added little to existing American stocks of such skills.[17] Thus most hand loom weavers going to the United States in the 1820s were unlikely to carry any but obsolete skills.

In the United States, where the number of power looms increased from under two thousand in 1820 to perhaps thirty-three thousand in 1831, power loom weavers were in great demand, especially by the mid-1820s, as Major Thomas Moody of the Royal Engineers, back from New York, reported to a Commons committee in 1826–1827.[18] But wages for power loom weavers below overseer level may not have been much better than in Manchester. In 1834 Lowell mill girls earned $1.90 a week (exclusive of boarding), on average, for running two or three power looms.[19]

If half to two-thirds of all textile immigrants—and 74 to 80 percent of textile operatives—between the 1770s and 1820s were hand loom weavers possessing no new industrial skills, what trades were likely to take these new skills across the Atlantic,

and how many workers were involved? Among operatives, no emigrant skills in the early 1770s can be certainly identified with the new textile technology. Over the five years 1809–1813, cotton carders (4), wool carders (1), spinners (19), cotton spinners (38), mule spinners (1), and wool spinners (7), all males over fourteen, most likely were familiar with the new technology. By the 1820s the range of industrial skills broadened. During the eight years 1824–1831, carders (17 male, 2 female), cotton carders (5 male), spinners (182 male, 71 female), cotton spinners (118 male, 6 female), wool spinners (5 male, 1 female), cotton twisters (1 male) and reelers (1 female), warpers (7 male, 3 female), dressers and sizers (11 male), machine printers (1 male), and calico printers (188 male, 18 female) were probably familiar with the new industrial techniques, though some of these trades still contained craft workers, especially calico printing.

Among managers, the picture was rather unfavorable toward American prospects. None of the dozen clothiers of the early 1770s could have had much, if any, technical familiarity with the new Arkwright technology being developed in the English cotton industry. At most they might have been acquainted with an early version of the carding machine or the jenny. Of the 79 managers who arrived in the United States from 1809 through 1813 and stayed with their trades, only 14 were certainly from the cotton industry; another 34 came from the woolen industry. From 1824 through 1831, no more than 14 cotton manufacturers' names appeared on the passenger lists. Yet this was a misleading occupational description, for at least six of them were merely factory operatives and hand loom weavers.[20] In addition there were 189 clothiers, including one female. The 28 manufacturers who emigrated in the 1809–1813 period and the 148 who arrived between 1824 and 1831 may well have come from the British textile trades. The term *manufacture* had been synonymous with the greatest of Britain's pre-eighteenth-century industries, the woolen manufacture. But it is impossible to know how many under this appellation came from the cotton and woolen industries in the 1820s.

The immigration of machine makers offered small comfort to American entrepreneurs hunting for British industrial skills, until the 1820s. In the 1770s the 49 machine makers classified as textile included 12 turners and 29 clockmakers or watchmakers. Of the 38 general machine makers then arriving, 25 were whitesmiths. On the eve of the War of 1812 the rate of immigration among machine makers was lower than the previous forty years. From 1809 through 1813, however, there was a larger number of machine-making trades. By the 1820s the rate of immigration of textile and general machine makers increased twofold and tenfold, respectively, over 1809–1813 rates.

The extent of the increases in the immigration of industrial workers between 1809–1813 and the 1820s may be magnified and distorted by several factors: disturbed Anglo-American relations prior to the War of 1812 may have reduced earlier levels of immigration, and the 1812 returns conceal attrition and ignore female workers. In the 1820s females amounted to 19 percent of spinners and 6.5 percent of finishers. But even if the 1809–1813 figures are increased by these margins in these trade categories, the difference between the 1820s profile and the earlier ones was slight. Only the unknown attrition rate of 1809–1813 could alter the picture much.

That picture is of a low-volume immigration carrying new industrial skills to the United States but with signs that the rate was appreciably rising over the first three decades of the nineteenth century. Among cotton and woolen operatives definitely associated with the new industrial technology, the average annual immigration rate rose from 14 (excluding any allowance for females) or 18 (making that allowance), plus a margin for attrition, to 77 between 1809–1813 and the 1820s. Among manufacturers with new industrial experience, the average annual number of immigrants apparently stayed around 2 to 3 for cotton managers (unless manufacturers are counted, in which case they rose from 5.25 in the 1809–1813 period to 20 in the 1820s), but increased for woolen managers from 7 before 1813 to nearly 24 in the 1820s. Textile machine makers' immigration rates rose from 7 to nearly 20 per year, and general machine makers' immigration rates shot up over the same period from around 10 to over 100 per year.

Next the amount of experience embodied in these textile workers must be established. Did they tend to be young and therefore inexperienced or rather older and therefore more experienced, and did the degree of experience vary over time and between trades? Table 8.3 shows that the immigration of British textile workers was a movement of young workers. Immigration in the 1770s was most characteristically youthful, with over 18 percent in their teens and another 52 percent in their twenties, no doubt because most were going to the New World as indentured servants and therefore could not take families.[21] In the unsettled years before 1812, teenage immigration fell off but then picked up again in the 1820s, when nearly 9 percent

8.3. Textile Immigrants, by Age and Trade, 1770s–1831

Age	1773–1775	1809–1813	1824–1831
Operatives			
14 and under	3.79%	0%	1.40%
15–19	18.01	3.31	7.77
20–29	51.66	59.68	53.96
30–39	15.88	22.36	20.76
40 and over	10.66	14.65	16.11
(N)	(422)	(635)	(3,632)
Managers			
14 and under			
15–19			5.74
20–29	58.33	58.23	42.08
30–39	30.00	22.78	23.22
40 and over	11.67	18.99	28.96
(N)	(12)	(79)	(366)
Machine makers			
14 and under			1.20
15–19	11.49	4.35	5.49
20–29	62.07	56.52	52.75
30–39	18.39	21.74	24.67
40 and over	8.05	17.39	15.89
(N)	(87)	(92)	(1,001)

of the immigrants were in their teens. Over the period as a whole, however, a rising proportion in all trade categories was over forty years of age.

A major variation in the dominant pattern of youthfulness occurred among the managers, who from the 1770s to the 1820s were an older group. By the 1820s over a quarter of the managers were above forty years of age. Such men took with them to America at least twenty years' experience, assuming that they were brought up in their trades.

In all of the textile trades—operatives, managers, and machine makers—the proportions of immigrants in their twenties remained high, at levels that might suggest a preponderance of Irish and Scots among the textile workers. The twenties cohort among the textile workers in the 1809–1813 period accounted for 60 percent of textile immigrants and in the 1824–1831 period nearly 54 percent. In contrast Charlotte Erickson found that scarcely 36 percent of English and Welsh immigrants aged over fourteen were in their twenties in 1831. But in that year 53.3 percent of the Scots and 56.4 percent of Irish immigrants to the United States were in their twenties.[22]

The age structure of immigrants carrying industrial skills is disclosed in table 8.4. Among operatives, cotton spinners and calico printers tended to be older than the average textile operative, with larger than average cohorts over forty and smaller than average cohorts in their twenties. On the other hand, plain spinners, 28 percent of whom were female in the 1820s, had larger than average cohorts in their twenties and thirties. Certainly there was no consistent tendency for the new industrial workers to be in their twenties. Such a tendency might have pointed to the occupational pattern traced by

8.4. Industrial Textile Workers, by Age and Trade, 1809–1831

	1809–1813				1824–1831			
Age	Spinners	Cotton Spinners	Calico Printers	All Operatives	Spinners	Cotton Spinners	Calico Printers	All Operatives
Operatives								
14 and under	0%	0%	0%	0%	1.98%	2.42%	4.85%	1.40%
15–19		5.26		3.31	7.11	5.64	3.89	7.77
20–29	63.16	47.37	40	59.68	54.94	49.19	43.69	53.96
30–39	26.32	21.05	60	22.36	22.13	24.19	22.81	20.76
40 and over	10.52	26.32		14.65	13.84	18.56	24.76	16.11
(N)	(19)	(38)	(5)	(635)	(253)	(124)	(206)	(3,632)
	Manufacturers	Cotton Manufacturers	Woolen Manufacturers	All Managers	Manufacturers	Cotton Manufacturers	Woolen Manufacturers	All Managers
Managers								
14 and under								
15–19					4.73		7.41	5.74
20–29	57.14	50	61.76	58.23	39.19	50	40.74	42.08
30–39	21.43	35.7	17.65	22.78	27.70	14.28	20.63	23.22
40 and over	21.43	14.3	20.59	18.99	28.38	35.72	31.22	28.96
(N)	(28)	(14)	(34)	(79)	(148)	(14)	(189)	(366)

Anderson for the 1850s, when most males who started in the cotton industry were forced to leave in their late teens in search of higher wages.[23]

Among industrial managers too there was no simple pattern distinguishing industrial from non-industrial experience. Manufacturers tended to be older than average, as were woolen manufacturers by the 1820s. By this time nearly a third of the woolen managers were over forty at time of immigration. Yet woolen managers contained a higher proportion of teenagers than the whole manager group. The age structure of cotton manufacturers is much less reliable in view of the tiny number of immigrants in the trade.

The implications of these age structures for the U.S. textile industries are difficult to determine in aggregate. At least it is certain that the majority of textile immigrants arrived from Britain with the experience of teenage work behind them. Most were in their twenties and thirties and presumably open to new ideas and work methods. Among managers, where long experience and the chance to accumulate capital was of most value, there was an above average proportion of immigrants (over a quarter) above the age of forty by the 1820s.

Prospects of permanency and stability among immigrant workers depended considerably on the extent of their family responsibilities. The accompaniment of a family restrained an immigrant from making more rapid or riskier changes of occupation or location. Conversely, unaccompanied immigration permitted higher degrees of occupational and geographical instability.

Marital status was adumbrated by the immigrants' age structure. Trades in which workers were younger tended to be trades in which they were

8.5. Percentage of Single or Unaccompanied Workers among British Immigrants in Textile Trades Entering United States, 1770s–1831

	1773–1775	1809–1813	1824–1831
Operatives			
Preparatory	91.67%	70%	61.84%
Spinners	100	68.57	62.06
Weavers	88.62	62.48	59.55
Finishers	100	63.33	60.59
Industrial operatives			
Spinners	[a]	68.42	56.13
Cotton spinners	[a]	63.16	70.97
Calico printers	[a]	20	60.19
Managers			
Industrial manufacturers	[a]	67.86	56.08
Cotton manufacturers	[a]	28.57	57.14
Woolen manufacturers	[a]	70.59	58.73
Machine makers			
General	97.37	60.34	68.28
Textile	97.96	73.53	58.17
Total for all trades	91.36	63.15	61.15
(N)	(521)	(806)	(4,999)

a. Not calculated; numbers very small for period 1773–1775.

single or at least unaccompanied. This was especially true for the 1770s, when over 73 percent of emigrant workers were in their teens and twenties while 91 percent were single or unaccompanied, again reflecting the presence of indentured servants (table 8.5).

In the early nineteenth century the proportion of textile workers emigrating without families came down to just over 60 percent, which was still much higher than the proportion for all British immigrants in 1831. For this year Charlotte Erickson found 23 percent of English and Welsh immigrants, 34 percent of Scots, and 40 percent of Irish immigrants arriving without families.[24] The fall in the number of single workers among the textile immigrants strongly suggests that prospects of secure work in the U.S. textile industries in the 1820s were encouraging more British textile workers to take their families with them, either initially or on a subsequent visit to the country.

Between 1809 and 1831, manager groups had consistently lower proportions (59 percent in 1809–1813 and 57 percent in the 1820s) of unaccompanied individuals than did operatives or machine makers. This is not a surprising figure. Higher proportions of managers were over forty years of age. And managers commanding higher levels of skill and sometimes possessing capital had stronger reasons for regarding emigration as a final or prolonged departure from Britain.

Machine makers' immigration exhibited two contrasting trends in the early nineteenth century. General machine makers were increasingly inclined to be single, and textile machine makers increasingly tended to take their families with them. Reasons for the difference are not readily apparent, for

both should have seen the expansion of U.S. textile manufacturing and the consequent demand for capital goods as strengthening their American opportunities. Why the general machine makers apparently did not is puzzling. Of course, shifts toward family emigration in the 1809–1831 period would appear greater because of the uncertain conditions before 1812.

Among operative and managerial trades with experience of the new industrial forms, there were signs that American industrial expansion was encouraging British textile workers to regard emigration as a permanent or long-term measure. Spinners increasingly took their families with them (44 percent in the 1820s) as did manufacturers (44 percent) and woolen manufacturers (42 percent in the 1820s). Small numbers of cotton manufacturers make it difficult to discern a trend for that group.

In short, prospects for stability among immigrant textile workers were rising between the War of 1812 and the 1820s. Greater than average and increasing proportions of immigrants accompanied by family were registered by operative spinners, manufacturers, woolen manufacturers, and textile machine makers, groups with favorable expectations in both the American and home labor markets, as Erickson pointed out.[25] Prospects for occupational and geographical stability among these trades, which formed the bulk of those likely to carry the new industrial technology (assuming that the manufacturers were frequently cotton manufacturers), looked brighter in the 1820s than they had two decades earlier. The geographic distribution of these immigrants and, far more, their numbers in relation to the expansion of the U.S. textile industries would determine the relative value of this improving stability.

American destinations of emigrant textile workers from Britain underwent a major shift between the 1770s and the 1820s. Whereas 70 to 80 percent of textile immigrants arrived in ports south of Philadelphia in the 1770s, by the 1820s that proportion was going to New York (tables 8.6–8.8). The explanations for the shift are clear. By the 1820s the indentured servant traffic was at an end, cotton and woolen mills were multiplying in the Northeast, and American foreign trade was increasingly funneled through New York, with its easy access to the Atlantic, and the Great Lakes and Midwest after the Erie Canal opened in 1825.[26]

In the three sources used to plot immigration profiles, the War of 1812 returns of enemy aliens are unique in showing where the immigrants were settling, as opposed to disembarking. Tables 8.6–8.8 show that New York–New Jersey retained 40 percent of operatives and managers but only 23 percent of machine makers. Most attractive to the machine makers was the Pennsylvania–Delaware region, where 43 percent of them settled. There too went 40 percent of the managers.

An examination of all immigrant textile workers from Britain registered during the War of 1812 (1,349 workers) showed New England short of managers and machine makers, compared with the Philadelphia region. The former had 23 percent of managers and 17 percent of machine makers, and Pennsylvania–Delaware had 32 percent of man-

8.6. Geographic Locations of Operatives among British Immigrants in Textile Trades, 1770s–1831

Port of Entry or Place of Residence	1773–1775		1809–1813		1824–1831[a]	
	No.	%	No.	%	No.	%
New England			145	22.83	312	8.59
New York and New Jersey	127	30.09	207 / 58 } 265	41.73	2,586	71.20
Pennsylvania and Delaware	59	13.98	103 / 48 } 151	23.78	734	20.21
Maryland	166	39.34	63	9.92		
South	70	16.59	6	0.94		
West			5	0.79		
Total	422	100	635	99.9	3,632	100

Note: The figures provided are according to point of entry for the 1770s and 1820s and for actual location during the War of 1812.

a. The passenger lists for Baltimore and the other Atlantic ports have not been used because the total volume of immigration through them was under 5 percent of that through Boston, New York, and Philadelphia, judging by the passenger lists on National Archives microfilm.

8.7. Geographic Locations of Managers among British Immigrants in Textile Trades, 1770s–1831

Port of Entry or Place of Residence	1773–1775		1809–1813		1824–1831	
	No.	%	No.	%	No.	%
New England			16	20.25	24	6.56
New York and New Jersey	1	8.33	26 5 } 31	39.24	289	78.96
Pennsylvania and Delaware	7	58.33	21 10 } 31	39.24	53	14.48
Maryland	2	16.67	1	1.26		
South	2	16.67				
West						
Total	12	100	79	100	366	100

See note to table 8.6.

8.8 Geographic Locations of Machine Makers among British Immigrants in Textile Trades, 1770s–1831

Port of Entry or Place of Residence	1773–1775		1809–1813		1824–1831	
	No.	%	No.	%	No.	%
New England	1	1.15	9	9.78	78	7.79
New York and New Jersey	3	3.45	13 8 } 21	22.83	744	74.32
Pennsylvania and Delaware	13	14.94	23 17 } 40	43.48	179	17.88
Maryland	54	62.07	21	22.83		
South	16	18.39	1	1.09		
West						
Total	87	100	92	100.01	1,001	99.9

See note to table 8.6.

8.9. Locations of All British Immigrants in Textile Trades during the War of 1812

| Place of Residence | Operatives | | | | | Managers | | | Machine Makers | | | Grand Total |
	Preparatory Workers	Spinners	Weavers	Finishers	Total	Woolen Managers	Cotton Managers	Total (Including all fibers)	General	Textile	Total	
New England		21.74%	19.73%	15.62%	19.3%	16.42%	40%	22.56%	17.60%	15%	16.76%	19.35%
New York	33.33%	17.39	35.96	25	33.5	40.29	36	30.49	23.20	16.67	21.08	31.43
New Jersey	22.22	18.48	7.38	6.25	8.6	5.98	8	6.71	7.20	6.67	7.03	8.15
Pennsylvania	22.22	16.30	13.20	12.25	13.6	16.42	8	21.34	9.60	36.67	18.38	15.20
Delaware	5.55	9.78	7.99	10.94	8.3	19.40	4	10.98	16	1.67	11.35	9.04
Maryland	11.13	10.87	10.41	21.88	11.2	1.49		4.88	17.60	18.33	17.84	11.34
South	5.55	2.17	2.06	6.25	2.4				4.80	3.32	4.32	2.37
West		3.27	3.27	1.56	3.1		4	3.05	4	1.67	3.24	3.11
(N)	(18)	(92)	(826)	(64)	(1,000)	(67)	(25)	(164)	(125)	(60)	(185)	(1,349)

See note to table 8.6.

agers (fifty-three men) and 30 percent of machine makers (fifty-five men) (table 8.9). Before concluding that this imbalance of British immigrant labor was related to New England's technical progress and Philadelphia's relative decline, it must be recalled that the New England returns are incomplete; those for New Hampshire are wholly missing, those for Connecticut partly missing, and those for Rhode Island probably partly missing.

A closer look at the 1812–1814 locations of workers conveying the new industrial skills shows that they were widely scattered through the seaboard states north of Maryland, with New York possessing the greatest share in most trades. A third of the 92 spinners were in New York and New Jersey. Nearly two-fifths of the 25 cotton managers were in New York, a fifth in Rhode Island, and two each in Massachusetts, Connecticut, New Jersey, and

Pennsylvania. About 40 percent of the 67 woolen managers were in New York, nearly 20 percent in Delaware, nearly 17 percent in Pennsylvania, and 12 percent in Massachusetts. Of the 125 general machine makers, 23 percent were in New York, 16 percent in Delaware, nearly 18 percent in Maryland, and 10 percent each in Pennsylvania and Massachusetts. The 60 textile machine makers were scattered; Pennsylvania led with 37 percent, double the shares of those in the next two places, Maryland and New York.

Within states there was a tendency for immigrants to cluster. In Philadelphia County, Pennsylvania, for example, there were 5 cotton spinners, 4 spinners, and 1 carder-spinner. At Columbiaville on the Hudson, twenty-five miles south of Albany, New York, a colony of British cotton workers was formed around a core of machine makers: 2 millwrights, 1 machine maker, 2 cotton machine makers, 1 cotton spinner, and 1 fustian cutter. Good mill sites and proximity to markets, as well as trade and kin networks among the immigrants, clearly conditioned the geographic dispersion of the immigrant textile workers. Only a series of local studies, beyond the scope of this chapter, could discover the significance of these small immigrant groups. At this time it can only be concluded that during the War of 1812, when the cotton and woolen industries in the United States rapidly expanded, British immigrants possessing new industrial skills were widely scattered through America's manufacturing districts.[27]

A closer scrutiny of the industrial skills carried to the United States in the 1820s revealed one significant geographic variation (table 8.10): a relatively high proportion of finishing workers among those emigrating through Boston. Whereas an average of 8 percent of textile immigrants entered the United States through Boston, nearly 24 percent of the finishers did. The explanation lies in the integration of calico printing in New England's cotton mills.

Immigration data disclose that at least 74 percent of the British textile operatives in America during the early nineteenth century were hand loom weavers. Only a small proportion of immigrants brought new industrial skills with them, but the proportion was slowly rising, reaching nearly 18 percent of operatives in the 1820s. Numbers of woolen manufacturers and general machine makers rose three and ten times to twenty-four and one hundred per year, respectively. Yet an overall 60 percent of workers in their teens and twenties, and a similar proportion unaccompanied by family, suggests a

8.10. Trades of Textile Immigrants from Britain by Port of Entry, 1824–1831

	New York	Phil-adel-phia	Boston	Total Number
Operatives				
Preparatory	67.10%	19.74%	13.16%	76
Spinners	72.36	18.03	9.60	427
Weavers	71.84	22.34	5.83	2,695
Finishers	66.82	9.23	23.96	434
Total	71.20	20.21	8.59	3,632
Managers	78.96	14.48	6.56	366
Machine makers				
General	75.12	18.40	6.49	848
Textile	69.93	15.03	15.03	153
Total	74.33	17.88	7.79	1,001
Total	72.39	19.32	8.28	
(N)	(3,619)	(966)	(414)	(4,999)

high level of transience. The geographic distribution of the textile immigrants, as indicated by ports of entry, showed a shift from the south Atlantic ports to New York between the 1770s and the 1820s, largely reflecting the rise of the port of New York as the gateway to the West. During the War of 1812, textile immigrants were widely scattered through the Atlantic seaboard states between Maryland and New England. Apart from the general preference for New York, only one other significant trend emerges in geographic distribution during the 1820s: the preference of finishing workers for Boston, and hence the New England bleacheries and printeries.

9
Aggregate British Textile Immigration and the Growth of America's Textile Industries in the Early Nineteenth Century

The majority of textile immigrants to the United States possessed obsolete skills, predominantly hand loom weaving. Relatively small numbers of workers with the new industrial skills—dozens rather than hundreds—reached the United States from Britain each year in the early nineteenth century, and an unknown number left their trades when in America. To demonstrate conclusively whether the influx of British textile workers was sufficient for the needs of the expanding American cotton and woolen industries, statistics of American industrial expansion are required. These have been assembled and comparisons made with concurrent levels of immigration without attempting to distinguish between textile workers with or without the new industrial skills.

During the years 1809 to 1813, 151 cotton and woolen companies were incorporated by the various states of the Union. If all companies went into operation, immigration from Britain could supply 4.2 operatives, 0.52 managers, and 0.61 machine makers per company. The figure for operatives favored American manufacturing prospects; not so the ratios of newly arrived managers and machine makers. All together, 79 managers and 92 machine makers arrived from Britain between 1809 and 1813, leaving shortfalls of 72 and 59, respectively. Arguably, earlier arrivals could come forward to meet these needs. Also, there are gaps in the War of 1812 records, not the least being the absence of women and children under fourteen. And some incorporated companies may not have functioned at all. Against these points must be balanced the number of unincorporated companies (including all companies in Rhode Island) coming into operation;

and, far more important, the fact that the immigrant statistics are gross figures, concealing the minority of immigrants carrying the new industrial skills.[1]

For the 1820s the case against a major British contribution to the early growth of American cotton and woolen manufacturing is much stronger still. Table 9.1 makes a gross comparison between the increase in U.S. cotton and woolen industry employment between 1820 and 1831–1832 and numbers of immigrant textile workers of all types between 1824 and 1831. The unavoidable absence of immigration figures for 1820–1824 (due to the prohibitory laws) is mitigated by two considerations. The early 1820s were years of low emigration from Britain, not least because the economy and the cotton industry in Britain were expanding. And in the gross immigration figures used, all textile workers have been counted, including the nonindustrial and those from outside cotton and woolen manufacturing. Another defect in the immigration figures used is that they do not include children under fifteen. Those employable in a textile mill, above the age of eight perhaps, might represent an additional 10 percent, judging by Erickson's figures for 1831.[2] The youngsters could work only in operative grades, however. Also omitted are the woolen immigrants of 1832 and those from the missing passenger lists. The last might add 11 percent to the immigration figures, since 31 of 288 months of lists are missing.

The reliability of the American census material might also be questioned. The U.S. census of manufactures is probably not complete, but as Margaret Walsh has noted, the early censuses are an approximation of reality on a national and regional scale, though not on a local level.[3] The major omission from the McLane Report of 1832 is the returns from Maryland, and this leads to an understating of the employment figures in the woolen industry and in machine making that year. The other weakness of the 1820 census relates to its timing. At this point the American cotton and woolen industries were emerging from depression. Consequently the many disused or undermanned mills recorded in the census returns confirm the 1820 position of unemployment in the textile industries, a position that causes some exaggeration of employment expansion between 1820 and 1831.

Compensating for all of these possible sources of distortion is difficult, if not impossible, and I have not attempted it here. Indeed it is hardly necessary, so great are the shortfalls between immigration and industrial expansion. New adult (over the age of fourteen) male immigration in the period 1824–1831 fell short of increased employment in the expanding American cotton and woolen industries by just over 14,000 workers. On a liberal estimate, it is safe to say that British textile immigrants could hardly have filled half the new jobs from current immigration. More conservatively, using the figures in table 9.1, they appear capable of filling only 25 percent of new jobs for male workers in the expanding American cotton and woolen industries of the 1820s, and perhaps half of all new jobs for all workers, if each male immigrant brought an employable family of four or five persons.

Similarly, at the managerial and machine-making levels new immigration from the British textile trades could not meet the labor demands created by expansion in the American textile industries in the 1820s (table 9.2). Whereas 356 new American cotton firms or mills were started between 1820 and 1831,

9.1. Growth in Employment in U.S. Cotton and Woolen Mills Compared with Volume of Immigration from British Textile Trades

Employment	1820	1831–1832	Increase from 1820 to 1831–1832
Cotton			
Adult males	2,637	18,539	15,902
Women and children	9,610	43,618	34,008
Total	12,247[a]	62,157[b]	49,910
Wool			
Adult males	1,892	4,749	2,857
Women and children	1,665	5,709	4,044
Total	3,557[a]	10,458[c]	6,901

	Aggregate	Adult Males
Increase in U.S. cotton and woolen industries' work force, 1820–1832	56,811	18,759
Immigration from British textile trades, 1824–1831[d]	4,999	4,724
Difference	51,812	14,035

a. USNA, Census of Manufactures, MS Returns, 1820. See Appendix D. Numbers here omit "adults" and "adults and children."
b. Friends of Domestic Industry, *Report*, p. 16.
c. *McLane Report* (1833). No returns for Maryland are given, and data for New York are incomplete.
d. U.S. Passenger Lists for New York, Philadelphia, and Boston, 1824–1831.

only 163 managers arrived from Britain between 1824 and 1831, and this number of immigrants included the 149 manufacturers. If cotton manufacturers only are counted, then 342 new firms or mills lacked British managers. In the woolen industry the gap was narrower: 81 firms or mills short of immigrant managers.

American textile-machine-making firms also expanded their employment faster than the rate at which immigrant machine makers arrived from Britain (table 9.3). Although female employment in America's capital goods industry contracted, male employment rapidly expanded, outpacing immigration to the extent of 1,600 jobs.

Without attempting to make corrections for the errors inherent in the sources, it appears that British immigration might supply about 46 percent of the new management positions opening up in the U.S. cotton industry in the 1820s (and possibly only 10 percent if cotton manufacturers only among the immigrants are counted), and 70 percent of the new management positions in the U.S. woolen industry. In American machine-making trades, immigration might meet 40 percent of the labor demand occasioned by industrial expansion. Even if allowance is made for older immigration for the years 1815–1823 (to include experience with Britain's latest power loom technology), at the same rate as later in the 1820s (a hypothetical and generous level), immigration would still have hardly supplied half the labor force required by the expanding American textile industries of the 1820s. Conservatively, well over half of America's cotton mills and something under half of its woolen mills would have been running without the assistance of immigrant managers and operatives. During the

9.2 Increase in Number of U.S. Cotton and Woolen Spinning or Spinning and Weaving Mills Compared with Number of Immigrant Managers Arriving from Britain

Mill Type	Number of Firms[a] 1820	Number of Firms or Mills[b] 1831–1832	Increase from 1820 to 1831–1832	Number of Immigrant Managers from Britain[d]	Ratio between Number of New Firms or Mills, 1820–1832, and Immigrant Managers, 1824–1831
Cotton (including cotton and woolen mills)	439	795[b]	356	163[e]	1:0.46
Woolen	293	565[c]	272	191[f]	1:0.70

a. USNA, Census of Manufactures, MS Returns, 1820. Most firms ran only one mill at this date.
b. Friends of Domestic Industry, *Report,* p. 16. This specified mills rather than firms.
c. *McLane Report* (1833) for woolen mill numbers. There are no returns from Maryland. For New York the two hundred mills estimated by the commissioner have been counted rather than the fifty-four evaluated. Firms and mills, not always distinguished, were counted equally.
d. U.S. Passenger Lists for New York, Philadelphia, and Boston, 1824–1831.
e. Includes 149 nonspecific manufacturers.
f. Includes clothiers and woolen cloth and worsted manufacturers.

War of 1812 and perhaps even earlier, American textile manufacturers began to cross the threshold into economic and technical independence of skilled industrial immigrants.

Because aggregate numbers reveal little about the distribution and activities of groups of immigrants, regional- and firm-level studies are needed to define and explain further the impact of British textile immigrants. Given the preponderance of preindustrial skills among the textile immigrants, one possibility at the regional level is that a concentration of them would delay the adoption of new technology. The one area best known in the American textile industry for the persistence of preindustrial methods was Philadelphia, where hand loom weaving survived until the eve of the Civil War.

Philadelphia's circumstances therefore invite investigation.

As late as 1859, Freedley reckoned that Philadelphia (population 600,000) contained 6,000 hand looms working on cotton goods, carpeting and hosiery. Their operatives worked either at home or in small weaving sheds of twelve to thirty looms. While power loom factories in the city produced $13.2 million worth of goods per year, the hand loom cotton and woolen weavers (excluding hosiery) made $4.76 million worth of goods.[4]

Philadelphia must be viewed as part of the industrial district covering southeastern Pennsylvania and northern Delaware. Only in this regional context is the survival of hand loom weaving explicable. The differentiation between this region

9.3. Growth in Employment in U.S. Textile-Machine-Making Industry Compared with Number of Immigrant Machine Makers Arriving from Britain

Labor	1820[a]	1832[b]	Change between 1820 and 1832	Immigrant Machine Makers from Britain 1824–1831[c]	Difference Between Increase in U.S. Machine Makers' Work Force and Number of Immigrant Machine Makers
Men	508	3,138[d]	2,630	992	−1,638
Women and children	249	68	−181	9	
Total	757	3,206	2,449	1,001	−1,448

a. USNA, Census of Manufactures, MS Returns, 1820. Ironworks have been excluded unless they mention machine making.

b. *McLane Report* (1833). This has no returns from Maryland or states to the south and none from the west except Ohio. Ironworks that specifically recorded machine making have been included.

c. U.S. Passenger Lists for New York, Philadelphia, and Boston, 1824–1831.

d. Compare this figure with the number of 3,200 American machine makers given for 1831 in the Friends of Domestic Industry, *Report*, p. 16.

and the northern states began before 1820.[5] That year Pennsylvania and Delaware together had under 11 percent of the national fixed capital investment in industrial cotton manufacturing. Measured by capital investment in cotton manufacturing, Pennsylvania ranked sixth and Delaware ninth in the United States. The two states also had a small share of the country's large firms, those capitalized at over $75,000; only five of the thirty-six in this category belonged to Pennsylvania and Delaware.

The difference between the Pennsylvania-Delaware area and the rest of the Atlantic states north of the Potomac was most apparent in fixed capital investment in power looms. Although this cannot be measured in monetary terms because looms and spindles were not required to be costed separately in the 1820 census, numbers of power looms sufficiently suggest the situation. Of the 1,665 power looms (1,623 in cotton, 42 in cotton and woolen firms) recorded in the United States in 1820, under 3 percent were in Pennsylvania and Delaware combined, and this was as many as in remotest Vermont. In contrast, the six New England states contained 60 percent of the country's power looms, New York 20 percent, and Maryland 11 percent. The Philadelphia district could not even boast large-scale efforts at power loom weaving; its largest unit was one factory with 20 power looms, compared with units of over 50 looms in New Hampshire, Massachusetts, Rhode Island, New York's southern district, and Maryland. Advantages of scale were especially sought by the Boston Manufacturing Company, which had 175 power looms, and a Maryland company, the Patapsco Manufacturing Company, with 80, in 1820.

The region's specialization in hand loom weaving was apparent by 1820. While there were still numbers of hand loom weavers working full time or part time throughout the Union in households or in weaving sheds attached to cotton spinning mills, nowhere but in Philadelphia was there a geographic concentration of small, horizontally specialized cotton weaving firms. All 15 firms were located in Philadelphia city and county. Their total capitalization, puny compared with cotton spinning or spinning and weaving mills, was $19,843. Their average size in hand looms was 11.6 (14 firms). Most had between 4 and 12 looms each, though one had 22 and another 40 looms. Their operations relied on very low capitalizations. Whereas each workers, on average, in 9 cotton spinning mills in the Philadelphia region represented $990.45 of fixed capital investment and each in 11 cotton spinning and weaving mills $608.37, the workers in the cotton weaving firms each represented only $81.66. Conversely, weaving firms were the most labor intensive (table 9.4). Their high proportions of male workers, the highest in all five types of cotton manufacturing businesses in the region, and the extremely low proportion of child labor, are also noticeable.

City directories show a growing concentration of hand loom weavers within Philadelphia between 1790 and 1830. Table 9.5 demonstrates that weavers increased both as a proportion of all textile workers in the city and as a proportion of the city's population. As a proportion of all textile workers, hand loom weavers increased from around 50 percent in the 1790s to 60 percent in 1821 and nearly 70 percent in 1830. Over the period 1790–1830, their rate of increase was triple the rate of growth of the city's population. And these figures, because of the limitations of city directories, understate the case. Although the directories list nearly 700 weavers living within city boundaries, it was reported in 1827 that 4,500 persons in the city were weaving cotton goods on hand looms. If this was the case, hand loom weaving rapidly expanded in the 1820s.[6]

Some explanation of the phenomenon may be found in the sorts of hand loom weaving that took place in the city and the regional structure of the cotton industry. Many in the hand loom trades seem to have survived after 1815 by turning to fancy and specialty goods: goods needing drop box mechanisms to insert a variety of weft colors, like checks and plaids; fancy cloths, principally satin weaves and, by 1824, jacquard-woven goods; smallwares, like tapes, ribbons, saddle girths, and webbing, though these had been made on powered Dutch or ribbon looms since the 1750s in England; and specialties like fringe, coach lace, and military braid.[7] That Philadelphia's goods were not commonly manufactured was hinted by the Morrises, Lancashire hand loom weavers, who referred not only to the city's coarse weaving but also to "many other kinds," adding, "It is all very good to learn."[8]

Incomplete data on the region's factory organization in the late 1820s and early 1830s show a preponderance of spinning factories and finishing establishments but few power loom factories.

9.4. Deployment of Labor in Cotton Manufacturing Firms in Philadelphia and Delaware Counties, Pennsylvania, and New Castle County, Delaware, 1820

Mill Type	Average Fixed Capital Investment (K)	Labor (L) Aggregate			Average per Firm			Sum of Averages of Men, Women, and Children	Crude Capital Intensity[a]	Crude Labor Intensity[b]
		Men	Women	Children	Men	Women	Children			
Cotton Spinning (9 of 11 firms reported L; all K)	$21,581.82	32	40	114	3.5	4.4	13.89	21.79	990.45	0.0010096
Cotton Spinning and Weaving (11 firms reported L and K)	49,472.7	231	204	458	21	18.5	41.82	81.32	608.37	0.0016437
Cotton Weaving (15 firms reported L and K)	1,322.86	148	79	15	9.86	5.26	1	16.12	81.66	0.0617298
Cotton Finishing (1 firm)	8,000	6			6			6	1,333.3	0.00075
Cotton and Wool Spinning plus weaving (1 firm)	80,000	35	16	90	35	16	90	141	567.4	0.0017625

Source: USNA, Census of Manufactures, MS Returns, 1820.
a. Determined by the formula average K/average L.
b. Determined by the formula average L/average K.

9.5. Textile Workers in Philadelphia City, 1790–1830

| Year | Operatives | | | | | Finishers | Managers | | Machine Makers | | Total | City Population |
| | Preparatory Workers | Spinners | Weavers | | | | Textile | Total | Textile | Total | | |
			Carpet	Stocking	Total							
1793	4 (5.1)			16	42 (53.8)	10 (12.8)	2	3 (3.8)	15	19 (24.3)	78	28,500 (1790)
1801	2 (4.2)	14 (29.2)		5	20 (41.7)	7 (14.6)			2	5 (10.4)	48	41,000 (1800)
1810	5 (4.8)	2 (1.9)	1	10	59 (56.7)	20 (19.2)	3	3 (2.9)	4	15 (14.4)	104	54,000 (1810)
1821	5 (1.4)	7 (1.9)	18	15	223 (62.3)	37 (10.3)	30	30 (8.4)	7	56 (15.6)	358	63,000 (1820)
1830	11 (1.1)	18 (1.8)	28	11	700 (68.6)	67 (6.6)	65	107 (10.5)	23	117 (11.5)	1020	80,000 (1830)

Sources: City directories for 1793 (James Hardie); 1801 (Cornelius William Stafford); 1810 (James Robinson); 1821 (McCarty & Davis); 1830 (De Silver). Population statistics from Daly and Weinburg, *Genealogy*, p. 92.
Note: The figures in parentheses for all but the last column are percentages.
a. Excluding wire workers.

Philadelphia County had only two cotton weaving factories in 1831, both in its industrial region of Frankford. Southwest of the city, Delaware County had three weaving factories. At this point the Frankford area had eleven bleaching, dyeing, and printing works.[9] Closer investigation is needed to confirm this impression of subregional specialization. It does appear, however, that the Philadelphia hand loom weavers survived because their craft skills lay beyond the capacities of existing power loom technology. Their economic position as manufacturers of specialist goods was strengthened by the proximity of specialist finishers and of a large market in the city. The latter gave them marginally higher cloth selling prices (table 9.6).

The part played by immigrant workers in the Philadelphia region's lag is hard to assess because not enough is known about the backgrounds of individual weavers. Emigration and immigration data, though inadequate, suggest some trends. It is clear that Philadelphia was not a preferred destination for immigrant weavers, as might have been expected. Of the 334 emigrant British weavers between 1773 and 1775, only 13.2 percent were bound for Philadelphia, compared with 36 percent for New York and another 36 percent for Maryland. The 1812–1814 census of enemy aliens (appendix C) shows a settlement pattern remarkably close to the 1773–1775 immigration pattern. Of the 826 British weavers recorded in 1812–1814, 36 percent were in New York State, 13 percent in Pennsylvania, 11 percent in Maryland, and 9 percent in Rhode Island. A higher proportion went to Philadelphia from 1824 through 1831. Of the 2,695 weavers emigrating through the ports of New York, Philadelphia, and Boston, 71.8 percent went through the first, 22.3

percent through the second, and 5.8 percent through the third.[10] It is possible that these distributions of weavers were largely a function of sailing routes and that having arrived, the weavers would migrate within the United States. Indeed, a higher proportion of weavers going to New York than of those entering through Philadelphia may have gone west.

If Pennsylvania neither received nor contained the highest proportion of immigrant weavers, immigrants, through the trades they imported, could have held back cotton manufacturing by underrepresenting crucial industrial skills like managing and machine building.

The Pennsylvania-Delaware region was certainly not short of skilled industrial immigrant workers during the War of 1812 (table 8.9). Thirty percent of 185 machine makers and 32 percent of 164 textile managers among the British enemy aliens were located in the two states. In contrast, some 21 percent of the immigrant machine makers were in New York and nearly 17 percent in New England, though the data from New England are incomplete. Among the textile managers, Pennsylvania and Delaware had as many as New York and twice as many as Massachusetts. Figures for the 1820s' immigrants suggest that Philadelphia was somewhat ill favored. Of the 1,001 machine makers going through New York, Philadelphia, and Boston in this period, only 17.8 percent went to Philadelphia, compared with 74 percent to New York, from where they might easily move upstate or into New England. An even smaller proportion, 14.4 percent of the 366 textile manager group, went to Philadelphia at this time.

In one respect, however, the position of the Philadelphia hand loom weavers was reinforced by immigrant workers. The development of a number of printing and dyeing plants in the area, chiefly by Englishmen, permitted quality finishes to cotton goods and thereby added further to the cotton manufacturing specialties of the region.[11]

The impact of a group of immigrant British textile workers at the firm level is well illustrated in the records of the Dover (New Hampshire) Manufacturing Company. Not only are the firm's letterbooks and accounts for the mid-1820s unusually complete, but the company itself typified the large-scale, vertically integrated northern New England cotton manufacturing firm of the period, engaged in spinning, weaving, and printing. At Dover, therefore, British workers encountered and were assessed by best-practice American cotton manufacturing.

As the immigrants meshed with Waltham-type manufacturing and technology, their activities drew comment in the company letterbooks in two areas: technical performance and labor relations. The positive contributions of the Dover immigrants in introducing mechanized calico printing have already been examined; here their failings are taken into account. In the first place, the horizontally specialized structure of Lancashire cotton manufacturing before the 1820s and the English practice of expecting a man to stay with his trade inadequately fitted an immigrant to manage a large-scale spinning and weaving mill. After two years' dealings with immigrant textile workers, the Dover agent outlined the difficulty:

I do not think the importation of a manufacturer from England, the course most likely to serve our interests. In England the course of manufacturing is essentially different from ours. Not only are the machines and manner of operating them on a dif-

9.6. Manufacturers' Cotton Goods Prices in Philadelphia County and City and in Rhode Island, 1820

Cloth	Philadelphia County and City (Cents per Yard)		Rhode Island (Cents per Yard)	
Shirting	12	(1 firm)	10–15[a]	(6 firms)
Sheeting			15–20[a]	(2 firms)
Drilling	25	(1 firm)		
Stripes			14–17	(3 firms)
Gingham	17	(1 firm)	15	(1 firm)
Checks	20–21	(6 firms)	14–20	(3 firms)
Plaids	16–18	(3 firms)	14–21	(6 firms)
Bedticks	22–30	(5 firms)	17–30	(4 firms)
Chambray[b]				

Source: USNA, Census of Manufactures, 1820, MS Returns.
Note: Relatively few firms gave prices for their goods. The number of firms quoting prices has been added in parentheses.
a. Power looms used.
b. Manufactured but no rates given.

ferent plan; but they have their spinning in one place, their dressing in another, their weaving scattered about through various cottages and cellars, and their bleaching and printing somewhere else. One who has been accustomed to attend only one branch of the business would hardly be qualified to superintend an establishment here where all the various departments from the picking of the cotton to the baling of the printed goods are carried on within the same yard.[12]

Among the operatives, mule spinners had no place at Dover, where throstles prevailed, and English mule spinners were regarded as expensive and contrary.[13] Efforts to employ hand loom weavers met limited success.

In April 1825 the company, through an agent, hired three English weavers. At first they did not "understand" the American machinery, but on April 27 the American mill agent reported, "The English Weavers are all at work at the looms & make beautiful cloth. They are first rate weavers & will be very valuable in the next mill."[14] Over the next two years, however, the inappropriateness of English hand loom weaving, even in some overseeing capacity, for American factory work became apparent. The Dover mill agent told the company agent in Boston:

Of the three overseers who had been overlookers in England, and who were qualified to become immediately overseers with us, it may be said they understood individually on handlooms, and their overlooking amounted simply to their having had charge of three or four looms each in their houses or cellars. They had no knowledge of the discipline of a room, keeping the accounts or anything connected with it and not one of them would have been competent the charge.[15]

Labor relations problems, judging by the Dover company's experience with its printers, largely arose from the very different industrial attitudes and behavior evinced by immigrant workers. Secretive attitudes came with a few immigrants like the Thorps who believed they possessed rare skills and knowledge.[16] The English printery superintendent, Bogle, fearing competition or exposure, refused to communicate with Porter when the American chemist, on his return from England, subjected his recruit's traditional learning to scientific questioning. Whether such secretiveness was a defensive English posture stiffened by Porter's questioning or whether it cloaked massive ignorance none of the Dover management could fathom. Certainly the extent of Bogle's secretiveness appalled his fellow English workers. At last Porter told Shimmin, "Mr. Bogle's services are I consider not only of no use in the concern but absolutely highly pernicious to your interest."[17] His refusal to explain the best method of purifying zinc sulphate led to several thousand yards of calico being more or less damaged. Eased out and compensated only with passage money, Bogle made overtures to Samuel Batchelder, superintendent of the Hamilton Manufacturing Company at Lowell, who was then planning a printery. But before he made a move, Bogle died of cholera.[18] Greenhalgh, the other supervisory-level worker, more adroitly negotiated a higher salary and a longer contract in return for abandoning "certain secrets in the trade."[19]

For a few immigrants, drink impaired work performance. The Dover manager told his treasurer in spring 1825 that the newly arrived carders (number not mentioned) and spinner made him "a little apprehensive from their frequent state of intoxication."[20] Possibly this reflected no more than American surprise at the relatively widespread intensity of British drinking habits. Certainly British consumption of beer and spirits was rising in the 1820s.[21] Yet the intemperance of a few immigrants became too much even by their compatriots' standards. In two instances at least, the Dover immigrants paid the passage money for excessive drinkers (a mule spinner in 1825 and a calico printer in 1828) to go back to England. In December 1827 the company manager reported to his Boston treasurer, "Thomas Lonsdale one of our machine printers, to whose family we are making advances in Manchester, has become very unsteady so much so that we consider it our duty to discontinue the appropriations."[22] The following March he reported that Lonsdale had left Dover for Boston and noted that "his friends [fellow immigrants] have paid in to us sixteen dollars and have become answerable for his board & passage money back. They wish him to be supplied with comfortable board but no liquor or pocket money to supply himself."[23] There is no indication whether alcohol was a problem for Lonsdale and others before they went to America. The possibility that a new environment weakened traditional sanctions on overdrinking is more than likely. Whatever the cause, intemperance was incompatible with the new work patterns of the factory.[24] And this was explicitly recognized by the immigrants who signed a certificate supporting the return of the mule spinner to England:

We the subscribers fellow passengers with James Leard from England & engaged in the service of the Dover Manufacturing Company by Mr. Swan at the same time Leard was do hereby declare that the said company have in all respects complied with the engagements of Mr. Swan so far as we are concerned & that Mr. Leard was a man who from habits of intemperance & ignorance of manufacturing except mule spinning was not calculated to promote the interest of any manufacturing company.
Signed Philip Scanlan. Wm. Stuart. Ed. Jos. Quinn.[25]

Other more widely shared aspects of immigrant behavior worried managers. Chief was the propensity to unionize. Although the first group of immigrant workers (carder, spinner, and weavers) made no attempt to do so, Greenhalgh's printers formed a combination when they learned in late December 1826 that Porter was hiring printers in England at lower rates than those prevailing at Dover. The situation, evidently an extension of the tensions in contemporary Lancashire, began to escalate. The printers who worked for piece rates wanted uniform wage rates and also refused to work alongside those who stayed outside their "trade or clan," terming such nonconformists *nob sticks*.[26] By March 1827 it was agreed that the new men (one of whom was a blackleg) should be paid 9 shillings a day (the exchange rate recognized in June 1827 was 2 cents to the English penny, or $4.80 to the pound sterling).[27] But the following month the arrival of two more printers, Robinson and Hough, also recruited at reduced rates by Porter, revived discontent among the immigrants. The body of printers voted to strike until the rates in the new immigrants' contracts were leveled up with the prevailing Dover rates. Over the two-day strike, the workers "engaged in gunning & drunkeness and other diversions" and told Williams that the Dover company would be blacklisted in England.[28] Williams informed the company treasurer of the threat, suggesting that they liase with the Lowell and Taunton managements. Simultaneously Williams discovered that the Dover printers were determined "to controul the printing establishments in this Country and their *belief* that they can do so."[29] With relief he heard a week later "that the Taunton and Lowell gentlemen are disposed to set their faces against combinations of the workmen. I doubt whether our men [will] make another attempt at present."[30]

But in May the immigrants' union pressured the agent for more housing so that families could be brought from England. Williams strongly objected; he was satisfied with the existing accommodations: boardinghouses for about twenty female boarders each and a few cottages for skilled men, reflecting the basic structure of the docile Waltham-system work force. To modify residence patterns in the direction of English models would radically change the composition of the work force, giving greater concessions to immigrants and their families, and so leave Williams increasingly exposed to militants.[31] Williams dragged his heels in negotiating with the printers about housing and additionally offered lower piece rates ("at Lancashire prices") when contracts expired in July, all of which further embittered relations between the American management and the English printers.[32] Some immigrants threatened to return to England or seek work elsewhere in the United States. Bridge, the assistant agent, wanted to fire all of the immigrants, but at least half the company's twenty-eight printing department workers were English immigrants in July 1827.[33]

Two of Porter's printers left, as did Greenhalgh (whom Williams suspected of being the leader behind the troubles). This apparently eased the situation.[34] As a precaution Williams asked the Boston treasurer for "Lowell prices and rules as we now wish to establish new Rules at this location which will afford some surety against sudden ruptures from trivial causes."[35] The coordination of rates and rules among the Waltham-system companies spelled the end for immigrants' trade unions. By late July the printers came to terms with Williams, agreeing to divide a reduced amount of work equally between themselves.[36] The following month cutters' wages were reduced from 13s. 6d. a day to $13.33 a week, or 10s. a day (for a six-day week).[37]

According to the company payrolls, average wages in the Dover printing department in July 1827 were lower than average male earnings in every other department except the blacksmiths' shop (table 9.7). This may have been due to labor troubles and diverse piece rates, and possibly some mysterious bookkeeping. Among the calico printers, ten who are known to have been English were paid either 28 cents or 1s. 6d. per thirty yards of cloth printed. The highest amount earned by one of them, Thomas Reed, was $16.30. But this was rather less than he should have received if the exchange rate was two cents to the English penny: he printed 403 yards at 28 cents per 30 yards and 1,178 yards at 1s. 6d. per 30 yards, for which he should have been paid $17.90. On this basis, the other printers were underpaid. John Ramsbottom, for example, received $15.59 but ought to have received $19.88 according to the rates and amount he printed. And the English block cutters, also paid by the piece, were underpaid: for 23 jobs at 13s. 6d. a job they should have gained $74.52 but in fact received $51.75. Either some deception was involved,

9.7. Labor and Wages at the Dover Manufacturing Company for Four Weeks Ending July 28, 1827

Department	Number of Workers		Average Weekly Earnings		Highest Weekly Earnings for Males
	Males	Females	Males	Females	
Carding room	5	23	$ 4.99	$1.35	$11.63
Spinning room	2	37	7.69	1.92	10.25
Dressing room	1	7	10.42	3.08	10.42
Weaving room	2	41	7.35	2.85	11.75
Bleachery	6		4.47		9.38
Printing	28		3.21		12.94
Machinery repairers	3		8.05		12.75
Blacksmiths	2		2.36		2.58
Manufacturing expenses	3	3	6.48	2.76	10.38
Boardinghouse keepers		6		1.55	
Total	52	117	9.77	2.44	

Source: Baker MSS, Dover MC, Payroll, April 1826–January 1829.

the exchange rate was different from that specified by the assistant agent in June 1827, or, most likely, the piece rates were more complex than the payroll suggests. In the eyes of the management, the block cutters unquestionably earned far too much. On July 26, 1827, the mill agent complained to the company treasurer: "We are loaded with expensive men in this department which nothing can release but getting into full work with machinery and other convenience for expediting the work in this branch with as little additional labour as possible."[38] A more explicit statement of the pressure of variable costs could hardly be made. It was presumably made on the basis of that month's payroll when four block cutters and designers, three certainly English, averaged $12.94 a week.

To understand the attitudes of the immigrant calico printers at Dover, the labor situation in contemporary Lancashire must be recalled. Combinations of calico printers, it was said in 1816, had a long history in the Bolton area, and two years later the Lancashire calico printers joined spinners and weavers in striking against wage reductions. By the early 1820s Lancashire calico printers had a strong feeling of solidarity with other workers against employers' moves to lower wages, a feeling that could be mobilized legally after the repeal of the combination laws in 1824. Up to the early 1820s the calico printers were not directly threatened by wage reductions. In Manchester over the troubled decade 1810–1819, their average wage remained steady at 26 shillings a week. After rising to 31s. 5d. in 1822, the Manchester calico printers' wages fell, reaching 17s. 6d. a week in 1825. The elements of this background of working-class organization and agitation, coupled with falling wages, the immigrant printers naturally projected into their parallel American situation.[39]

Mass immigration from Britain's textile trades in the early nineteenth century thus was characterized by a preponderance of preindustrial skills carried by young workers unaccompanied by family. When compared to the 1820s' expansion in the work force of the American cotton and woolen industries, it could apparently supply little more than a quarter of American needs for male workers and perhaps half of American needs for workers of both sexes and all ages. New immigration could meet only about 46 percent of the actual expansion in cotton mill management positions, 70 percent of that increase in woolen mill management positions, and about 40 percent of the rise in machine-making jobs. But the mass immigration of Britain's preindustrial workers cannot be blamed for the persistence of hand loom weaving in the Philadelphia area. This district was not especially favored by British weavers or especially avoided by British industrial immigrants. Rather hand loom weaving continued in this district because the market led small firms to concentrate on textile products mostly beyond the capacity of power loom technology, up to the 1830s at least.

Placed at the other end of the spectrum of American textile technology—a large-scale, vertically integrated northern New England cotton mill—British textile immigrants as a group experienced difficulties in adjusting to best-practice American textile manufacturing. Their skills of mule spinning, hand loom weaving, and mill managing proved inappropriate or inadequate for technical arrangements in the American mill, and their work practices of secretiveness, liberal drinking, and unionism clashed

with American management attitudes and structures nearly as much as in England. In America the
insecurity of the immigrants' position possibly explains why the unionists stopped short of physical
violence, though other factors may have been at
work, such as the restraining influence of a Methodist leader or the prospects of alternative occupations. At any rate, British textile immigrants as
a group, both quantitatively and in some respects
qualitatively, fell well short of requirements in the
American cotton and woolen industries of the
1820s, judging by national, regional, and local
studies.

IV
American Modifications to the Imported Technologies

Introduction

By 1830, perhaps a half or more of the American cotton and woolen concerns were expanding without the managerial or technical skills of men with recent training and experience in Britain's textile centers. The threshold to independence from immigrant labor was crossed in several ways. As first-generation industrial immigrants trained their own families and other workers, or at least allowed American workers to learn from observing the new technology in operation, the new skills and knowledge passed into the American labor force. Secretiveness among immigrants slowed the process marginally, as did conflict in the technical backgrounds of immigrants in woolen manufacturing. But visits to Britain by Americans and the publication of technical works helped to reduce reliance on the immigrant worker, confining his role to the practice of nonmechanized techniques and the introduction of British industrial best practice. Americans could also import crucial machine parts to compensate for their lack of specialized machine makers. Card clothing, rollers, spindles, reeds, and printing cylinder shells, to name but a few known examples, were all imported in this period.[1] A third major alternative open to American entrepreneurs lay in substituting American capital for immigrant labor, an important aspect of the larger process of the American adaptation of the imported technologies to suit their own economic and social circumstances.

Although the fiery destruction of the U.S. Patent Office in 1836 precludes the sort of exploration of patent evidence that was conducted with British inventions earlier, it does seem that Americans made few major modifications to the imported textile technologies before 1812. This was hardly surprising in view of the limited success of factory production before 1807. However, in those early years, one invention of incalculable value for factory production was made. The card clothing machine, developed partly in response to the demands of household manufacturers and partly in response to the needs of the early cotton spinning mills, was as important as Whitney's cotton gin (omitted here because it belongs to the primary sector of the industry). By 1790, Americans possessed Kay's two machines for bending wire teeth and piercing the leather in readiness for setting. Over the next decade or so, they fully automated and combined these two machines into one card clothing machine. The machine for producing sheet card clothing was patented by Amos Whittemore in America in 1797 and in England in 1799. Despite the credit that these patents accrued for the Cambridge, Massachusetts, gunsmith and card maker, Whittemore was not unaided in the necessary developmental work. Earlier and imprecisely known efforts to improve the card clothing machine were made by American mechanics and instrument makers like Oliver Evans, Ebenezer Chittendon, Nathan Cobb, and Joseph Pope. Whittemore was especially indebted to Eleazer Smith, a Walpole mechanic.[2]

During the War of Independence, when supplies of card clothing from Britain dried up, Smith built a machine for making card teeth for Jeremiah Wilkinson of Cumberland, Rhode Island. He next served another card maker, Jonathan Hale of Framingham, Massachusetts, and later moved to Boston to work for a company formed by Giles Richards and William and Amos Whittemore. With them in the 1780s Smith made several improvements to the card clothing machines. Then he returned to Walpole to work on a nail-making machine, work that took him into the employment of Jacob Perkins of Newburyport in the early 1790s. Eventually in 1795–1796 he rejoined Amos Whittemore and his associates with the assignment of perfecting the card clothing machine. Unknown to Smith, Whittemore was secretly working on another model into which he was building Smith's improvements. To quote an earlier trade history: "His [Smith's] machine was gradually nearing completion. . . . He had succeeded in making it prick the leather, make the teeth and set them in *straight*, and was about to apply his ideas in putting on the second bend to the teeth, when he heard of the patent granted to Amos Whittemore, in 1797, who had forestalled him in this last contrivance and given the machine an automatic completeness."[3] It is hard to tell from primary sources just how much of Whittemore's patent was plagiarized from Smith. Patent litigation, which dragged on in the First Circuit Court at Boston between 1810 and 1815, shows that the Whittemores' extension of their 1797 patent, in 1809, aroused deep and widespread resentment among their competitors and led to a challenge of Amos Whittemore's claim to originality.[4] The defendant in the case, William F. Cutter, contested that the card clothing machine of 1797 "was invented by Messrs. Elezer Smith, Lemuel Cox, Giles Richards, John McClench, Gersham Cutter, Abel Sherman, Samuel Condon, Ezekiel Cutter & by divers other persons, & was compiled, from their inventions, by said Amos Whittemore, who added, at most, very few parts of his own."[5] Among the individuals Cutter named was a former Whittemore partner (Richards) and a former Whittemore employee (Eleazer Smith). Later in the trial Cutter produced fresh evidence against

Whittemore's claim to invention: "I can prove by the testimony of Abel Stowell of Worcester . . . and of Eleazer Smith of Walpole . . . that the said Eleazer Smith is the true and sole inventor of the improvement in the manufacture of cards for which the said Amos Whittemore obtained the patent upon which this action is founded."[6] The defendant's assertions may be questioned. Cutter and other card manufacturers had a vested interest in breaking Whittemore's extended patent monopoly. And if Giles Richards significantly contributed to the invention, why did he suppress his claims for fourteen years? On the other hand, it is more than likely that Whittemore learned much from others and especially from Smith, a clever mechanic lacking capital resources and business skills.

Another improvement in the card clothing machine was made at Leicester, a center for hand card making in central Massachusetts. One of its card clothing manufacturers, the Quaker Pliny Earle, supplied machine card clothing to the Worcester Manufacturing Company in 1789 and a few months later received Almy & Brown's order for the same, prior to Slater's arrival in Rhode Island.[7] After joining Almy & Brown, Slater wanted fillet card clothing, in the form of an endless two- to three-inch wide strip, for spiraling around his doffer cylinder. For this new purpose, diagonally set or twill teeth were required. To achieve the teeth arrangement Earle was forced to prick the fillet leather by hand, an experience that led him to invent a machine for the purpose. It was patented in 1803, and the following year Earle advertised the sale of patent rights in a New York newspaper.[8] Like Whittemore, Earle was forced to defend his patent,

and in 1814–1815 contested five infringements in the First Circuit Court.[9]

The economic importance of the new card clothing machines cannot be overrated. Without them, American cotton and woolen manufacturers would have had to rely either on imports or on traditional labor-intensive methods. The former would have crippled the industry during the War of 1812, a period that in fact witnessed dramatic expansion. Although it cannot be absolutely quantified, reliance on traditional methods would have increased an expensive item in machine carding capital equipment. Card clothing manufactured by the new machines was a third of the total cost of a new set of carding machines. In 1814 Duplanty, McCall & Co. on the Brandywine paid Woodcock & Smith of Leicester $1,065.19 (including $15 transportation costs) for the clothing for ten carding machines purchased from a Brandywine firm at $200 each.[10] The clothing might last between three and seven years—fine-yarn production required the cards to be in much better shape than did coarse yarn production—so it also represented a heavy depreciation or maintenance cost, depending on how this was regarded.[11] While this relative cost of machine-made card clothing was doubtless inflated by the existing patent monopoly, it is almost certain that traditional hand setting (such as persisted in Yorkshire) would have pushed up the card clothing component of carding capital costs above $33\frac{1}{3}$ percent. In America labor was generally reckoned to be more expensive than in England, and if the older methods were more economical, then Whittemore and Earle should have had no trouble from patent infringers.

10
Modifications to the Imported Cotton Technology I: The Shaping of Waltham-System Innovations

That Americans fundamentally reshaped their imported British cotton manufacturing technology during the two decades after the War of 1812 is not open to debate, but the reasons for it are.[1] At the heart of one current historical controversy are rival explanations for the American pursuit of standardized production lines, greater mechanization, and interchangeability—the American system of manufacturing on which English visitors commented in the 1850s.[2] Habakkuk reopened the question by arguing that labor shortage was the dominant economic influence, because the cost of labor relative to capital was higher in the United States than in Britain and because the labor supply (especially of unskilled labor) was inelastic, a situation he attributed to the abundance of land in America.[3] On the other hand, Temin, after looking at interest rates, contended that the higher price of capital in the United States induced the use of less, rather than more, capital per worker than in Britain.[4] More recently, Lebergott pinpointed social mobility as the cause of high labor costs, which, because they were known more accurately than other costs, motivated efforts to save labor.[5]

Here a more specific and inductive line than that followed by most other participants in the Anglo-American technology debate has been hewed. My starting point is the technology itself. As Salter noted, new and commercially adopted hardware is frequently evidence of "a new production function which is superior to its predecessor in the sense that less of one or more factors of production is required to produce a given output, the input of other factors remaining unchanged."[6] Assuming that the technology itself can help to explain the influences that shape it, American innovations in cotton manufacturing technology are now reexamined in the light of the most frequently cited formative pressures: the initial power loom technology, market conditions, and factor influences.

Francis Cabot Lowell, Patrick Tracy Jackson, and their ten fellow stockholders whose combined investment of $100,000 launched the Boston Manufacturing Company (BMC) in 1813 committed themselves at the outset of their enterprise to weaving by power, a decision with a number of significant repercussions.[7] It meant that Americans shared research and development costs with British developers, for the power loom imported by Lowell was imperfect. To reduce these costs and get the loom into commercial production quickly, Lowell and his engineer, Paul Moody, opted for a cruder technology than was concurrently developed in Britain. In turn this decision confined manufacture to lower-quality goods. But power loom weaving also involved the establishment of a whole technological system, not merely the construction of one new machine, and thereby pointed the way to vertical integration. This, Stigler observed, was the typical organizational form followed in young industries. The costs of a firm's functions or technical processes are interrelated, and only the expansion of an industry (arising from widening markets) increases the magnitude of functions to such an extent that increasing returns permit a firm to specialize in one function.[8]

Lowell's motives in choosing vertical integration can only be surmised. A neighboring firm, like the Waltham Cotton and Wool Manufacturing Company (which started cotton spinning in 1810 and had two thousand spindles running in 1814), could supply the low-number yarns needed for his power loom.[9] Presumably Lowell's mercantile experience taught him the importance of exercising maximum control over business operations. Reliance on another firm for yarns—inevitable while his power loom model was undergoing trials between February and September 1813—promised delivery delays and disputes and diminished profit margins after production began. And the commercial and technical separation of spinning and weaving precluded opportunities to make the innovations that would widen profit margins; for example, Moody's filling frame patent of 1821 introduced conical bobbins that had to match both the spinning frame and the power loom shuttle, and therefore assumed organizational control over and technical familiarity with spinning and weaving.[10]

Power loom technology also necessitated higher operating speeds in the prior processes of warping and dressing.[11] Unless these were achieved, power loom weaving would be uneconomic. In other words the pressure of what Rosenberg has called complementarities, or technical accommodations to overcome bottlenecks, concentrated inventive effort on warping and dressing innovations. And the need for higher processing speeds pointed to the logic of flow production, already practiced in American flour mills. In the BMC's Waltham cotton mill, flow production was organized upward, with cards on the ground floor, spinning frames on a middle floor, and looms on the top floor, leaving the basement for the machine shop.[12]

The commitment to power loom weaving involved the employment of new levels of indivisible fixed and variable (overhead) costs. In the face of these, average unit costs could be lowered only by increasing output. This step incurred additional raw materials, fuel, wear and tear on equipment, and labor. In short, the power loom pushed the Waltham entrepreneurs toward large-scale operations, which in turn created pressure to save variable costs. Of these variable costs, raw cotton prices did not fall until 1818, the cost of water power probably did not change much, and wear and tear on equipment was a function of the prices of wood, iron, coal, power, and labor.[13] If labor were the most easily measurable charge in manufacturing, as Habakkuk and Lebergott thought possible, higher labor productivity through technical change was an obvious goal to pursue.

Markets at each end of the manufacturing process also played a part in shaping Waltham innovations. The comparative advantage of American manufacturers with regard to the cotton supply is best seen in the difference between U.S. export and Liverpool prices for Upland cotton. In 1815 the American price of Upland staple was 10.5 pence a pound, compared with the Liverpool average price of 20.8d. a pound.[14] Upland prices at Liverpool remained above 15d. a pound until 1819, when a long decline began. P. T. Jackson's first purchase of cotton for the BMC, in December 1814, cost him 27 cents a pound, equal to 16.2d. a pound, at a time when no American cotton in Liverpool was under 23d. a pound. By August 1815 Jackson was paying only 19 cents a pound for his cotton, equal to 11.4d., when Upland cotton in Liverpool ranged between 18 and 25.5d. a pound.[15] Jackson publicly reported in 1831, "It is well known that the quality of cotton which is used in this country is much finer than that in general use in England."[16]

American access to finer-quality cottons had im-portant consequences for innovation. With better qualities, cruder production machinery was possible. For example, fine cotton spun into low-number yarns would make warps better able to withstand the strains imposed by the harsh motions of the Waltham loom. The point is corroborated by the experience of the Ware Manufacturing Company, an expanding Massachusetts cotton and woolen manufacturing firm. In 1828 the agent told the treasurer, "It will not be possible for us to get along with the fine goods No. 40 without using one half Sea Island Cotton, as the double cards will not card well enough for New Orleans."[17] Sea Island cotton was the finest, and here it was allowing the use of a novel double card, presumably the one invented by the company's machinist, Colonel Anthony Olney, in 1826.[18] In other words the cotton supply situation reduced the technical problems faced in developing a crude technology.

Product market influences tended to reinforce the decision to make coarse cloth. While the market was appreciably growing—the U.S. population rose by 33 percent between 1809 and 1819, to 9,379,000, but its access to British imports was restricted by fractured Anglo-American relations from 1807 to 1814—it remained diffuse. In 1820 over 90 percent of the population still lived in rural settlements of less than twenty-five hundred inhabitants each, and 25 percent lived in the Mississippi valley.[19] In the West, life was close to subsistence level. An Englishman's glimpse of conditions in Illinois as late as 1833 suggests a market demand for the most basic clothing and housing needs:

Indeed you would smile at the extreme simplicity of dress in these parts; we look rather singular by

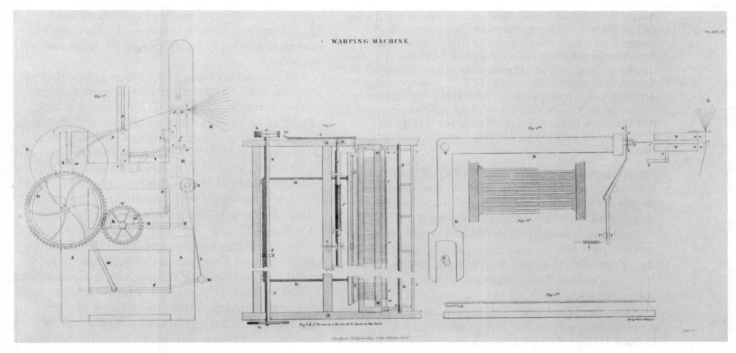

10.1 American cotton warping frame, showing stop motion; model of 1840, similar to the warper patented by Paul Moody for the Boston Manufacturing Company of Waltham, Massachusetts, in 1816. The stop motion, described in fig. 4, is of the drop wire and rocker type. When a yarn (B) breaks, its drop wire falls and jams the tumbler (s,t,s) on top of the oscillating rocker (R). This depresses the lever (u ,v) at its rocker end and thereby releases the spring bolt (r). The bolt then smashes into the belt-change fork, shifting it from the live to the dead pulley on the main power shaft (see fig. 1). (Source: Montgomery, *Practical Detail.* Courtesy Merrimack Valley Textile Museum.)

being dressed in broad cloth. Stockings seem quite out of fashion as a common article of dress, and shoes also when they are quite free from Company in summer weather, even the women think it a luxury to go barefoot . . . the houses as you know all of wood, and excepting cousin D.P.'s, not showing anything of what we call in England comfortable accommodations, but in the wilderness people must not expect what we call comforts, for it is quite necessary to abolish all artificial wants as much as possible.[20]

Rural consumers, who formed America's mass market in the early nineteenth century, therefore looked for cheapness and durability, the qualities that distinguished Waltham goods, in textiles.[21]

Appleton claimed that the BMC's goods were "an entirely new article; a fabric, more useful for common purposes, than any then in use."[22] The better quality of cotton, low-yarn number, uniformity in spinning, and, above all, the high twist in the yarn explain why Waltham goods rapidly displaced what Appleton harshly called "the inferior trash of India."[23] High twist made an especially hard yarn and therefore a tough piece of cloth.[24] In both price and quality the power loom woven goods outclassed their only rivals, the poor-quality imports from India and household handwoven cloths.

Production of these goods had two technical effects. Mass markets allowed long production runs of standardized cloths and therefore of simpler standardized technology.[25] For example, one size of yarn, like the number 14 made at Waltham, dispensed with the need for frequent changes in carding, roving, and yarn size and thereby removed the necessity of change gear devices in the double speeder that Moody developed for the company. Second, a higher number of twists per inch in-

creased the ratio of spindles to looms, unless spindle speeds could be increased. Since power was cheap, this promoted the search for faster spindles.[26]

The competitiveness of Waltham goods became invincible after the tariff of 1816 revalued all imported cottons up to 25 cents a square yard and levied an ad valorem duty on all cotton cloths. This raised the price of Indian cottons from 9 cents to 31¼ cents a yard, while Waltham sheeting was at 25 to 30 cents a yard.[27] So successful was the Waltham product that in 1828 Manchester manufacturers, Appleton reported, were "making tolerable imitations" for export, "with stamps of our most noted factories."[28]

The Waltham entrepreneurs extended commercial protection by means of the American patent system. Reconsidering Taussig's infant industry argument, Paul David noted that tariffs are justified primarily when the initiation of a new and unfamiliar industrial pursuit involves a period of learning by doing which seems so costly that even the "most alert and foresighted entrepreneurs might well 'hesitate' to finance [the] initial interlude of commercial losses."[29] Commercial losses would be expected "if production methods developed by costly trial and error could readily be copied, at negligible expense, by subsequent entrants to the industry, or applied by firms in entirely different industries."[30] In this light, technical innovation became an extension of the tariff. While the latter safeguarded all American manufacturers from foreign competitors, the former, by a series of patented inventions, guarded the Waltham company's research and development costs, and to some extent profits, from depredations by subsequent American entrants to the industry. Between 1793 and 1836, patenting in America was both cheap and easily secured, involving payment of a $30 fee and formal registration, patentees being left to defend their claims to invention and monopoly in the courts.

The BMC used patent protection extensively by selling patent rights and by defending those rights in the courts. Their first patent, logically, was for the power loom. Taken out by Lowell and Jackson on February 23, 1815, it was assigned to the company after Lowell's death in 1817. What it covered is unknown. It may have been for the weft stop motion said to have been installed in the loom at its experimental stage, or it may have been for the friction arrangement for stopping the loom when the shuttle jammed in the shed, the design that strongly impressed Job Manchester.[31] Whatever it covered, the company paid $5,000 for it to Jackson and Lowell's executors.[32]

The BMC hedged its technology with a substantial degree of patent protection. Between 1816 and 1821, Paul Moody, the company engineer, added another nine patents to the power loom one, and on April 14, 1817, patent rights for a loom temple were bought from James Stimpson. At least two of Moody's patents were revisions of earlier ones declared faulty in patent litigation. Between them, these patents covered a bobbin winding machine (in fact a warping frame), a dresser, a double speeder, and a filling frame.[33]

Having secured this protection, the company proceeded to recoup its research and development costs.[34] Patent rights to build and use the Waltham loom were $15 per loom; for the double speeder, warper, and dresser, $300 each; and for the filling

frame, $2 per spindle. Sales of licenses between 1817 and 1823 totaled $8,354; loom licenses yielded the largest sum ($3,555). Patent protection was also exercised in the sale of patent machines, which were substantially more costly than those of rival machine builders. Waltham looms, for example, sold at $125 and cost $90 to build, compared with Rhode Island looms which sold at $70. The BMC made net profits totaling $33,190 on their sale of machinery from 1817 through 1823. In addition, their sale of patent licenses, including general manufacturing rights, netted $34,354, which shows just how important a source of income licensing became.

Inevitably patent infringements occurred. Between 1820 and 1822, Moody and the BMC fought at least eight actions over patent invasion in the circuit courts of Massachusetts and Rhode Island.[35] Only one resulted in victory for the BMC, and then the company was awarded less than a third of its claim for damages of $6,000. This was the company's case heard in the Massachusetts court between October 1820 and May 1821 against three of its former employees, Jonathan Fiske and Ephraim and Jacob Stevens, who were convicted of illegally building Moody's dresser, the patent for which he assigned to the BMC on January 14, 1819.

Between Moody's loss of the defense of his double speeder patent (April 3, 1819) in the First Circuit Court in October 1820 because of faulty patent drafting and his registration of a revised patent on January 19, 1821, the alleged infringer, Jonathan Fiske, took out six double speeder patents of his own on December 7–8, 1820. Their existence complicated the situation. But after a special district court held at Boston on November 6, 1821, in which another litigant (Ebenezer Wild) attacked Moody's claim to invention in both his double speeder and filling frame patents, Moody withdrew (as did Wild). The following year Moody also declined to contest any further three double speeder patent actions pending in a Rhode Island district court. What lay behind the BMC's retreat from its patent defense can only be guessed at now. Possibly litigation looked increasingly less profitable than out-of-court settlements. What is clear from Gibb's compilation of BMC machinery profits is that the company's sale of machinery and patent rights slumped immediately after 1822, a time of industrial expansion as well as the company's withdrawal from the courts. Did this signify the decline of patent protection and the end of learning by doing for the Waltham company? Whatever the reason for the BMC's abandonment of its patent actions in 1822, the evidence supports the view that patents were an important method of securing infant-industry protection under the tariff by enabling the pilot company to recoup some of its research and development costs against competition from subsequent entrants to the industry.

Product market conditions also induced an innovation in marketing techniques at Waltham. But the emergence of the textile company selling house has been studied elsewhere and is not one of the technical innovations with which this study is primarily concerned.[36]

Commodity and product market conditions largely influenced the technical innovations made at Waltham by allowing the use of cruder technology and by reinforcing the decision to produce low-quality goods for mass markets. The company's leadership in these markets led Waltham entrepreneurs to secure infant-industry and pilot-firm pro-

tection through the tariff and through patented inventions.

Since capital in America was more expensive than in Britain, American innovations appearing before 1830 ought to have been capital-saving.[37] Strassmann and Habakkuk detected some such tendency in the rapid obsolescence, flimsy construction, and faster running speeds of American machinery.[38] Williamson explicitly linked rapid machinery replacement to expectations of obsolescence, the use of wood, and high interest rates.[39] To what extent did Waltham innovations prior to 1823 exhibit an effort to save capital? Answers may be sought in the price, construction, and design of Waltham equipment.

Waltham equipment was expensive compared with other American textile machinery. The Waltham warping frame, for example, cost $200 to build and sold at $600. The BMC's total machinery sales from 1817 through 1823 grossed $95,164. Since costs of construction totaled $61,974, the company's profits, and an added fixed capital charge to consumers of the machinery, amounted to nearly 54 percent more than the cost price.[40] Writing in 1821, a New Hampshire entrepreneur told his brother, "The Waltham plan, taking it all together, is the most perfect system of all other known in this country. . . . If any thing is said about the expence of building Waltham machinery it may be understood that it is of the most expensive kind."[41] Waltham machinery prices were also much higher than current Lancashire ones. Exact comparisons are difficult to make because some values were firms' stock valuations and some machine makers' selling prices; further, accessories and costs of installation were not always clearly included or excluded. And on the English side, the only known machinery valuations relate to mule spinning, not used in the Waltham system. Machinery prices given in table 10.1 come from McConnel & Kennedy, a fine-cotton spinning firm in Manchester, and therefore represent some of the top English machinery prices in 1812 (the highest valuations for the years 1812–1820). The prices of Waltham pickers and drawing frames were two and three times higher than the Manchester equivalent. Surprisingly, Waltham throstle spindles were between two and twelve times higher in price than Manchester mule spindles (some of McConnel & Kennedy's mules were valued at £1 a spindle and some, in their New Mill, at £66 15s. for a 300-spindle model). Lowell stretcher spindles in 1830 were eleven times more expensive than English ones in 1812—also an astonishing differential, due possibly to the relative cheapness of English mule spindles. The difference would be very slightly narrowed by adopting a slightly higher ($4.85:£1) exchange rate, but exchange rates tended to fluctuate while Britain was off the gold standard prior to 1820.[42]

The design and construction of Waltham equipment yields equivocal evidence about capital-intensive or -saving trends. Wooden framing, used in all machines until the early 1820s at least, suggests capital saving.[43] But only the dead spindle (introduced in 1815) and the filling frame (patented in 1819) could be regarded as especially capital-saving. The former allowed higher spindle speeds and so reduced the number of spindles needed for high-twist yarns. The filling frame replaced two machines with one. But in doing so the amount of cast iron required was increased, with cast-iron

cones supported by a cast-iron frame superimposed on the wooden framing.

Other Waltham innovations certainly evidenced more capital intensity in their construction, either because they embodied additional mechanisms to save labor or because they utilized an increasing number of metal parts. The warper had a stop motion. The dressing frame had a stop motion, a clock to measure yardage, and an additional drying fan. The double speeder utilized four instead of two cone pulleys (as in the English model) and a series of complex mechanisms, not least a wagon carriage which was basically wooden but had a number of metal accessories. The filling frame had a stop motion too.[44] The increasing use of metal is seen in a record of the 296 different forgings (wrought-iron and steel parts) made in the BMC machine shop for the equipment of the Merrimack Manufacturing Company's second mill, built at Lowell in 1823:[45]

Card	44 different forgings
Drawing frame	15
Double speeder (patented)	49
Throstle	43
Stretcher	48
Warper (patented)	38
Dresser (patented)	37
Loom (patented)	22.

This list does not tell how much metal was used in the various machines, but it does dispel any notions of a totally wooden American technology.

Finally, some evidence contradicts the idea that rapid obsolescence affected Waltham innovations as much as other American improvements in the period before 1840. A number of Waltham designs, and presumably machines, lasted over twenty

10.1. English and Waltham Cotton Machinery Prices, 1812–1830

Equipment	English Machinery Prices, 1812[a] ($4 = £1)	Waltham Machinery Prices	
		1817–1823[b]	1830[c]
Blowing machine	$260		
Picker		$400	$737.50 (large picker)
Card (single)	$160 (cast iron)	$175	$168.75 (half the price of a double card)
Drawing frame	$96 (6 heads)	$300	$187.50
Roving frame stretcher	$400 (150 spindles, $2.66 per spindle)		$1,091.72 (stretcher[d] of 36 spindles) ($30.32 per spindle)
Spindle and flyframe	$160 (24 spindles)		
Speeder		$2,000	$943.75 (24 spindles)
Spinning equipment: Throstle			
Warp throstle		$8 per spindle	$8.68 per spindle
Filling throstle		$10–12 per spindle	
Mule (300 spindle)	$1,200 ($4 per spindle)		
Bobbin machine		$175	
Warper		$600	$156.25
Dresser		$1,000	$537.50
Power loom	$40 (Glasgow power loom)[e] $52 (fustian power loom)[e]	$125	$80

a. McConnel & Kennedy, Manchester, England: Manchester University Library, McConnel & Kennedy Inventory, 1812–1827, pp. 12–17, 262. The prices quoted are the firm's valuations of the stock.

b. BMC, Waltham, Mass.: Gibb, *Saco-Lowell Shops*, p. 47. These are selling prices.

c. Proprietors of Locks & Canals, Lowell, Mass.: Baker MSS, Merrimack Directors' Minutes, 1822–1843, pp. 82–85. These are selling prices.

d. The Merrimack MC did not include a valuation for its stretchers in 1830 so this figure comes from the Hamilton MC, another Lowell corporation. See Baker MSS, Hamilton MC, Minutes of Proprietors and Directors (1824–1864), 1830 valuation of stock presented to stockholders' meeting, February 2, 1831, "Machinery in Mill C." These are new machinery prices.

e. PP (Commons), 1824 (51), 5:309, 383.

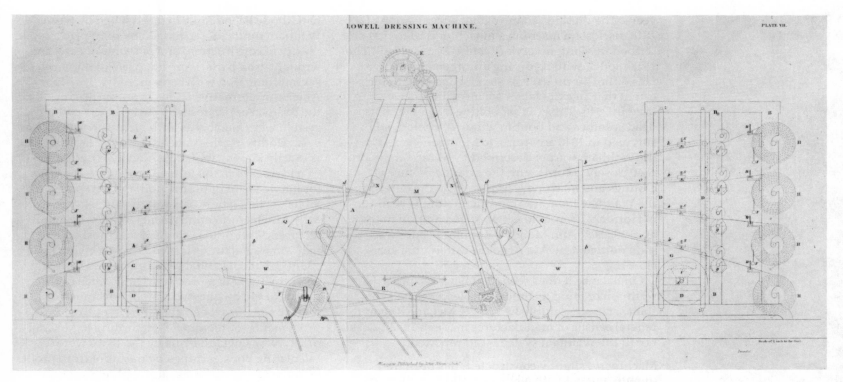

LOWELL DRESSING MACHINE. PLATE VII.

10.2 American dressing machine for sizing cotton warps; Lowell model of 1840 similar to the dresser patented by Paul Moody for the Boston Manufacturing Company in 1818. Sizing rollers (y,y,y,y) impart size to warp ends unwound from the section beams (H,H,H,H) which have already been prepared on the warping frame. Brushes (s,s,s,s) and hot-air fans (L,Q) dry off the size. Then the warp ends from all eight section beams are combined to form a power-loom warp (E). (Source: Montgomery, *Practical Detail*. Courtesy Merrimack Valley Textile Museum.)

years. The dressing frame patent specification of 1818 discloses a machine almost identical to that described in Montgomery's *Practical Detail* (1840). The 1818 model, with its drying arrangements of two fans or hot air pipes, was superseded only to the extent that an extra fan or steam cylinders were used in 1840. The Waltham filling frame, with its dead spindles and bobbin winding motion, still survived in 1840 according to Montgomery. And the Waltham warper described by Montgomery in 1840 closely accords with the extant evidence of castings and machinery orders.[46] On the other hand the Waltham loom was modified with the addition of Horrocks's variable batten speed motion in the mid-1820s.[47] And Moody's double speeder was soon outdated by Asa Arnold's differential gear (1823).

On balance it does seem that Bezaleel Taft's description of rapid technical obsolescence and constant scrapping, published in the McLane Report (a federal census of manufactures made in 1832), does not fit the Waltham system. Taft reported:

Machinery, slightly constructed, has been frequently nursed with all the attention required by ricketty children—a continual tax on the business for which it was designed. In other cases, the use of expensive and well-constructed machinery has been entirely superseded by the introduction of new improvements, by which a saving in labour was to be obtained beyond the value of the old machinery, or the expense of the new. In this way, the garrets and outhouses of most of our manufactories have been crowded with discarded machinery, to make room for that of more approved character.[48]

Taft, an Uxbridge, Massachusetts, manufacturer, lived close to the Rhode Island line and probably had Rhode Island mills in mind when he wrote. Waltham innovations, judged by their price, complexity of construction, and amounts of cast- and wrought-iron parts, were not particularly capital-saving compared with contemporary British or American alternatives. This is not surprising since the Boston Associates who backed the Waltham system were merchants with plenty of capital to invest. Modifying this conclusion is the fact that little is exactly known about machine lives before 1830.

With regard to the influence of labor, the other formative factor of production, the first point to establish is that labor costs were accurately known by early nineteenth-century American textile manufacturers. Taft in 1832 claimed that most of his fellow manufacturers lacked "that precision in accounts, that could enable the owners to ascertain with any thing like precision, how much of the profits of the business, or rather how much of their funds, have been absorbed in outfits, improvements, or repairs."[49] According to McGouldrick, New England firms by the 1830s were calculating book earnings by means of cloth profit: manufacturing earnings minus (chiefly) cotton, labor, and repair expenses. Cloth profit did not include interest or rental income and was always gross of depreciation expenses.[50] Assuming that the BMC used the same accounting methods, the Waltham entrepreneurs must have been most aware of rising labor costs in their first years when rapid expansion magnified labor expenses.[51] The timing and nature of the Waltham innovations firmly suggest that they were designed to reduce labor costs.

The timing of the Waltham innovations can be pinpointed from patent litigation evidence. Patrick

Tracy Jackson, founder-director and company agent (executive mill manager) and Lowell's brother-in-law, testified in 1821 that

in 1815 Mr. Moody invented and made a machine for warping. The next year he made important and useful improvements in the English Dressing machine for sizing yarn. . . . He was not assisted in making them by Mr. Lowell or myself. In 1815 we were informed by Mr. Thomas Faulkner that machines were used in England for roving Cotton, called Bobbin & Fly frames, which he thought better for the purpose than the stretcher we then used. Mr. Moody (with the consent of Mr. Lowell & myself) built one as nearly like those used in England as he could, from the information given him by Mr. Faulkner and what we could obtain from other sources. This frame was put in operation in the summer of 1816.[52]

The roving frame was not satisfactory, so Moody, with Lowell's and Englishman Faulkner's help in making calculations (and Faulkner's in making drawings), built a second machine, which they called the double speeder and which went into operation in December 1816.[53] Work on developing the filling frame, Moody himself testified, began in fall 1817, and the machine was patented in 1819.[54] In addition the Boston newspapers for March 1815 announced that the BMC was installing improved spinning frames that required "only five persons to do the same quantity of work that requires at least twenty persons to do, and saves much of the cotton which is unavoidably wasted in the usual method, and requires much less power, less space, and the cotton comes out much more equal."[55] The description fits the characteristics of the dead spindle, which was associated with the Waltham system and which figured in Moody's filling frame patents.

One version of the dead spindle in 1832 could reach eighty-four hundred rotations per minute compared with forty-eight hundred with the old throstle.[56] This raised the spindle-for-spindle productivity of the throstle frame above that of the mule and so facilitated at Waltham the displacement of mules by throstles, in the form of stretchers and warp and filling frames. In short, all of the major Waltham innovations, including the power loom, were developed, perfected, and put into commercial operation between February 1813 and early 1818. Although spanning wartime expansion and postwar depression for the rest of the American cotton industry, these years saw the progressive growth of the BMC, resting on its market penetration and tariff and patent protection.

The timing of these innovations can now be related to the company's growth. Production in the first factory at Waltham commenced in January 1815, when the nearby Metcalf Machine Shop delivered the last of its BMC order: ten carding machines, five 60-spindle throstle frames, two 192-spindle mules, four reels, two winding blocks, and sundry accessories. Power looms were built in the BMC's own machine shop, which was occupied in November 1814. Seven wide and fourteen narrow power looms were in operation in February 1815. The power loom improvements and the dead spindle (installed by January 6, 1815) were therefore reached before production began. Consequently considerations internal to the technology (for example, whether the mechanisms ran smoothly and in time with each other) and generally perceived directions of cost saving must have shaped the innovations; reducing labor costs was one of the latter.[57]

10.3 Twisting frame of sixteen spindles, attributed to the Merrimack Manufacturing Company, Lowell, Massachusetts, ca. 1830s. Below the wooden cross-member under the drop wires (top of photograph) can be seen the rocker bar. When a drop wire, released by a broken yarn, impedes the bar, the lightly restrained spring bolt flies against the belt-change fork, halting the machine as explained in illustration 10.1. (At Slater Mill Historic Site; photo by author.)

The remainder of the Waltham innovations came after production began in January 1815 when commercial pressures could be more accurately assessed. Table 10.2 shows that improvements coincided with a period of rapid expansion. First there was expansion within the original factory, which more than doubled its spindle and power loom capacity between January 1815 and August 1816. During these months, modifications to the warping frame, the dresser, and the double speeder were made. Then in late 1817, when plans for a second mill (started early in 1818, and equipped with 3,584 spindles) were under review, Moody began developing the throstle filling frame.[58] It was introduced before summer 1818, when Fiske (who made its pattern) and the Stevens brothers (who finished its gears) were dismissed from the company and went off to make their own machines in alleged violation of Moody's patent rights.[59] Obviously the warper and dresser were needed to keep up with the pace set by the power looms. To some extent the double speeder and filling frame had the same object, though their combined effect was to reduce or eliminate the need for the very skilled stretcher and mule spinners.

Patent evidence helps resolve the question of whether labor-saving considerations explain the BMC's post-1814 innovations, rather than its adoption of the alternative of adding to the existing stock of old models, derived from Britain. Specifications sometimes record the merits of an invention in economic as well as technical terms. And it is certain that Moody's patents represented innovations as well as inventions, that they were commercially successful, and that they were not merely an engineer's fantasies. Jackson in his 1821 affidavit

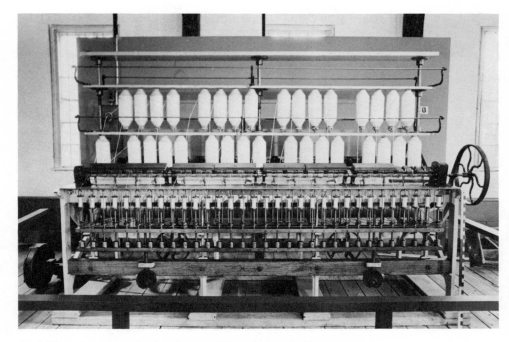

10.4 Throstle-type spinning frame of sixty-four spindles built by the Locks and Canals Company of Lowell, Massachusetts, ca. 1825–1835. (At Slater Mill Historic Site; photo by author.)

10.2 The Timing of Waltham Innovations, 1813–1817

Date	BMC's Capacity	Innovation
1813		Power loom improvements; patented February 23, 1815
1814		Dead spindle built by Metcalf Machine Shop
1815	First mill finished	
January 6	300 throstle spindles; 384 mule spindles[a]	
February	300 throstle spindles; 384 mule spindles; 21 power looms[a]	Warping frame developed this date according to P. T. Jackson; patented March 9, 1816
1816		
August 31	1,084 throstle spindles; 384 mule spindles; 54 power looms; 3 warping and 3 dressing frames[b]	Dressing frame developed this date according to P. T. Jackson, patented January 17, 1818
		Double speeder I built summer 1816, according to P. T. Jackson
December		Double speeder II put into operation December 1816; patented April 3, 1819
1817		
February 21	1,084 throstle spindles; 936 mule spindles[b]	
April 14		Loom temples of J. Stimpson assigned to BMC[a]
Late summer, autumn	Construction of second mill begun; 3,584 spindles in this mill[b]	Filling frame developed late summer 1817 according to Moody; patented May 6, 1819

a. Mailloux, *Boston Manufacturing Company*, pp. 68, 80, 85, 89.
b. Bagnall, *Textile Industries*, 2:2014, 2025.

made this point clear: "Mr. Lowell & myself had, during his lifetime, the immediate direction of the business of the company, and Mr. Moody made no alterations or improvements in any part of the works without consulting one or both of us, explaining the alteration he wished to make, and obtaining our consent thereto."[60] The engineer was firmly subservient to the entrepreneur. Moody had to justify his technical improvements on grounds of profitability, as well as mechanical ingenuity. Furthermore, Moody's contract inhibited any indulgence of fanciful inventions, for it obliged him to hand over all his improvements to the company; the mechanic's private interests were subordinated to the business interest of the corporation.[61] In this respect Strassman understates the Waltham innovations when he alludes to them as "no more than a few important innovations among many."[62] Their technical success was also commercial success and not just for one firm but for all the promotions of the Boston Associates: at Waltham, Lowell, and elsewhere.

The specification of the warping frame, the first improvement following the commencement of production, has not survived. Other evidence and Montgomery's 1840 description show that one of its singular features was a stop motion. The function of the warper was to take about two hundred to four hundred warp yarns from a rack of bobbins, lay them in parallel, and then wind them onto a section beam. The warps from several beams were then combined and sized in the dresser and wound onto a warp beam, which, when full, would be fitted into the power loom. If one yarn broke during warping, the section beam had to be stopped immediately and turned back; the severed end had to

be found and then pieced before starting the machine again. Unusual vigilance was essential in the warping frame operative. Above a certain speed, however, even the most alert operative would fail to detect a breakage until many feet of warp were wound onto the section beam. Either the machine would have to be run slowly and more machines used, or else the machine could run faster with the assistance of some form of automatic shutoff device. The stop motion therefore represented a saving in both capital and labor.[63]

Moody's modifications between 1816 and 1818 to Johnson's English dressing frame (patented 1804) comprised several improvements according to the patent specification. Moody first altered the position of the section beams, setting them above (instead of behind) each other. By this means, the warps from each section were sized and dried separately, which accelerated drying and thereby strengthened the warp yarns so that "they do not break so often." The new position of the section beams made their warp yarns more accessible also: "The ends can be mended up more easily than in the said Johnson's machine." Second, Moody employed two drying fans instead of one and made provision for drying by hot air. Third, Moody's dresser incorporated a measuring device coupled with a stop motion. When thirty-two yards of warp (the length of a Waltham warp) were dressed with size, the machine halted, and the warp beam could be removed to the power loom. With his improvements Moody claimed, "I could dress double the quantity in the same time, that could be done on the English machine."[64]

Dresser tending was still hard work, for not only was there no stop motion to signal a broken end (sticky yarns would adhere to the wire stop motion guides) but also the operative had to regulate the pace of the machine according to the rate of drying. In 1827 the Ware Manufacturing Company agent complained,

There is no work in the factory considered so laborious for females as that of attending the Dresser. We have already had upwards of 20 Girls in our new Dressing Room, out of which not more than four have become sufficiently case hardened to remain there. The others either discouraged before they got experience or worn out after they had become efficient, & were obliged to quit. This keeps us more or less in raw hands which can only be remedied by time.[65]

His double speeder, Moody later claimed, was invented because the first English-type bobbin and fly frame, built from Faulkner's descriptions, "could not be made to lay the roping equally on the bobbins."[66] Briefly the challenge was to slow down the speed of the bobbins as the roving increased their outer circumference and simultaneously to slow down the up-and-down motion of the spindle rail. Moody solved the problem in his double speeder by using two pairs of cone pulleys, each pair with its "separate regulator": one to change the bobbin speed and the other pair to slow progressively the speed of the spindle rail traverse. The "desired result" was "winding the roping on the bobbin with equal tension at all times with perfect precision."[67] Because imperfect winding at the roving stage would constantly cause breakages, the new mechanisms saved labor costs in avoiding inordinate lengths of piecing time. They also saved labor in another important respect. As Jackson recalled, some form of bobbin and fly frame was preferable in Moody's opinion because it would replace the stretcher that the BMC first used.[68] Stretchers, at first a type of mule, required very skilled operatives.[69] The Waltham double speeder therefore saved on the wage bill at this level of processing.

Last of Moody's patented inventions was the filling frame, inspired according to Appleton by a negotiating session with Silas Shepard of Taunton, Massachusetts. Shepard quibbled over the license fees he proposed charging for his patent winding machine, whereupon Moody threatened to dispense with a separate winding machine and spin the filling yarn directly on the bobbins.[70] Such a machine was built, Moody attested, "towards the fall of 1817 on account of the expence and wastage of winding the filling from the bobbins on the quills."[71] The wastage was surely not in cotton alone but even more in labor costs, for the yarn from a throstle frame was previously rewound onto shuttle bobbins in the spiral cop form that would permit unwinding inside the shuttle without the bobbins being rotated. Now Moody combined two machines in one and again saved labor costs, as well as substantial capital costs in this instance. The purpose of his invention was made clear in Moody's patent:

The main object of my invention is to give to the spindle rail such an alternate ascending and descending motion that the thread as it is spun in the usual way, may be wound upon the bobbins in

10.5 Rotary temple for power loom, patented by Ira Draper of Weston, Massachusetts, in 1816; figures 8–10 show the temple and figure 2 shows the temple in position in the loom (left-hand side). (U.S. patent drawing 2608; courtesy U.S. National Archives.)

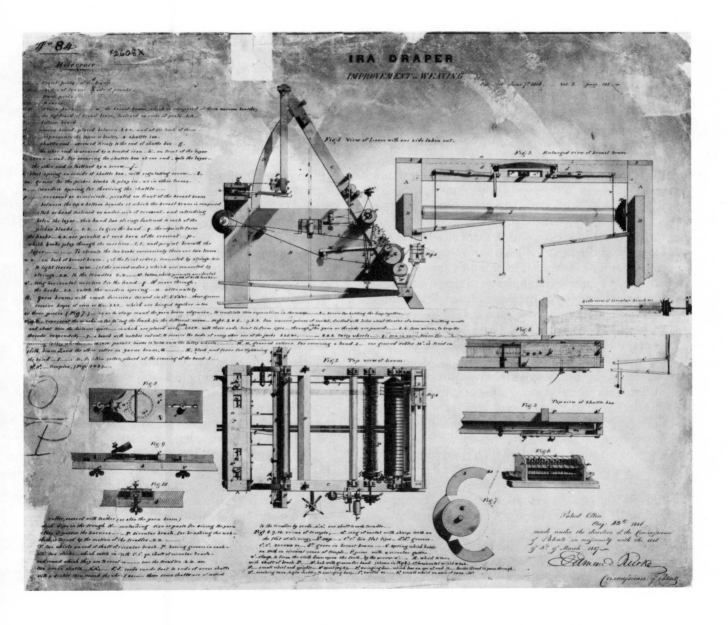

IRA DRAPER

IMPROVEMENT in WEAVING

such a manner that the bobbins may be put into the shuttles for weaving, thus combining in one operation the two several processes heretofore used, of spinning, and afterwards winding the thread upon quills for the shuttles.[72]

Moody also added a bolt and spring stop motion, which stopped the filling frame when the bobbins were full, another instance of automaticity. Of equal, if not greater, importance was the value of the filling frame in allowing the BMC to abandon its mule spinners, who previously made the filling yarn.

In these early years of production, the BMC acquired at least one invention from outside its own machine shop. In April 1817 it purchased patent rights for loom temples invented by James Stimpson of nearby Dorchester. Their merit was to increase the speed of the power loom.[73] Without them, the loom had to be stopped once every six inches of cloth for the temples to be reset (temples kept the cloth stretched laterally in the loom). The Waltham loom was set at forty-four picks per inch; assuming its shuttle speed was seventy picks per minute (above hand loom rates of fifty to sixty per minute), then it made ninety-five inches of cloth an hour (the Patapsco Co. of Baltimore had power looms that made only sixty inches of three-quarter shirting an hour in 1820), which therefore involved about a dozen stoppages an hour and a possible loss of up to ten minutes in every hour's weaving time.[74] Hence a perpetual or self-acting temple significantly increased loom speed and also the number of looms an operative could tend, though even with self-acting temples (the most successful being Ira Draper's of 1816, to which the BMC turned in 1824) American operatives were reported to man-

age only two looms each in 1840, no more than in England.[75]

In sum the patent evidence shows that the Waltham innovations made savings in capital and, especially, in labor costs. Both kinds of savings were achieved by higher running speeds (dead spindle, warper, dresser, and loom temple), the displacement of skilled labor (double speeder and filling frame), the integration of two processes into one (filling frame), and greater automaticity through stop motions (power loom, warper, dresser, filling frame). Stop motions were apparently widespread in American cotton manufacturing equipment in this early period. In the 1821 revision of the 1819 filling frame patent, Moody brushed aside the mechanical details of the stop motion because "there are so many modes of doing this, equally good and well-known, that I have not thought it worth while to describe it minutely."[76]

An analysis of the earnings and composition of the Waltham labor force indicates the ways in which the pressure to save labor costs was felt by the Waltham entrepreneurs. The earliest extant BMC payroll starts on May 3, 1817, unfortunately after Moody's innovations were introduced; nevertheless it is indicative of the labor situation faced by the company in its formative years. Earnings for employees starting between May 3 and 17, 1817, and paid off on or before August 30 are shown in table 10.3. It reveals a number of interesting points. Men in the manufacturing departments overall received average weekly earnings nearly triple the overall average female earnings. Even the lowest-paid man, a card hand, earned nearly twice as much as any woman, except the double speeder and dresser operatives and the drawing-in girls (who

10.3. Earnings in the BMC Mill, Waltham, May 3–August 30, 1817

Job	Males		Females	
	Mean Weekly Pay	No. of Workers	Mean Weekly Pay	No. of Workers
Card room				
Superintendent	$11.853	1		
Picker	7.886	1		
Roper	15.452	1		
Double speeder			$7.393	1
Hand	6.388	3	2.783	12
Spinning room				
Superintendent	9.088	1		
Mule spinner	20.267	2		
Hand			2.247	13
Dressing Room				
Superintendent	10.269	1		
Winding frame			3.484	6
Warping frame			3.443	3
Dresser			4.185	3
Drawing in			5.285	2
Weaving room				
Superintendent	10.742	1		
Weaver			3.516	30
Cloth picker			2.605	4
Nonmanufacturing chores				
Sweeper in weaving room			2.483	2
Watchman	7.523	2		
Teamster	8.845	1		
Overall for the above	10.831	14	3.742	76
Machine shop				
Superintendent	11.882	1		
Machine maker	10.730	1		
Machinist	8.721	21		
Carpenter	8.772	2		
Laborer	6.221	4		
Overall machine shop	9.265	29		
Overall mill departments, excluding roper and mule spinner	9.074	11		

Source: Baker MSS, BMC, vol. 80 (Wages, 1817–1818) ff. 1, 13, 14, 24, 34, 40, 53, 66, 81, 101, 117.
Note: The above covers only workers starting in the period May 3–17.

manually threaded the ends of each warp through the heddle eyes of the loom harnesses and splits in the reeds). Most employees seem to have been paid piece rates at this time, though this is not wholly clear. In 1820 the company reported that some of its hands were paid by the piece and some by the day.[77] Mailloux, looking at the wage books for 1817–1820, concluded that female spinners received fixed weekly wages and most other workers received piece rates.[78] But this was not true for the months analyzed here. Whatever the method of remuneration, the consistently high wages commanded by males led to the BMC's well-known strategy of replacing them with females wherever possible. And some possibilities for doing this were created by changing the technology.

The table also shows that some males—the roper and the two mule spinners—earned far more than the overall average of $10.83 a week. The roper must have been the stretcher operator noted above. Both the mule-type stretcher and the mule demanded high levels of skill from their operatives, but this does not explain the high wages they commanded. The real reason for these high wages was the short and inelastic supply of mule spinners resulting from Britain's early-nineteenth-century shift into high-quality markets, as reflected in its technological innovation. Not until the late 1820s and early 1830s, according to Montgomery, did British cotton manufacturers turn back to coarse yarn and goods manufacture, a change attributed to the power loom.[79] Slightly puzzling is the fact that fine mule spinners in Lancashire were not earning as much as $20 or £5 a week at this time.[80] Fine mule spinners in Manchester in 1817 averaged 32s. a week, just over $6; Thomas Houldsworth's fine spinners

grossed 90s. a week in 1814 but their net earnings were 60s. Because the cost of living was generally assumed to be lower in America than in England, emigration ought to have appealed to mule spinners.[81]

Evidently inelasticity in the supply of mule spinners in Massachusetts combined with the demand for low-number weft yarns (spun more quickly than fine yarns) and piece rate payments to give the mule spinners twice the earnings of mill department superintendents. No wonder plans for expansion late in 1817 impelled Moody to work on a filling frame, so that the mule spinners could be displaced and a bottleneck in production and labor costs eliminated. Noticeably the double speeder operative was already earning less than half of the roper's wage; stretchers for producing coarse roving were on the way out in the Waltham system. Once the mule and stretcher spinners were retired, overall average male earnings in the mill came into line with overall average male earnings in the machine shop.

It comes as no surprise to discover that two and possibly all three of the male operative spinners were British immigrants. Cowan, the roper or stretcher spinner, came from Manchester and was still employed by the BMC in 1821.[82] Joseph Robinson, one of the mule spinners, emigrated at the age of ten (presumably with his family) in 1803 and by 1812 was a cotton spinner working at Watertown, Massachusetts, with a wife and three children. In 1817 his average weekly earnings were $16.31.[83] Two immigrants named William Burgess (the name of the other BMC mule spinner) were registered during the War of 1812 as enemy aliens. One was a machinist in Delaware who emigrated in 1809 at the age of twenty-nine. The other was a cotton weaver at Watertown who emigrated in 1806 at the age of twenty. The latter, with his operative trade and residence close to Waltham, was probably the BMC mule spinner who in summer 1817 averaged $24.22 a week.[84] If he was that employee, his switch from weaving to mule spinning underscores the acuteness of the labor supply situation.

In comparing male earnings in mill and machine shop, another interesting trend becomes apparent. Without the mule spinners, the company was achieving labor savings through the production of capital-intensive, labor-saving equipment, with twenty-nine mechanics averaging $9.26 a week and eleven male mill workers averaging $9.07 a week.

While female wage rates were low enough to cut the mill wage bill by around 66 percent, they were still high enough to attract girls from New England farms.[85] Reliance on females presented other problems in reshaping the imported technology. One of them was high turnover rates; as Gitelman noted, the women were secondary wage earners with a transient attachment to industrial employment.[86] Only six of the thirteen female spinners and eleven of the thirty female weavers working for the company in early May 1817 were still holding the same jobs in the company three years later in March 1820. This is a crude turnover rate and conceals the movement of workers within the factory. For example, at least one of the departing spinners (Mary Stetson) graduated to the company's weaving department, and one of the departing weavers (Hannah Stimpson) moved up into the drawing-in section, in both cases for higher wages.[87] Dublin, in the most intensive study of the early Waltham system's labor force produced so far, concludes from a survey

of the Hamilton Manufacturing Company's work force in 1836 that 26.4 percent of the females on the payroll entered or left the company over a five-week period. As a result over 20 percent of the employees in July 1836 were "spare hands" learning a new mill job, and another 20 percent were engaged in training the newcomers to their respective departments.[88] In New Hampshire the Dover agent in 1827 reported that "the greatest difficulties which this establishment labours under, is to induce the help to continue. When the girls have been here two or three months they immediately start off for [the] lower works, Somersworth, New Market, or Lowell; help from Dover Factory get a place in some new mills. In this way we have to learn two or three sets of hands in the course of a year." In the face of this, a situation that had perplexed him for nearly two years, Williams went on to recommend "leaving off some of the most worn out machinery" and making "some little improvement and repairs on the rest."[89] If the parent company in the Waltham system experienced turnover rates anything like this, then the BMC machine shop was surely challenged to make equipment that required little skill to operate, thereby reducing operatives' learning times, and had as many automatic devices as possible to compensate for operative inexperience during training times, which averaged two and a half months in 1836.[90]

In a similar way, work sharing (the temporary substitution of one operative for another, to cover for sickness or holidays) may have played some part in the drive toward mechanization and automaticity. Dublin showed that work sharing was a feature of the female Lowell work force in 1836; when it started is unknown.[91] However, my first ideas about the influence of the quality of the BMC's female work force still await verification.[92] Sources consulted do not positively show that intelligent and educated farm girls were more easily distracted or physically weaker than available male workers.

The question of how workers adjusted to the newly imported technology requires a study of its own. A superficial assessment suggests that the Waltham-system mills more effectively displaced traditional habits and customs with a factory code of respectability and industriousness than did their English counterparts. The choice of middle-class teenage girls provided, probably, the most respectable work force then available to America. Concentrating the girls in boardinghouses run by determined and irreproachable matrons evidently diminished the incidence of irregular attendance, drinking, bad language, and idleness that faced English mill masters.[93] Being teenage girls, at a formative stage of life, and living in a community where youth and femininity played subordinate social roles, they could be trained to think in terms of working by the clock rather than by the job more easily than could a middle-aged craftsman. Ware, and more recently Gersuny, showed that the Waltham-system girls served under punitive moral regulations that dismissed workers for minor or personal deviations, including levity, hysteria, captiousness, and discontentedness.[94] Such harshness evidently aimed to guard the respectability of the operatives and to inculcate a discipline that exalted the work of the factory. The aggravations the girls experienced in adjusting to the moral demands of factory authority have yet to be explored beyond mill payrolls. Piecework and the threat of fines or dismissal enforced this new authority,

though it might be pointed out that if the New England girls went into the early mills for noneconomic or secondarily economic reasons, then fines and even the loss of a job would be much less disastrous than in England or in states south of Massachusetts where the prevalence of family labor meant that all mill earnings counted toward the subsistence of the family. The remoteness but high concern of parental authority and the inconsistency between respectability and brutalizing corporal punishment probably meant that the Waltham-system managers would have to rely on moral and economic inducements and sanctions rather more than did English-type mill managers.

The physical aspect of working conditions in the Waltham system also seems to have required less operative adjustment than it did in English factories. Ware concluded that "the saving feature of mill work [in the boardinghouse mills] was that it was not strenuous."[95] Of course degrees of mental taxation and physical exertion and hazard varied between departments in the mill. Dresser tending was reckoned to be the most wearisome. And discharges for poor work performance at the Hamilton Manufacturing Company in Lowell from 1826 through 1838 most heavily involved weavers.[96] In various ways, Waltham technology lightened operative chores, although this goal was not always uppermost in the minds of the Waltham innovators. Technically it was possible for women to operate light mules, as Archibald Buchanan's women had done in Scotland in the mid-1790s.[97] Instead, and largely because of operative turnover rates, the Waltham system replaced its stretcher and spinning mules with throstles, which were much easier to tend in making minimal demands on physical strength and coordination. Numerous stop motions compensated for operative inattention. The double speeder and the filling frame reduced the number of piecing operations per unit of work done and hence the physical activity of the operative. The self-acting temple further reduced operative interference with the moving parts of the power loom. In short, while Waltham innovations primarily were intended to allow the substitution of cheaper, unskilled, female labor for more expensive, skilled, male, and immigrant labor, they had the wider effect of relieving the work of most cotton factory operatives. The temptation then facing mill managers was either to increase the speed of the production line or to augment the number of machines allotted to each operative, thus nullifying the operatives' gains.

11
Modifications to the Imported Cotton Technology II: The Shaping of Rhode Island–System Innovations

Whereas Waltham innovations emerged, initially at least, from the machine shop of one firm, Rhode Island innovations appeared from a number of machine shops both inside the state of Rhode Island and in states farther south where the Rhode Island system of manufacturing (smaller-scale mule spinning factories producing a variety of goods with family and child labor) was copied.

Rhode Island entrepreneurs secured a power loom rather different from that developed at Waltham. Incorporating the Horrocks's variable batten speed motion, it allowed manufacturers to weave middling as well as low numbers of yarn, and varied qualities of cotton goods gave higher returns from shorter production runs. Consequently Rhode Island men felt less economic need than the Waltham management to achieve large-scale production. Thus of the seventeen Rhode Island firms with power looms in 1820, only four had capitalizations of $50,000 or more, and five had capitalizations of less than $20,000.[1] The largest of these firms had a fixed capital of $200,000, compared with the BMC's $400,000, and the smallest power loom weaving firm had a fixed investment of $6,000. Each of them combined spinning and weaving, again confirming Stigler's point that interrelated functions lead to vertical integration in the early history of an industry.[2]

Besides the initial power loom technology, product market influences also played a part in shaping Rhode Island innovations. Of the seventeen Rhode Island power loom weaving firms in 1820, nine firms with a total of 192 looms made four or more types of cloth, and five firms with a total of 89 looms made one cloth only (shirting in four and sheeting in one).[3] The cloths most favored by the seventeen firms were shirting (thirteen firms), stripes (eight firms), sheeting (seven firms), and plaids (seven firms). Bed ticks (four firms), checks (three firms), ginghams (two firms), and chambrays (two firms) were also made by these power loom companies. Plaids, checks, and ginghams were certainly not woven on power looms because they needed a drop box mechanism, then confined to hand looms; shirtings and sheetings certainly were, and stripes might have been quite easily but apparently were not.[4]

While diversity and quality of product distinguished the aspirations of Rhode Island manufacturers, they were unable to pursue these lines immediately after the War of 1812. Touring Rhode Island in the summer of 1816, Lowell and Appleton were told, "There was not a spindle running in Pawtucket, except for a few in Slater's old mill, making yarns."[5] The tariff of 1816 failed to give Rhode Island manufacturers the protection in quality markets that they wanted, though the tariffs of 1824 and 1828 helped a little by raising the minimum valuation of imports to 30 cents (1824) and then 35 cents (1828) a yard, and ad valorem duties from 6.25 cents (1816) to 7.5 cents (1824) and then 8.75 cents (1828) a yard.[6] The effect of this was to reduce, but only marginally and intermittently, the values of British imports of calicoes (figure 11.1).

Despite the persistent and high levels of British imports of colored calicoes, Rhode Island manufacturers continued to emulate their English counterparts by moving up-market. When calico printing was introduced into New England in the 1820s, Rhode Island mills, attracted by the prospect of higher unit profits, began making fine print

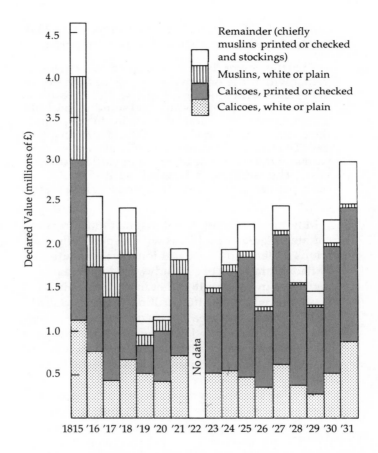

11.1 Exports of British Cotton Manufactures to the United States, 1815–1831. Source: P.R.O., Customs 9, vols. 3–18.

cloths for Fall River and Taunton printeries. In 1831 the Arkwright Cotton Mills of Warwick reported:

In 1820 the power-loom was adopted, which, together with other improvements, caused the cloth to be made with much less expense. In 1823 built an additional cotton mill, with 2,500 spindles, and 100 looms, to make fine cloth of 50 skeins to the pound, the two mills making about 2,000 yds of cloth per day. The fine cloth, yard wide, sold at 25 cents per yard and the coarser at 18 cents per yard for several years. The same goods, now selling at 15 and 11 cents per yard, have been declining for the last six years.[7]

Moving up-market meant that manufacturers encountered greater problems with respect to fluctuations in demand due to fashion. Farther south the Philadelphia agent of a Brandywine mill told its managers in January 1819 that in Philadelphia it was believed that "cotton drilling pantaloons *shall* be fashionable next season, & consequently, as not much has been ordered out from England, there is a fair opening to make a handsome profit on the article, if it is made sufficiently good & is ready *in time*."[8]

Another example of the influence of the product market comes from the Taunton Manufacturing Company in southern Massachusetts. On October 26, 1827, its directors voted to procure "regular notice and patterns of the styles of prints intended for the spring and autumn sales, so as not to be anticipated in the market."[9] The essence of success in such markets was the technical ability to change quickly from the manufacture of one type of yarn or goods to another. And this was a major product market influence in the shaping of Rhode Island innovations.

Attention focused on the roving frame, in which it was desirable to move one step beyond Moody: to achieve the double speeder's capacity to slow down bobbin rotation and spindle rail traverse in unison but also to introduce a mechanism by which the grist (thickness) of the roving could be easily changed. The Waltham speeder was designed to produce only one count of roving. An elegant solution to the problem came from Aza Arnold, a Rhode Island mechanic who served his apprenticeship with Samuel Slater.[10] His own account of this important invention, the differential gear which survives not only in roving frames but in the back axle of nearly every motor vehicle, is worth quoting at length:

I invented the compound motion (so called) as early as 1818, and while repairing a Hines' speeder in 1820 it occurred to me, that by using it, I could simplify the speeder, and produce a much more useful machine, by taking the exact difference between one graduated motion, and one uniform motion, and using it for a second graduated motion. I consulted the most able mechanicians of my acquaintance to see if they could tell of any method by which it could be accomplished; David Wilkinson, Larned Pitcher and others thought it impracticable. Ira Gay comprehended the correctness of the problem but thought it could not be mechanically used. I put the speeders in operation in 1822 and my patent was issued in Jan. 1823 and a notable prejudice was kept up by Waltham and Lowell manufacturers: the Waltham speeder was exclusively used at Waltham and Lowell until I had constructed the Great Falls factory at Somersworth N.H. which actually produced 30 per cent more goods per week than the Lowell or the Waltham factories had produced of equal quality. . . . this brought down the directors of the Lowell corporations to our place, to enquire into the cause of so great a difference. It

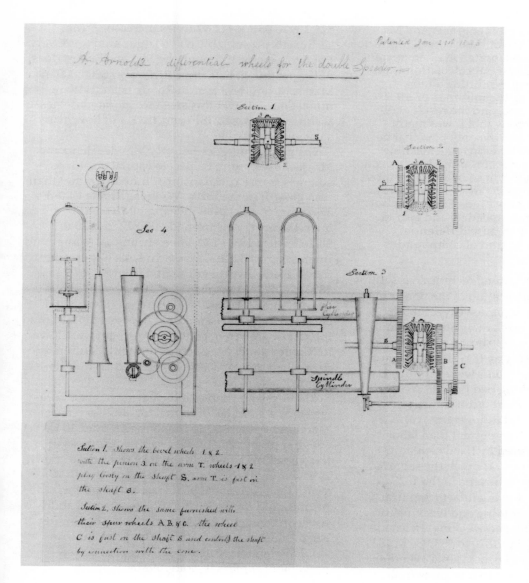

11.2 Roving frame differential gear, for retarding bobbin speed during winding; patented in 1823 by Aza Arnold of North Providence, Rhode Island. Close-up reviews of the differential are in sections 1 and 2. (Source: Zachariah Allen papers, Rhode Island Historical Society.)

brought also Mr. Moody, their engineer and Mr. Geo. Brownell, the foreman of their Lowell Machine shop. They also sent the celebrated mathematician, Mr. Warren Colburn to see if our calculation was right. I had the pleasure of exhibiting all the machinery and explaining the minutiae of a R. Island invention, a third time, and the result was, that Mr. Colburn told Moody that it was mathematically correct, and it was the only plan that he had seen or ever heard of by which the machine could be adjusted to all sizes of roping. I have been informed by Mr. Brownell that soon after this interview they commenced making my kind of gears at Lowell and not only built my kind of speeders but also took up their Waltham speeders and geared them over & converted them into differential speeders by putting in my kind of compound motion.[11]

Another innovation to extend technical versatility in production was the adaptation of the power loom to make a twill as well as a plain weave. Job Manchester, a Rhode Island mechanic, altered Gilmore's loom of 1817–1818 by using "sliding cams" to activate four harnesses as early as 1819. In 1825 over a hundred of Manchester's looms were running in Pawtucket alone, all making bed ticks.[12]

Of the two more important factor influences bearing upon Rhode Island innovations, capital may well have been more important than labor. Average firm size, as measured by fixed capital investment, in Rhode Island was much the same as average firm size in Massachusetts in the 1820s. Whereas the average cotton manufacturing firm (engaged in spinning, spinning and weaving, or wool and cotton spinning and weaving) in Rhode Island grew from $24,378 in 1820 to $53,986 in 1831 (a 2.2 times increase), in Massachusetts over the same period the average cotton firm grew from

$26,129 to $51,564 (1.97 times increase).[13] Massachusetts presumably had a larger number of very big firms by 1831, though the McLane Report does not show this (it lists three firms over $200,000 in Massachusetts and four in Rhode Island). Ware lists nineteen firms over this size incorporated in Massachusetts between 1815 and 1831, but how many survived is not clear.[14]

The great difference between Rhode Island and Massachusetts with respect to capital resources emerges from a comparison of industry rather than firm growth in the 1820s. Whereas Rhode Island's aggregate fixed capital investment in cotton manufacturing rose from $1,730,822 in 1820 to $6,262,340 in 1831, a 12.4 percent annual (compound interest) growth rate, Massachusetts's aggregate fixed capital investment soared from $1,567,745 in 1820 to $12,891,000 in 1831, a 21.1 percent annual growth rate.[15] At this level there was less capital available in Rhode Island than in Massachusetts, and the rate of growth in capital investment was slower in Rhode Island. Massachusetts manufacturers evidently had greater access to capital than did those in Rhode Island.[16]

Rhode Island innovations may have been influenced by a shortage of capital for another reason. Ware found that Rhode Island men tended to invest a higher proportion of their capital resources in fixed assets; a non-Waltham firm invested as much as seven-eights of its capital in property, while the Waltham managers kept at least a third of their capital in liquid forms.[17] This suggests capital abundance in Rhode Island, but only at the beginning of a firm's history. Thereafter a Rhode Island firm would be unable to spare much capital on improvements unless it ploughed back profits. Thus,

Rhode Island firms possibly looked for capital-saving improvements, small modifications to existing machines, rather than new machines.

A contrary trend emerges from an impressionistic glimpse of Rhode Island's capital goods production during the 1820s, which indicates that a significant number of manufacturers were switching from wooden to cast-iron frames. Isaac Peace Hazard, older son of the Quaker manufacturer Rowland Hazard, observed in 1827 that "nothing but cast Iron Frames will do at present for our Cotton Manufacturers and every thing of the nicest kind. I visited Slatersville, Blackstone, Woonsocket & some other Manufacturing establishments . . . and found all were encreasing their works."[18] Both trends help to explain why a higher proportion of a cotton firm's resources was invested in capital equipment in Rhode Island than in Massachusetts. And this growing market presumably led to increased competition among Rhode Island machine makers. Certainly Rhode Island machinery was cheaper than that obtainable in Massachusetts from both the Waltham-system machine shops and from smaller machine makers (table 11.1).

The relative capital shortage in Rhode Island must also have induced simpler and more productive capital equipment. The three roving speeders that distinguished the Rhode Island system, though not invented in the state, may well have derived from this pressure. The Taunton speeder or tube roving frame was invented by George Danforth of Norton, Massachusetts, close to the Rhode Island line, and "put in shape" in the Taunton Manufacturing Company factory by the superintendent, Silas Shepard. Danforth patented his machine, sharing its returns with the Taunton managers, on

11.1. Machinery Prices in Massachusetts and Rhode Island, 1817–1830

Equipment	Massachusetts			Rhode Island: Wilkinson Machine Shop, Pawtucket 1827[d]
	Waltham-System Prices		Ware Prices	
	1817–1823[a]	1830[b]	1826–1827[c]	
Picker	$ 400	$737.50 (large)		$250
Card, single	175	168.75	$230 (30-inch sheets)	110
Drawing frame	300	187.50	225	180
Speeder	2,000 (20 sp.)	943.75 (24 sp.)	700 (20 sp.) (Taunton)	700
Spinning machinery				
Warp throstle	8	8.68		
Filling throstle	10–12			
Mule (per spindle)		4[e]		3
Bobbin machine	175			65 (spooling machine)
Warper	600	156.25	150	625 (warper and dresser)
Dresser	1,000	537.50	550	
Power loom	125	80	125 (wide) 70 (narrow)	85
Loom temple			1 (per loom)	

a. BMC machine shop selling prices in Gibb, *Saco-Lowell Shops*, p. 47.
b. Proprietors of Locks & Canals, Lowell, selling prices, in Baker MSS, Merrimack Manufacturing Co., Directors' Minutes, 1822–1843, pp. 82–85.
c. Ware MC, Ware, Mass., machine shop selling prices, in Merrimack Valley Textile Museum MSS, Ware Manufacturing Co., Agent's Letterbook, letters of July 14, July 25, August 26, September 6, November 4, 1826, February 21, 1827.
d. David Wilkinson's selling prices, Wilkinson Machine Shop, Pawtucket, R.I., in Baker MSS, Slater Collection, Providence Iron Foundry Minutes, 1819–1832, p. 6.
e. Nashua MC, N.H., in Baker MSS, Nashua MC Letterbook 1825–1835, p. 114 (1827). This was a self-actor mule spindle selling price.

September 2, 1824.[19] Its first variant, the plate speeder, was being built by the machine makers Godwin, Rogers & Co. of Paterson, New Jersey, a year later, for a factory in the lower Hudson valley.[20] Another variant, the Eclipse speeder, was patented by Gilbert Brewster in 1829. An inventive mechanic who moved from Rhode Island to upstate New York in the early 1820s, Brewster started work on a speeder as early as 1824.[21] The chief merit of these speeders was that they produced a vast amount of coarse roving very quickly; one tube of the Taunton speeder was five times faster than one spindle of the English bobbin and fly frame.[22] In addition they were much cheaper than the Waltham double speeder. Whereas the latter sold at $100 a spindle in 1823 and nearly $40 a spindle in 1830, the Taunton speeder retailed at $35 a spindle in 1826–1827, a sum that included a patent license fee of $10 per spindle (see table 11.1).

The composition and costs of labor in Rhode Island also affected the way in which the imported British technology was reshaped. The Rhode Island system of cotton manufacturing relied on child labor much as in Britain. In 1820 nearly three-fifths of Rhode Island cotton workers were children. This proportion diminished between 1820 and 1831, and increasingly adult female workers largely took the place of the children (table 11.2). Whereas Massachusetts manufacturers turned increasingly to women (the 1831 figure of 80 percent must include teenage girls), Rhode Island manufacturers used a combination of women and children, and the balance between them was moving toward women. By 1831 there were roughly equal proportions of women and children in the Rhode Island cotton mills.

11.2. Composition of Cotton Manufacturing Work Forces in Rhode Island and Massachusetts, 1820–1831

	Men	Women	Children	Total
Rhode Island				
1820[a]	18%	25%	57%	2,412
1831[b]	20%	39%	41%	8,500
Massachusetts				
1820[a]	19%	32%	49%	2,225
1831[b]	20%	80%		13,343

Note: Massachusetts workers of unknown sex or age in 1820 have been redistributed according to the known proportions of men, women, and children in the state's cotton industry at that date.

a. U.S. Census of Manufactures, 1820, MS Returns (see appendix D). Includes cotton spinning, cotton spinning and weaving, and cotton and wool spinning and weaving mills.

b. Friends of Domestic Industry, *Report*, p. 16.

A significantly lower proportion of male workers was employed in Rhode Island and Massachusetts than in the contemporary Lancashire cotton industries, as table 11.3 demonstrates, and this too must have had some bearing on the Rhode Island innovations.

The relative costs of Rhode Island labor are much more difficult to establish for the cotton industry in aggregate. While an estimate of aggregate earnings and of the aggregate consumption of cotton in weight was made in 1831, it contained no corresponding estimate of the value of the output of cotton goods. If this is disregarded, it seems that $1 of labor in the Rhode Island cotton industry processed about 17 percent less than $1 of labor in the Massachusetts cotton industry per week in 1831.[23] Manufacturers, unaware of aggregate labor costs, were sensible only of their own firms' wage bills. A comparison of weekly wages in 1831 shows that wage rates were lower in Rhode Island than in Massachusetts. For Massachusetts, the average rate for males was $7.00 and for females $2.25. In Rhode Island, the rate for males was $5.25, for women $2.20, and for children $1.50. These lower wage rates were partly explained by the employment of family units of workers, as in Britain in the late eighteenth century. In one of Slater's mills in 1816, for example, Ware found that a work force of sixty-eight was made up of a dozen single adults and the members of thirteen families.[24] The preponderance of young people meant not only cheap labor but also very unskilled and physically limited labor, more so than the Waltham girls' labor. Theoretically the presence of large numbers of children should have created an interest in extending mechanization and automaticity, as it did at Waltham. But this does not appear

11.3. Composition of Anglo-American Cotton Manufacturing Work Forces, Early 1830s

	Cotton Workers in Lancashire and Cheshire, 1833[a]	U.S. Cotton Workers, 1831[b]	
		Massachusetts	Rhode Island
Men	37%	20%	20%
Women	41%	80%	39%
Children	22%		41%
Total	48,645	13,343	8,500

Note: "Children" refers to males and females under age fourteen in Lancashire and under age twelve in Rhode Island.
a. Factories Commission Report cited by Collier, *Family Economy*, pp. 67–69.
b. Friends of Domestic Industry, *Report*, p. 16.

to have occurred, insofar as successful and enduring innovations indicate. The most obvious explanation is that the Waltham improvements spread into Rhode Island and states farther south. Moody, after all, defended his double speeder in the Rhode Island courts and in 1821 spoke of the widespread use of stop motions in American textile technology.

At the level of individual skills, mule spinning and dresser tending were the mechanical processes in which labor costs were relatively high. Dresser tending was especially arduous for women. In 1827 it was reported, "In Rhode Island they employ almost exclusively males instead of females as Dressers."[25] Dresser rates were therefore probably higher in Rhode Island than in Massachusetts. Speed in dresser tending was essential in keeping up the flow of production. In the Ware Manufacturing Company in central Massachusetts near Worcester, an inexperienced dresser hand would not average two cuts (pieces) of warp a day, compared with an experienced hand who could manage seven cuts a day, and in the case of the Ware

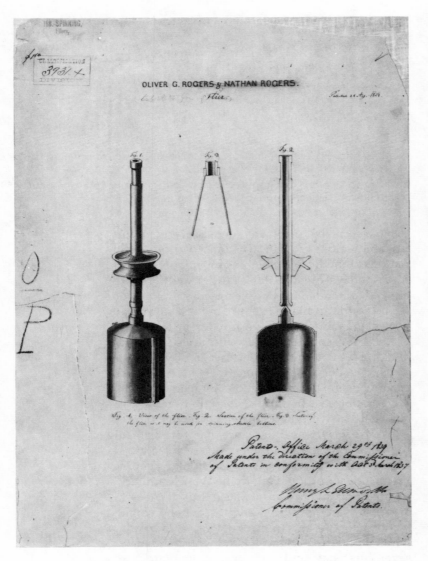

11.3 American antecedents of the cap spindle: bell and trumpet flyers patented in 1824 by Oliver G. and Nathan Rogers of New Hartford, New York. The patent specification describes the flyer "principally in its cylindrical form": the center of the cylinder resting on a spindle and "small tubes or hooks either on the outside or inside for conducting the thread or roping to the bottom from whence it passes on to the bobbin." (U.S. patent drawing 3931; courtesy U.S. National Archives.)

company an inexperienced dresser hand meant that power looms were idle (in 1827).[26] Labor-compensating equipment was therefore especially needed in dressing. Two other dressers, besides the Waltham model, were developed in the smaller mills in southern Massachusetts (and probably within the Rhode Island system), though little is known about them. Colonel Anthony Olney, machinist of the Ware company, informed a prospective textile industry machinery purchaser, "You speak of the Taunton Dresser. I think you would not use one of them after trying ours. We have one ourselves, but do not use it, & cannot therefore recommend it to others."[27]

Mule spinning was the other skill for which labor costs were high. It is very difficult to quantify these costs—though Caroline Ware says that mule spinners, like overseers, dyers, and calico printers, earned $1.50 to $2 a day—because mule spinners were paid by piece rates, which depended on the size of mule, the fineness of the yarn spun, and the amount of yarn spun.[28] At any rate a self-actor promised not only to displace high wage earners but also to produce the middling and fine yarns needed for print cloth.

Thus while Richard Roberts was busy in Manchester perfecting his self-actor between 1825 and 1830, Ira Gay contrived an American version. He constructed his first model while in partnership with Larned Pitcher, a Pawtucket machine maker, in 1823 and 1824. When Gay moved from Rhode Island to New Hampshire in 1824 to become machinist at the newly formed Nashua Manufacturing Company, he took his self-actor plans with him, and by February 1827 the company had seven of his self-actor mules running.[29] The machine spread to some extent. In Rhode Island Pitcher and Brown built Gay's mules, and in Oneida County, New York, Oliver and Nathan Rogers agreed to pay Gay twenty cents a spindle for every self-actor mule spindle they sold, at $4 a spindle.[30] Four years later, Gay sold his self-actor licenses at 17 cents a spindle, but his machine never really caught on.[31] Self-actor mule spinning in fact did not spread in New England until the 1840s when Roberts's mule and another American self-actor, William Mason's, were accepted.[32]

Precisely what sort of influence lay behind the improved throstle spindles developed outside the Waltham system is hard to tell. Spindle improvement usually meant reaching higher speeds. This meant greater output per operative hour and also, especially for high-twist yarns, greater output per spindle. Higher spindle speeds thus saved capital and labor. They also gave the throstle a new advantage in relation to the mule. Several spindles were developed before really important breakthroughs came. For example, a warp throstle spindle capable of five thousand revolutions per minute was patented in 1819 by William B. Leonard of the Matteawan Company machine shop in the lower

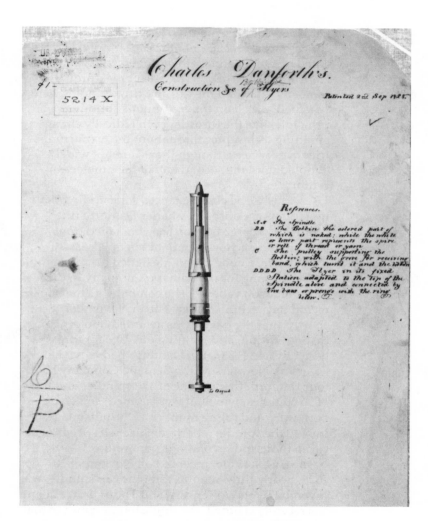

11.4 The cap spindle, patented on September 2, 1828, by Charles Danforth of Ramapo, Rockland County, New York. (U.S. patent drawing 5214; courtesy U.S. National Archives.)

Hudson valley.[33] And in 1824 a trumpet flyer, in appearance a precursor of the cap spindle, was patented by Oliver and Nathan Rogers, then at New Hartford, New York.[34] The great breakthroughs came in 1828 when the cap and ring spindles were patented. Based on the continuous spinning principle, like the throstle, both radically changed spinning techniques throughout the textile industries everywhere, though they were not widely adopted until the second half of the nineteenth century.

Curiously, the circumstances surrounding both inventions are obscure. Charles Danforth (the brother of George Danforth), who hailed from Norton, Massachusetts, about ten miles east of Providence, Rhode Island, designed his cap spindle while working for the Ramapo Company in the lower Hudson valley.[35] The company's time books record its machinists busy late in November 1828, shortly after the patent was awarded, molding, casting, fitting, and reaming out caps for throstles.[36] By late 1833, after Danforth had moved to Paterson, New Jersey, the Ramapo Company had four Danforth frames, one of 120 spindles and three of 96 spindles, which, it was claimed, could spin both warp and filling number 30 "to advantage."[37] In 1840 the spindle reportedly reached speeds of seven thousand revolutions per minute.[38]

The ring spindle, patented on November 20, 1828, nearly three months after the cap spindle, was invented by the American John Thorp, not, as Penn has recently claimed, the English immigrant of that name. Thorp's biography was carefully researched fifty years ago but the circumstances that led him to his revolutionary design remain unknown.[39] Despite its potential for high speeds at less power,

the ring spindle met early teething troubles (possibly in machining a perfectly balanced spindle), and not until the 1840s did it start to become widely accepted. It is of some slight interest that a Manchester engineer, Archibald Thomson, in 1809 patented a ring for inserting twist at high speed; in contrast to Thorp's device this ring rotated, the yarn being guided onto the spindle through a hole in the circumference of the ring. It was not promoted in England, possibly because of the cotton industry's shift to mule spinning.[40]

Several broad points can be made about American innovations in cotton manufacturing in these years of prolific inventiveness between 1814 and the early 1830s. In the Waltham system, innovations came from a single machine shop serving a very large firm in the first decade of its history. Heavy investment, astute management, market protection, and technical skill all played a part in securing the commercial success of these Waltham inventions. But the direction of these innovations was primarily aimed at reducing labor costs because these variable costs threatened to rise most sharply as the firm expanded, and failure to check them would reduce profit margins. To achieve reductions in labor costs, the technology had to be adapted so that cheap, unskilled, female labor could be substituted for expensive, skilled, male labor.

Rhode Island–system innovations, coming from a number of different machine shops, emerged more haphazardly. Their timing coincided with the expansion of the industry in the state during the 1820s. Compared with Waltham mills, Rhode Island manufacturers experienced some degree of capital scarcity (implying higher interest rates in that state) and therefore looked for improvements that would save costs in this direction too. At the same time labor may have been as expensive as in Massachusetts (perhaps even more), so there was some pressure to save labor costs. Certainly this was true of specific operative tasks like mule spinning and dresser tending. Rhode Island machine shops may have been more innovative than the larger Waltham-system shops for another reason. Business uncertainties and the smaller capital resources of independent mechanics gave them a preference for short-term leases. For example, in 1830 Aza Arnold, the machinist, considered the possibility of his partner, the manufacturer Samuel Greene, taking a two-year lease on a Rhode Island cotton factory; at most the lease could be extended to five years.[41] Under such a short lease, the pressure to sell products very likely drove machine makers toward technical improvements.

Most, if not all, of these early American innovations saved both labor and capital. Higher speeds cut labor and capital costs because more cotton was processed per unit of labor and per unit of capital. Manufacturers of capital equipment were conscious of the multiple merits of their inventions when it came to selling. The agent of the Ware Manufacturing Company in 1826 claimed that the self-acting temple invented by his machinist, Anthony Olney, "bids fair (from the solidity of its construction, its simplicity, its little liability to be out of order, its cheapness, the beautiful selvage it produces, & the additional quantity of cloth pr. day which can be wove by it) to outdo every thing of the kind which has been heretofore created. It will cost near one dollar for each loom."[42]

Waltham and Rhode Island system innovations did not remain confined to their respective man-

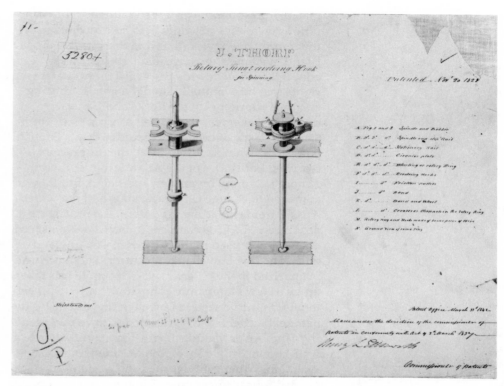

11.5 The ring spindle, patented on November 20, 1828, by John Thorp of Providence, Rhode Island. Because they attained much higher speeds, the cap and ring spindles displaced the traditional U-shaped flyer in the second half of the nineteenth century. (U.S. patent drawing 5280; courtesy U.S. National Archives.)

ufacturing systems. The large Waltham-type corporations were especially quick to take advantage of the improvements coming from the smaller Rhode Island machine shops. Arnold's speeder was adopted with such wide disregard for his patent rights that he eventually sued the greatest infringers, the Proprietors of the Locks and Canals, machine makers for the Lowell corporations, for $30,000. Partly deceived by Houldsworth's English patent for Arnold's differential, the proprietors searched both sides of the Atlantic to disprove Arnold's claim to invention, and failed, but they managed to change U.S. patent law and prevented Arnold from extending his patent. Eventually Arnold received $3,500 compensation.[43] Taunton speeders were also used in the northern New England mills, and one Lowell manufacturer, Thomas Hurd, plagiarized them.[44] Gay's self-actor mule was taken up by a New Hampshire company.[45] A twill power loom was introduced to Lowell in the mid-1820s by Samuel Batchelder, who claimed to have invented it.[46] The Dover Manufacturing Company wrote to a well-known Providence, Rhode Island, machine maker for paper cylinder rolls and a calender needed for calico printing.[47] In the other direction, Moody's actions in the Rhode Island courts show that Rhode Island men did not hesitate to borrow Waltham technology when they believed it appropriate to their own situation.

Technology was frequently transferred between the two regions and systems through the capital-goods sector. Rhode Island mechanics like Ira Gay and Aza Arnold moved into northern New England to set up machinery, gaining posts as mill managers or machinists.[48] When the partnership of Fiske and the Stevens brothers split up, sometime after the

patent battle with the BMC, Fiske went to Dover, New Hampshire, as machinist for the Dover Manufacturing Company (later the great Cocheco Company); one or both of the Stevens brothers moved to the Matteawan Company in the Hudson valley.[49] Occasionally mills shared wooden patterns for castings. In 1829 the Newmarket Manufacturing Company gave extraordinary permission to the Taunton Manufacturing Company to use its card and dresser patterns, then at General Leach's furnace (at Lowell or Foxborough, Massachusetts).[50]

Nevertheless the two systems of manufacturing cotton preserved their distinctive roles in the early nineteenth century. Waltham managers produced for lower-quality mass markets, while Rhode Islanders sought higher profits through diversity and higher quality of product. And although their respective innovations primarily appear to have saved labor costs, they also achieved capital savings. Habakkuk's explanation of labor scarcity due to cheap American land may be generally accepted, but in the case of cotton mule spinners, Britain's product market strategy toward fine yarn and goods production induced an inelasticity in supply, which had far-reaching consequences for American spinning technology.

Zevin calculated that these innovations caused "cloth production to expand at the fairly evolutionary pace of 5 percent to 6 percent per year."[51] Their impact can also be gauged by comparing American with British coarse cotton spinning and weaving mills in the data provided by Montgomery in 1840—the yardstick of "throughput," as later nineteenth-century managers called it.[52] From this it is evident that a 128-loom American mill in two weeks produced 83 percent more cotton yarn, by weight, and 14 percent more cotton goods, by square yardage, than did a 128-loom English mill. Montgomery also found that labor costs were 19 percent higher per yard of goods in the United States than in Britain. When all costs were considered, the American mill was 3 percent more profitable than the comparable British one.[53] Once American labor costs fell, as they would with rising immigration, America's textile technology would give it a substantial advantage in manufacturing coarse cotton yarn and goods unless British manufacturers adopted American innovations.

12
American Modifications to the Imported Woolen Technology

Although the household manufacture of woolens in the United States continued well into the second half of the nineteenth century, surviving longest in remote rural districts, it was quickly displaced by factory production wherever transportation networks or concentrations of population widened the market. In 1810 a conservative estimate was that two-thirds of the clothing worn by Americans was the product of family manufactures. Over 90 percent of this household production came from states north of the Mason-Dixon line.[1]

By the 1820s household cloth producers in the eastern states were largely confined to two manufacturing functions: hand loom weaving for woolen factories that had not yet concentrated their hand looms in workshops or switched to satinet power looms, and spinning and weaving in rural areas still beyond major transportation routes, as in parts of northern New England and New York or western Pennsylvania. In the western states of Ohio, Kentucky, Tennessee, Indiana, and Illinois, household production could include carding as well as spinning and weaving. Increasingly, however, carding and fulling mills were moving into sparsely populated and frontier districts to support family spinning and weaving. This is clear from table 12.1, which shows a prevalence of carding mills and spinning wheel makers in the five western states where in 1820 there were under 2,000 woolen mill spindles and around 18 percent of the country's 9.6 million population.[2]

In the late 1830s about 60 percent of the wool manufactured in the East was processed in factories and the rest in households, while in western and rural areas the proportions were about equal. What

12.1. Cloth and Capital-Goods Manufacturing Activities Supporting Household Woolen Production in the Western States, 1820

State	Carding Mills	Carding and Fulling Mills	Fulling Mills	Spinning Wheel Makers
Kentucky	20		4	1
Indiana	10	1	2	13
Ohio	50	9	26	22
Tennessee	10			22
Illinois	1			
Total	91	10	32	58

Source: USNA, Census of Manufactures, 1820, MS returns.

proportion of household cloth manufactures went to market is hard to tell. Partly this varied with the state of the national economy. A fresh impetus to family manufactures was given in the period 1807–1815, when deteriorating Anglo-American relations diminished foreign cloth imports. But the flood of importations accompanying the end of the War of 1812 stemmed the distribution of homespun. Whatever the proportion of household cloths that went to market, it is certain that after 1815 most was bartered locally, frequently in exchange for manufactured goods from retail storekeepers. Coarse in quality and uncertain in supply, household cloths could hardly compete against factory products in wider markets. As Cole succinctly put it, "Household production for household use was the rule."[3]

In these circumstances, attempts were made before 1830 to improve the performance of household cloth-making equipment, particularly spinning im-

plements, presumably because the output of the domestic spinning wheel was so low. A modern hand spinner reckons to make 340 yards of low-count (2 run) woolen yarn an hour on the wool or spindle wheel (not including fiber preparation, chiefly carding)—some 4,080 yards in twelve hours.[4] If her wool were machine carded, an eighteenth-century New England housewife could spin six skeins (3,360 yards) a day.[5] In contrast, a Yorkshire slubber in the 1790s, on a billy of unknown spindleage, could make eight and a half skeins (each of 1,520 yards), or 12,920 yards a day.[6] He was, of course, helped by child pieceners, and made slubbing, not yarn. This rough comparison indicates the margin of difference between hand spinning and early machine spinning in the woolen industry. It was therefore not surprising that the period of strained Anglo-American relations saw an increase in spinning wheel patents, some of which modified factory machines, especially the billy and jenny, for use in the home.

One was that of Oliver Barrett, Jr., of Schaghticoke near Troy, New York, who advertised his "domestic roving and spinning machine" in spring 1812, emphasizing its simple construction, ease of repair, convenient size (no longer than a bed), simplicity of operation ("a child may learn in a short time to spin from 8 to 12 runs [about 12,800 to 19,200 yards] in a day, with a machine of 12 spindles"), and high productivity and low labor costs (under 25 percent of the labor costs incurred on the spindle wheel). Barrett was careful to conceal the design of his machine.[7]

Not so the Reverend Dr. Burgess Allison, of Philadelphia and later of Burlington, New Jersey, who took out three domestic spinning patents between 1812 and 1814.[8] One of them, published in Thomas Cooper's *Emporium of Arts & Sciences* in 1813, showed a novel but clumsy drafting system, with the piecing apron and the yarn bar clasp extending from a swinging frame pivoted above the spindles. Besides simplicity of construction and operation and small size, the machine was recommended as "very portable." Its amalgam of billy and jenny functions must also have appealed to household spinners.[9]

All together, fourteen American spinning wheel patents were taken out between 1790 and 1812 inclusive, ten in the years 1810–1812.[10] Only one seems to have become a successful innovation. This was Amos Miner's "accelerating wheel-head for spinning wool," patented November 16, 1803. By placing an additional pulley between the driving wheel and the spindle in the wool spinning wheel, it more than doubled spindle speeds and, Miner claimed, doubled the output of household spinners. Its success was related to the new carding mill technology: "It appears that machine-rolls being much easier drawn out than those carded by hand, did not receive a sufficient twist, which difficulty is now obviated."[11]

Miner, who lived in a rural area (Marcellus, near Syracuse, in upstate New York), found that his simple but effective incremental improvement in the spinning wheel was very popular among household spinners. Demand was great enough for him to go into partnership with two others in 1809. They set up a small machine shop with ten lathes, a boring machine, and two saws, all water powered, and the following year had twenty workers producing between six thousand and nine thousand accelerating heads a year. Miner took out a second

patent (April 11, 1810) as protection against pirates, one of whom he feared was in every New England town, so widespread were his wheel heads.[12]

While household manufactures derived some innovations from industrial woolen mill technology, like the jenny and its variants, they exerted little influence in the other direction, apart from creating a demand in rural areas for carding machines and cloth finishing equipment. Innovations in household woolen manufacturing technology continued as long as home production, and a spate of improvements were patented up to the 1890s.[13]

Unlike innovations in cotton manufacturing technology, those in the American woolen factory industry were not confined to one firm, one geographic region, or one particular system of manufacturing. Instead they appeared in different sections of the industry and at widely differing times.

The rotary cloth shear, the development of which was recorded in a series of patents between 1792 and 1812, came primarily from machine makers and clothiers in upstate New York, Vermont, and western Massachusetts. Yet it was not wholly a product of rural woolen manufacturing, because six of the twenty-three U.S. cloth shearing patents registered from 1792 through 1812 were taken out by men from Connecticut (four), Rhode Island (one), and New Jersey (one).[14] Of these six, George C. Kellogg of New Hartford, Connecticut, and William Stillman of Westerly, Rhode Island, made shearing frames that were recorded in commercial use in 1820.[15]

On the other hand, the card condenser was developed between 1824 and 1826, chiefly by manufacturers and machine makers in eastern Massachusetts. John Goulding, a machinist from

Dedham, Massachusetts, combined the ring doffer (patented by Ezekiel Hale of Haverhill, Massachusetts, February 18, 1825) and revolving twist tubes (patented by George Danforth in his Taunton speeder, September 2, 1824), which, with the addition of twisting keys in the tubes, enabled the carded woolen sliver to be completely and continually twisted. The condenser was a major innovation for it eliminated the use of the billy.[16] But even before Goulding patented his improvement in 1826, other woolen manufacturers in New England were making roving from the doffer (the last cylinder in the finisher card) instead of on the billy.[17]

American efforts to apply the cotton power loom to woolen cloth weaving were made before 1820. In that year, the federal census recorded forty-three power looms in seven woolen spinning and weaving mills in New Hampshire (one firm, four power looms), Massachusetts (two firms, fourteen power looms), Connecticut (two firms, ten power looms), New York, northern district (one firm, thirteen power looms), and New York, southern district (one firm, two power looms). (See appendix D for an analysis of the fixed capital investment in the American woolen industry in 1820.)

A powered jenny, the equivalent of the cotton mule and the last of the major American innovations in woolen manufacturing technology in the period, was developed between about 1814 and the mid-1820s by Gilbert Brewster, who moved over these years between Vermont, Connecticut, and upstate New York, all areas of wool growing and woolen manufacturing.[18] The history and descriptions of these innovations have been well surveyed by Arthur Cole and others, so the rest of this chapter concentrates on the broad economic influ-

ences that shaped the development and spread of these four innovations, an aspect that has not been extensively explored yet is crucial for an understanding of technological modification, the last stage in the transmission process.[19]

The first limitation imposed on the technology used by American woolen manufacturers was the relative inferiority of the American wool clip. Despite the efforts of merino enthusiasts in the first decade of the nineteenth century, it was still inferior to Spanish, Saxony, or the best English fleeces, though the overall quality of the American clip somewhat improved between 1800 and 1830.[20] Partridge, the immigrant dyer, in 1823 acknowledged, "America raises very little that is equal to the best of the second quality Spanish," which itself was then inferior to Saxony wool.[21] He also complained of disparities in the quality of the American wool supply just when, he believed, American woolen factories should be specializing in and therefore using only one or two qualities of wool, for cheaper or more expensive cloths, rather than a wide range of wools. Only continued attention to selective breeding could improve the situation, and this, Partridge correctly observed, would come with "extensive capitals, vested in the wool trade, to purchase domestic and foreign wool of every quality, and in sufficient quantity to supply the market. Agents, who are judges of the article, should be employed in Spain and Germany, to make such purchases as the present and increasing demand may require."[22]

This problem in the supply of wool made it difficult for American manufacturers to compete against the superfine and second cloths imported from Britain. Confined to the lower end of the mar-

12.1 Cloth shearing frame patented in 1792 by Samuel Griswold Dorr, clothier of Albany, New York: the first American attempt to mechanize, through rotary movement, cloth shearing. As shown in fig. 2, Dorr's American shearing frame consisted of a wheel of twelve "spring knives," fixed like spokes and set at an angle of about 45° to the horizontal. Under this wheel and on the same axle rode a second one carrying four "tangent knives," which lay almost flat upon the cloth. As the two wheels rotated above the cloth surface, they acted in "the manner of shears." (U.S. patent drawing 46X; courtesy U.S. National Archives.)

ket, Americans accumulated a manufacturing experience less concerned with quality than with quantity production, and for this they appropriately modified their imported technology. British reactions to the American innovations reinforce this point.

For a decade or more, the American rotary cloth shearing frame was regarded with suspicion in Britain, where a mechanized version of the old hand shears, Harmar's patent, was preferred. In 1817, nearly twenty-five years after the first American patent for a rotary cloth shear, a British patent was taken out for an improvement in the Harmar frame, "which are commonly made use of."[23] The popularity of this machine derived from the quality of its work. As Partridge noted in 1823:

There are many shearing machines in use in this country [United States], less than eight or ten having been patented in a few years; but none of them perform the work anything like so well as the first patented English machine, called Harmar's shearing frame, on which two shears work exactly as by hand, excepting that the bobs move by mechanical motion.[24]

The difficulty with American rotary cloth shearing frames was that they were likely to traverse faster than they cut, leaving the nap in ridges. These might be removed by subsequent "kerfs" or cuttings, "yet, to a nice observer, accustomed to the business, it can easily be discovered when the cloth is finished."[25] Presumably cheaper and lower-quality cloths could more easily be sold with this less-than-perfect finish than expensive superfine cloths intended for the most discriminating distributors and consumers.

The American condenser was also regarded with

12.2 Close-up of an early-nineteenth-century American cloth shear, showing its spiral cutting blades. This shear last ran in the factory of E. Dole & Co. of Campton, New Hampshire. (At Merrimack Valley Textile Museum; photo by author.)

hostility by the older generation of English cloth manufacturers because, in their opinion, it led to loss of quality. As late as the 1890s one Yorkshire manufacturer called the condenser "a curse" and attributed the decline of the West of England trade of superfine coatings and doeskins to "the 'will-o'-the-wisp' delusion of the Yankee 'condenser.'"[26]

Circumstances in product markets also focused attention on achieving innovations appropriate to the production of lower qualities of cloth. The woolen fabrics made by household producers offered an important market for cloth finishing technology. Cloth finishing—fulling, raising the nap, and then cropping it—was not easily performed in the home. After the mid-seventeenth century, it was usually carried on in the traditional fulling mill in America. In this context the influence of household manufactures can be seen in the distribution of shearing frames in 1820 (table 12.2). Of the 425 shearing frames reported, some 231 or 54 percent were in the more rural areas: Maine, New Hampshire, Vermont, the northern district of New York (counties north and west of the Catskill Mountains), the western district of Pennsylvania (counties west of the Appalachians), Virginia, Kentucky, Indiana, and Ohio. Of these 231 frames, 119 or 51.5 percent were located in cloth finishing (usually fulling) or carding and finishing mills, which predominated in regions of household woolen manufacturing. In Vermont, northern New York, and Ohio, the proportion in carding or carding and finishing mills was 32, 65, and 62 percent, respectively. Not all of the shearing frames recorded in the 1820 census were of the rotary type. One New Jersey firm reported "1 shearing mashene carrying 2 hand shears."[27] But the prevalence of the rotary shearing frame in the

12.2. Distribution of Machines and Mills with Cloth Shearing Machines, 1820

State	Mills	Shearing Frames	Shearing Frames in Cloth Finishing or Carding-and-Finishing Mills
Maine	5	12	1
New Hampshire	4	8	
Vermont	21	34	11
Massachusetts	15	38	
Rhode Island	4	5	1
Connecticut	15	33	
New York			
North	91	124	81
South	17	42	
New Jersey	12	20	1
Pennsylvania			
West	7	11	3
East	13	22	1
Delaware	2	14	
Maryland	14	20	1
Virginia	4	5	2
Kentucky	1	1	
Indiana	4	7	3
Ohio	22	29	18
Total	251	425	123

Source: USNA, Census of Manufactures, 1820, MS Returns.

226

patents and its popularity among shearing frame makers suggest that the vast majority of American cloth cropping machines in 1820 were of this type.

In product markets, import competition had significant effects on the course of technical innovation in the American woolen industry. It was more than a coincidence that eighteen of the twenty-three U.S. cloth shearing patents between 1792 and 1812 were taken out during the years 1807–1812, when Anglo-American trade was disrupted. Reduced foreign competition encouraged minimal as well as mechanical finishing techniques. As Cole showed, wartime protection led American manufacturers to start flannel production for the first time. Nathaniel Stevens of Andover, Massachusetts, switched his entire mill to flannel making in 1814, and ten years later 690,000 yards of flannel were annually produced within forty miles of Boston.[28] The nature of British competition after 1815 is shown in figure 12.3. The overall value of American woolens imported from Britain shrank progressively by 74 percent between 1815 and 1830. Much of this might be explained by the fall in British wool prices, which halved over these years, from twenty-two pence to ten pence a pound approximately.[29]

Especially obvious in figure 12.3 are the heavy falls in the values of imports of kerseymeres and flannels, which plummeted by 96 and 99 percent, respectively, between 1815 and 1830. A rising standard of living in America may explain part of this trend. Per-capita income in the United States, it is estimated, was rising by about 1.5 percent per year between 1800 and 1840, and this could allow some shift away from the cheapest woolen cloths.[30] More likely, the fall in British kerseymere and flannel imports resulted from tariff protection.

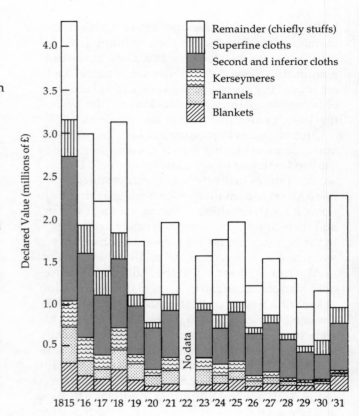

12.3 Exports of British woolen manufactures (including worsteds) to the United States, 1815- 1831. Source: P.R.O., Customs 9, vols. 3–18.

The effect of the tariff in woolen manufacturing was double-edged. While the tariffs of 1816, 1824, and 1828 raised the ad valorem duty on imported woolen cloths from 25 to 30 to 45 percent, respectively, they also increased the duties on imported raw wool from 15 to 20 to 50 percent, respectively.[31] On balance the tariff offered most protection to American woolen manufacturers who made cheap cloths from American-grown wool. That the tariff played a major role in reducing flannel imports was indicated in evidence printed in the McLane Report, published in 1833. It was noted that imported British flannels in 1829 sold at forty-six cents a yard, compared with New York–made flannels at twenty-eight cents a yard.[32] The agent of the Amesbury (Massachusetts) Flannel Factory affirmed, "Flannel which is made to a very considerable extent in this county [Essex] sells lower than they can be imported for with the present duty."[33]

Flannels, used for underclothing and shirts, and kerseymeres, used for coatings by the less prosperous sections of the population, were lightly finished. Kerseymeres received far less fulling, raising, and shearing than did broadcloth, and flannels were merely washed and pressed.[34] In other words their manufacture favored the development of innovations in woolen carding, spinning, and weaving, as well as finishing, to which British inventive activity in woolen manufacturing was largely devoted.

Behind the infant-industry protection of the tariff, improved transportation networks and the extension of factory production broke local monopolies and lowered prices, thereby sharpening pressures to adopt and improve the factory technology still further. Flannel production illustrates these changes in product market conditions. In 1820 a Vermont factory valued its flannels at $1 a yard, an upstate New York factory at 50 to 75 cents a yard, and a southern New York factory at 25 to 30 cents a yard.[35] These price variations may have concealed quality differences. By 1832 New York flannels in the Hudson valley and on Lake Ontario were down to 28 cents a yard, in contrast to those made in remote New Hampshire households, which their producers valued at $2 a yard.[36]

American development of a new woolen cloth, satinet, also contributed to the American reshaping of woolen technology. Distinguished by its cotton warp, satinet was made by a number of New England woolen manufacturers (some immigrants) from about 1808 on.[37] The cotton warp reduced manufacturing costs, as did the light finish that satinets required. The resultant moderately priced fabric gained ground for trouser cloth as knee breeches went out of fashion. By 1830, Cole guessed, nearly half of the country's woolen factory production was satinet.[38] From the viewpoint of technical innovation, the salient feature of the new cloth was its cotton warp. This suggested and allowed the application of the power loom to woolen manufacturing. In 1820 the seven American woolen spinning and weaving firms running the forty-three recorded woolen power looms were all using their water- or steam-powered looms to make satinet.[39]

Another product market influence assisted in the adoption of the new American innovations. Selling agents tried to forecast fashion and to match production to markets. In so doing they communicated information about techniques, as well as styles.[40] John Scholfield during the War of 1812 was advised by Rensselaer Havens, a New York merchant, on the proportions of various wools needed to make flannels, how much to raise them, and, for other cloths, how much to full, raise, shear, and press.[41] Havens sent patterns of fashionable cloths for Scholfield to copy: "Willing to throw in your way any information in my power, I have taken the liberty to inclose you some patterns of cords which are now & have for some time been fashionable for pantaloons."[42] In exchange Scholfield gave Havens information on the costs of manufacturing flannels so that Havens could "at all times judge how much I ought to ask for the goods" and adjust his profit margins accordingly.[43]

Another New England woolen firm, the Walcott Woolen Company (later the Hamilton Woolen Company) of Southbridge, Massachusetts, received similar advice from its Boston agents, Tiffany, Sayles & Hitchcock, in the late 1820s. The selling agents also passed on information about new machinery. For example, in June 1828 the selling agents wrote to James Walcott, the owner-manager, to say that looms (power, presumably) built by Hayden & Trumbull at Northampton for $125 each were the best available. In this opinion they were advised by a Colonel Shepard.[44]

Commodity and product market influences thus disposed American woolen manufacturers to pursue innovation in the technology of lower-quality cloth production. Compared with their British counterparts, they were therefore less exclusively interested in cloth finishing technology. And the activities of marketing agents helped to spread the innovations.

Factor influences also contributed to the reshaping of the imported woolen manufacturing technology. Higher interest rates in America than in Britain in the early nineteenth century, it might be supposed, would produce capital-saving innovations in the American woolen industry. Before making comparisons, it must be emphasized that the differentials between English and American machinery prices are hard to establish. There may well be hidden differences between the alternative machinery types; for example, the respective sizes and outputs of the various Harmar and Hovey cloth shearing frames, the prices of which have survived, are unknown. It is also possible that these prices conceal regional variations in capital costs. Yet the evidence in table 12.3 seems to suggest that Americans succeeded in reducing the cost of carding and cloth shearing equipment, compared with the West of England, between 1810 and the mid-1820s.

The lower American equipment prices may be related to two of the major innovations in American woolen manufacturing technology: the condenser and the cloth shearing frame. But this does not explain why Americans first sought capital savings in carding and shearing rather than, say, spinning and weaving equipment. The only indication from relative machinery costs emerges from the Trowbridge equipment prices. There, the billy, which Americans replaced with the condenser, was over three times more expensive per spindle than the jenny. Why there was such a difference in price is hard to understand because the billy was but a variant of the jenny and surely required hardly any extra amounts of raw materials to construct.

American machinery prices were relatively much higher than West of England prices for spinning and hand shearing equipment. Massachusetts jenny spindles were between three and six times more

12.3. English and American Woolen Machinery Prices, 1810–1830

Equipment	English Prices: J. & T. Clark, Trowbridge, Wilts.[a] 1811–1826 ($4 = £1)	American Prices (paid by manufacturers)		
		Du Pont Firms, Brandywine Valley, Del.[b] 1810–1816	Ware Manufacturing Co., Ware, Mass.[c] 1825–1830	Hamilton Woolen Co., Southbridge, Mass. 1828–1830[d]
Cards				
Scribbler or breaker	$600 (44 inch)	$518.79 ⎫ $825	$300	
Finisher		⎭		
Billy	$84 (60 spindles; $1.40 per spindle)			
Condenser			$50	$100 (plus $45 for installing)
Condenser and double card			$300	
Spinning machinery				
Jenny	$36 (90 spindles; 40¢ per spindle)	$100 (80 spindles; $1.25 per spindle)	$90–180 (90 spindles)	
Jack			$2 per spindle	$1.60–235 per spindle
Brewster machine		$1,500 (144–250 spindles)		
Looms				
Narrow			$40	
Wide or broad			$50	$95[e]–175[f] (power)
Cassimere			$90	
Satinet				$70–90 (power)
Finishing machinery				
Gig mill	$200	$150 (with shearing frame)		
Shearing equipment				
Hand shears	$16	$30–45		
Harmer frame	$147			
Rotary frames	$608 (Davis's, 1826)	$190 (Hovey's, 1816); $340 (Hovey's, 1817)	$65 (Hovey's)	$95 (Swift's; 33 inch blades)

a. Beckinsale, ed., *Trowbridge Woollen Industry*, pp. xix, xxiii, 112, 122, 130, 137, 161.

b. EMHL, Acc. 500, vol. 37, Woolen Journal, 1810–1815, entries for March 31, May 31, 1811, March 31, 1812, November 30, 1813, March 25, June 14, 1814; ibid., Acc. 500, Ledger 62, Victor & Charles Du Pont, Woolen Factory Ledger, 1814–1820, f. 39; ibid., Acc. 500, C. I. Du Pont & Co., box 26, bills (1816); Gibson, "Delaware Woolen Industry," p. 97.

c. MVTM MSS, Ware MC Agent's Letterbook, letters of July 14, November 15, 1826, August 7, 1827, March 10, June 29, 1829.

d. Baker MSS, Hamilton Woolen Co., Southbridge, Mass., bills of March 12, 22, 24, 1830; Tiffany, Sayles & Hitchcock to Samuel Hitchcock, May 10, 1830, Wheelock & Prentice to Co., July 16, October 25, 1830; William Stowell to Co., August 8, 1830; and Abraham H. Schenck to James Walcott, Jr., June 4, 1828, and Peter H. Schenck to James Walcott, Jr., August 23, 1828.

e. Howard's power loom, 1830.

f. Matteawan power loom, 1828.

expensive than those in the West of England. In part this may have been due to the difficulty of making the actual spindles. English jenny spindles, illegally imported into the United States on the eve of the War of 1812, sold in Philadelphia at $25 per hundred, representing about a third of the cost of an American jenny.[45]

American hand shears, between two and three times the price of English shears, may also have been relatively expensive because of difficulties in achieving comparable qualities of steel in blades and bowsprings. Prices paid for hand shears in the Philadelphia area by the Brandywine firm of du Pont, Bauduy & Company were much the same before and during the War of 1812: around $30 a pair. But, whether due to wartime shortage or difference in quality of product, the company paid 50 percent more in 1813 for two pairs of hand shears, purchased in Albany, New York, at $45.25 a pair.[46]

Regarding the influence of capital on innovation from the view of investment, it is clear that the four major American contributions to woolen manufacturing technology came in those states where fixed capital investment in the woolen industry was relatively high. The states in which the cloth shearing frame, the card condenser, the woolen power loom, and a self-actor woolen mule were developed were New York, Vermont, Massachusetts, and Connecticut. In 1820 all but Vermont were in the top five states or regions ranked according to their shares of the nation's fixed capital investment in woolen manufacturing. New York came first with nearly 27 percent of the national fixed capital investment, Connecticut second with 18 percent, Massachusetts fifth with 10 percent, and Vermont seventh with 4.6 percent. (See appendix D.) Aver-

age sizes of woolen manufacturing firms in these states confirm the evidence of machinery prices: nearly all of the innovations (the exception being Brewster's powered mule) emerged from and were suitable for relatively small production units. The average size of woolen firms (measured by fixed capital investment) in New York was $12,655; in Connecticut, $10,453; in Massachusetts, $8,673; and in Vermont, $5,668. The largest firms had capitalizations of $64,000 in New York, $80,000 in Connecticut, $30,000 in Massachusetts, and $15,000 in Vermont.[47]

Brewster's water-powered spinning machine, which cost $1,500 in 1816, could be afforded only by the larger woolen companies.[48] In Connecticut, for example, one firm with a fixed capital of $20,000, in Jewett City, had a 200-spindle Brewster machine, and two more Brewsters, with 250 spindles each, were run by the Pameacha Manufacturing Company of Middletown, a firm with a fixed capital of $80,000.[49] These machines were the largest Brewsters recorded in the 1820 census. Partridge, in his 1823 treatise, thoroughly approved the quality of the work turned out by the Brewster machine, particularly the uniform twist that it imparted to the yarn (in contrast to hand or jenny spinning), but he remarked on its high cost, "both in the first purchase and in the subsequent repairs."[50]

In the diffusion of American woolen innovations the manufacturers of capital goods played some part, but how important a part compared with selling agents is hard to tell. The brothers William and Eleazer Hovey, millwrights from New Hampshire, moved to Lebanon, New York, and began to build shearing frames for clothiers in western Massachusetts but before 1811 settled in Worcester in central Massachusetts. This county, as Riznik noticed, had one of the highest per-capita values of household-made textiles in the state, and this probably attracted the Hoveys.[51]

Mill machine shops also spread American innovations in woolen technology, judging by the experience of the Ware Manufacturing Company in Massachusetts. The company, which manufactured both cotton and wool, followed the policy of making machinery for other manufacturers in order to keep its own machine shop functioning to full capacity. Considering the terms of a contract with another firm, the Ware agent observed, "If we consent to a relinquishment of two thirds of the contract, it must be done in such a way as to make it for the interest of the 3 Rivers Co. not to erect a Machine Shop or a cupola furnace, but to contract with us for the other two factories whenever they shall be wanted."[52] And the Ware machine shop sought an edge over its competitors by offering a wide range of improvements, a strategy that compelled the company's machinist, Anthony Olney, to invent or to borrow.[53] Sale of the improvements achieved diffusion. In this manner the woolen condenser was brought to the Ware area. Four years after the event, the Ware agent recalled: "As near as we can ascertain the condenser was applied to our cards about the 4th April, 1825. Mr. Cook left here for Canton on the 14th March, was gone 7 days and in about a fortnight after his return put the condenser in operation."[54] He referred to Canton, Massachusetts, from whence Olney hailed. Significantly it was close to Dedham, where John Goulding, patentee of the most successful condenser, lived, though April 1825 was more than a year before Goulding took out either English or American

patents. The first Ware condenser may indeed have been no more than a slight variation of the ring doffer patented by Hale in February 1825. Whatever its technical nature, the Ware condenser was sold for $50 in summer 1826.[55] Three years later, Ware double woolen cards fitted with condensers sold at $300 each.[56] They were not as successful as Goulding's device, which in twenty years was adopted by a high proportion of woolen mills in the Union.[57]

Eastern machine makers, aware of the limited competition arising from poor transportation networks in the Midwest, moved across the Appalachians to serve the small carding or carding and finishing mills and larger integrated concerns that appeared there before 1820. One of these was Isaac Hodgson, who emigrated from Manchester, England, in 1811, one of five brothers who settled first in the Brandywine valley in Delaware.[58] In 1820 the census recorded Isaac at Woodbourn, Washington Township, Montgomery County, Ohio. With capital equipment valued at $5,000 (including four turning lathes, two bellows, and two anvils), he made "all kinds of Cotton & Woollen Mushinry."[59] Presumably he carried both British and American textile manufacturing technology into the Midwest.

Changes in labor costs partly reveal how that factor induced alterations in the imported woolen technology. American woolen manufacturers calculated their labor costs in two different ways. One was to reckon a worker's earnings over time (per day, week, or month). The other method was to estimate how much the labor in each processing stage cost for the production of a given weight of wool or length of cloth. British and American figures for the first method are given in table 12.4. It is clear that by the early 1820s, slubbers and mule spinners in

12.4. English and American Woolen Industry Weekly Wage Rates, 1814–1825

Job	English Rates[a]	American Rates[b]
Wool sorter		$ 6.00
Wool scourer		5.00
Picker	$0.84	
Slubber	2.50–3.80	4.26–6.27[c]
Billy boy	0.48–0.72	0.80–1.25[d]
Spinner		
Male	1.44–1.92	5.50–6.00
Female	1.96	4.55–8.76[c]
Weaver	1.92–3.84	4.52[e]
Dyer		5.00
Fuller	5.00	5.25–6.25
Gigman	5.00	
Shearer	2.40–5.00	5.50–6.00
Finisher		5.31–6.27[c]
Burler	1.20	
Overseer		9.00–9.25
Manager		20.00
Mechanic	6.60–11.64	7.50
Carpenter	5.82	6.00
Laborer	4.44	3.75–4.50

a. West of England wage rates, July 1825, copied from a London newspaper by Zechariah Allen, in RIHS, Allen, "Journal," p. 113.
b. EMHL, Acc. 500, Ledger 63, Victor & Charles du Pont, Woolen Factory Petit Ledger, 1817–1823, passim, except for the manager's salary. This was Clifford's in 1814.
c. Piece rates.
d. Apprentices.
e. Piece rates for one weaver, March–August 1819.

the American woolen industry earned twice as much as their counterparts in the West of England, and sometimes more than that. This figure neglects a time difference of three to seven years in the two sets of figures and does not take into account the respective costs of living, though the cost of living was reputedly lower in America than in England. American woolen weavers earned between a fifth and two and a half times more than in England, while finishers' earnings equaled top English earnings. Carpenters' and laborers' earnings were about the same in both countries, but mechanics earned up to 35 percent less in America than in the West of England, if Zachariah Allen's American rates are to be trusted.

These wage rates must be placed in the context of the development of British and American woolen technologies. By 1820 most American woolen mills had mechanical and powered cloth shearing machines. For this reason, American shearers' wages were not that much different from English rates. And the table distorts the true position because skilled shearmen were not widely employed to run the American frames. Partridge in the early 1820s claimed that a sixteen-year-old boy could run two of Harmar's frames operating four pairs of shearing blades.[60] The hostility of the shearmen delayed this technical development in England but not in America.

The main point to emerge from table 12.4 is the relatively high labor cost involved in slubbing and spinning in America. Attempts to reduce variable costs through technical innovation might be expected to begin with slubbing and spinning, especially after the War of 1812 when profit margins slumped. This in fact was the case. According to

Cole, attempts to find a substitute for the labor-intensive billy began in America as early as 1810.[61]

Confirmation that slubbing and spinning were regarded as expensive labor charges is suggested by the efforts of Victor and Charles du Pont to shift their woolen spinners from piece to monthly rates. In 1818 this Brandywine company had three male spinners (John Backhouse, James Lavars, and Henry Pierce), all on piece rates. Two years later Lavars was still spinning for piece rates, but the other two male spinners (now Isaac Peters and Henry Pierce) were on a rate of $22 per month, with deductions for days off.[62] It seems that the company was negotiating to pay time rather than piece rates, in hopes of simplifying wage scales and reducing labor costs.

The other way of calculating labor charges was applied by managers in England, as Julia Mann showed. Labor charges, as a proportion of manufacturing costs, varied according to the value of the wool used and cloth made and the amount of machinery employed. Before the introduction of machinery, labor charges comprised over 40 percent of all manufacturing costs and raw wool 38 percent. By 1798 labor costs were down to 33 percent of manufacturing a West of England superfine broadcloth. And in 1825, after the introduction of mechanical shearing, labor was reckoned to cost half the value of the wool.[63] Table 12.5 shows some American estimates. Unfortunately the three estimates are not perfectly comparable since they were for different cloth types. Nevertheless, the cloths were all of medium to low quality, so to this extent approximate comparisons can be made. The falling proportion of wool costs reflects improving wool supplies in the 1820s, with wool growing increasing

12.5. American Costs of Manufacturing Woolen Cloths, 1810–1830

Expenses	Pair of Woolen Blankets, 1810–1820[a]		Grey Narrow Kersey, 30 Yards, 1813–1814[b]		Grey Woolen Cloth, 100 Yards, 1828[c]	
Wool	69.07%		61.27%		57.12%	
Labor						
Cleaning picking			3.06%	(8.57%)	1.41%	(4.09%)
Carding and slubbing	9.94%	(34.61%)	7.35%	(20.58%)	2.36%	(6.85%)
Spinning	6.63%	(23.08%)	4.51%	(12.62%)	7.58%	(22.01%)
Warping			3.53%	(9.88%)	0.72%	(2.09%)
Weaving	6.63%	(23.08%)	8.82%	(24.69%)	6.53%	(18.96%)
Finishing	5.52%	(19.22%)	8.45%	(23.66%)	9.97%	(28.96%)
Sundries					5.86%	(17.02%)
Total	28.72%	(99.99%)	35.72%	(100%)	34.43%	(99.98%)
Materials (oil, glue, sacking, etc.)	2.21%		3.19%		8.45%	
Total cost	$9.05		$20.40		$27.575	

a. EMHL, Longwood MSS, Group 6, Box 1, n.d. but c. 1810–1820 (one of the du Pont woolen firms), undated bills and estimates.
b. Ibid. I have omitted wear and tear on $8,000 of machinery ($0.66) and interest on capital ($2.40).
c. MVTM MSS, Ware MC, Agent's Letterbook, S. H. Hewes (for agent) to James C. Dunn, September 26, 1828.

in Ohio as well as in Massachusetts and New York, and better transportation facilities.[64]

Most interesting are the differences between the respective components of the labor costs. Between 1814 and 1828, picking, carding, and slubbing charges fell from around 30 percent of labor costs to 11 percent. Warping and weaving declined from 34 to 21 percent. On the other hand, spinning costs increased from 13 to 22 percent. Without knowing cloth specifications—number of warp ends, picks per inch, yarn finenesses, and weave—it is impossible to regard these changes as precise quantifications; however, it is clear that the spread of the condenser was saving labor costs in carding. The relative increase in spinning charges reaffirms the point that in the 1820s spinning was the one manufacturing operation that could be expected to attract a good deal of inventive effort.

Evidently Americans succeeded in lowering labor and capital costs in their innovations, but this does not explain why American labor costs for woolen carding, spinning, and shearing were higher than British costs. One explanation, that the supply of immigrant skilled workers was short, was indeed the case (see appendixes B and C). Only three cloth dressers and three cloth workers emigrated to the American colonies from Britain between 1773 and 1775; only two shearmen, who stayed with their trade, emigrated from 1809 through 1813, making a total of three British shearmen registered during the War of 1812; and in the years 1824–1831 merely two cloth workers, three cloth finishers, three croppers, and thirty-four cloth dressers emigrated to the United States. These numbers were far too small for skilled immigrant labor to meet all of the needs of America's expanding woolen industry.

The situation was worse in the supply of woolen carders and spinners: none emigrated between 1773 and 1775; one woolen carder and seven woolen spinners emigrated and stayed with their trades from 1809 through 1813, making a total of two woolen carders and eight woolen spinners from Britain in the United States during the War of 1812; but in the 1824–1831 period, only one wool picker and six woolen spinners arrived from Britain.

It might be countered that immigrant woolen manufacturers like Isaac Bannister, manager of a Brandywine mill, were familiar with all processing techniques and therefore could teach settled Americans a range of trades. This was true enough, but the number of immigrant woolen manufacturers also fell far short of the growth in new American woolen firms. Thus, to a considerable extent, American innovations were a substitution of machines for skill, of capital for labor.

Still this explanation does not account for the low numbers of woolen carders, spinners, and finishers emigrating from Britain to the United States. Relative wages for slubbers and spinners evidently favored the United States in the period 1815–1825, and the failure of English wool spinners to emigrate to America remains somewhat puzzling. Perhaps workers found it more profitable to migrate into British cotton mule spinning than to emigrate into the U.S. woolen industry.

The cloth shearers' case is more readily explicable. Prior to 1810 the American woolen industry offered relatively limited opportunities for specialist shearmen; there were few manufacturing concerns, and their future was uncertain. Meantime in the 1790s shearing frames appeared in England, and over the following decade rotary machines were developed in the United States, where they were adjuncts to household manufactures. In response to the shearing frame, English shearmen turned to combinations and machine breaking, thereby acquiring a reputation for militancy that could hardly attract potential American employers. By 1810 neither England nor the United States held much promise for hand shearmen, with Harmar's frame in the former and a variety of rotary machines in the latter. Partridge summed up the situation in 1823: "Hand-work is so expensive, that in some countries, in England, it is entirely laid aside, and would be, in all of them, if the workmen would permit it."[65] Like George W. Powell, a shearman who emigrated to the United States in 1801 at the age of eleven, the hand croppers would have to broaden their skills or else leave their trade altogether. Powell, working for Victor and Charles du Pont between 1818 and 1823, adjusted to technological obsolescence by exercising other skills, such as warping, spooling, and pressing.[66]

One other formative influence shaped innovations in America's early woolen technology. This was what Rosenberg has called complementarities, technical accommodations that overcame bottlenecks and allowed other inventions to succeed.[67] This appears to have been why the condenser was so important. Not only did it cut labor costs, but also it produced uniformly drafted and twisted slubbing that would not break as easily as that which was pieced unevenly by hand. Goulding's first condenser patent (May 2, 1826) shows that he intended his condenser yarns, packaged on spools or long rolls, to be spun on a woolen mule or jack.[68] The following year, he took out a second American patent for condensing and spinning. For

infringements of this he sued Benjamin Bussey of Roxbury, Massachusetts (from whom Goulding rented his Dedham mill), for $5,000 in the First Circuit Court and won.[69] Patent litigation again confirmed the commercial success of a technical improvement; with condenser carding, power could be applied to spinning and, as Cole noted, spinning machines could be increased in size.[70] But this was not all. In making a more uniform yarn, the condenser card, in conjunction with the mule or jack, paved the way for the power loom weaving of woolen cloth. The problem lay in the relative weakness of woolen warps, explaining why woolen manufacturers first used the power loom to make satinet or mixed cloths with cotton warps and woolen weft or filling.

Goulding pioneered the improvement. He was able to apply the power loom to woolen weaving probably because he was already improving the uniformity of his yarns. According to the recollections of one of his employees, Daniel Bonney, Goulding started making broadcloth on a power loom in 1822. This antedated Goulding's condenser patent by four years, but the claim remains convincing because of what Bonney also recalled was happening with Goulding's carding equipment:

When I went to work in the factory at Dedham, in June, 1823, for Mr. Goulding, my work was attending cards, where the rolls were pieced with a machine, instead of children's piecing them by hand, as at Halifax (Mass.). This machine was used until they began to make the roping on the cards, which was, I think, in 1824. Mr. Calvin Whiting, of Dedham, was then an old man, who used to be at the factory almost every day, while Mr. Goulding was experimenting on the long roll. He, Mr. Whit-

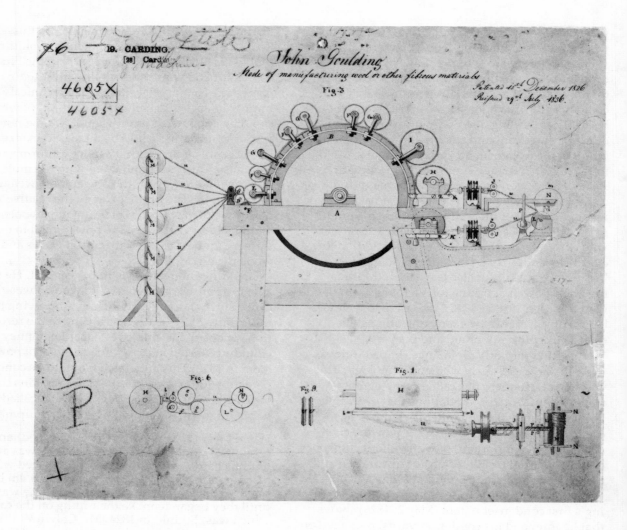

12.4 The device that enabled woolen carding machines to deliver a continuous slubbing: the woolen condenser patented in 1826 by John Goulding of Dedham, Massachusetts. Figure 7 shows the two ring doffer rollers (*H,H*) that removed the carded fibers in two endless strips. Figures 1 and 2 detail the twist tubes and keys that inserted twist in the strips and so completed slubbing formation. (U.S. patent drawing 4605; courtesy U.S. National Archives.)

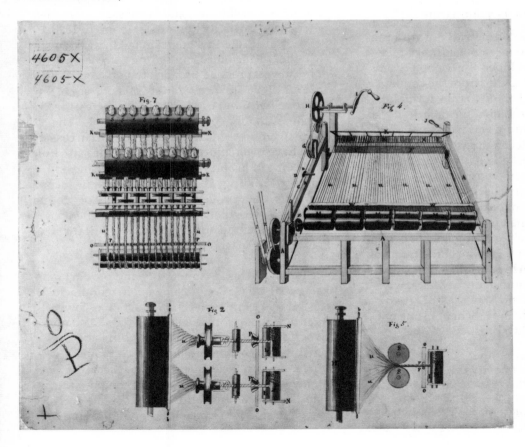

ing, had some carding machines which he had improved, so that the roping ran off into boxes, which were set behind the jennies and spun from, but the roping would tangle and break so that they could not do much with it. Mr. Edward Winslow, a very ingenious man, who came from Middleborough, Mass., worked for Mr. Goulding, and he it was who invented the method of winding the roping on to the long bobbin, by having the bobbin run on a drum, and also having drums on the jennies, to turn the roping off when they spun from it.[71]

Besides showing that Goulding was not working single-handed, this evidence reveals that Goulding's machine shop developed the friction drums (on which the spools of roving from the condenser cards were placed) that helped to transform the woolen jenny into the woolen mule or jack. It also demonstrates that Goulding was working on improvements in woolen technology on several fronts simultaneously, the most efficient way of dealing with problems of complementarities.

In this context it is worth noting that one of the more successful American broadcloth power looms of the 1820s was developed in a thriving factory and machine shop complex already engaged in cotton and woolen manufacturing and machine building. This was the Matteawan broadcloth loom patented by William B. Leonard, mill agent at Matteawan, near Fishkill, New York, on May, 23, 1827 (two patents). A year later Abraham H. Schenck, one of the Matteawan partners, told a prospective Massachusetts customer:

I think we have arrived at nearly the "ne plus ultra" for weaving Broad Cloth. We have one of this character in operation for these four or five months. The cloth is more even than we have ever had it by hand, and made at any desirable thickness at pleasure, is simple and durable. We are build[ing] three for ourselves, which fills us up. All our weaving is done by power. We have 28 Looms in operation and finishing 40 pieces a week. We use Amadon's [illegible word] Dressing [machine], which works highly satisfactory, and our weaving is second to none in this country.[72]

The Matteawan woolen power loom cost $175, excluding reeds, shuttles, and harnesses, more than twice the price of a cotton power loom. In 1828 Schenck attributed a 40 percent fall in the value (whether he meant cost or selling price is unclear) of his lowest grade of broadcloth between 1824–1825 and 1826–1827 to a variety of labor-saving improvements but especially to "the 'dressing machine' and the 'broad power loom.' "[73] The Matteawan's technical and commercial success must have owed much to the similarities and links between cotton and woolen manufacturing and machine building, as well as the company's search for technical improvements at all stages of production, a search evidenced by Abraham Schenck's tour of the manufacturing districts of Britain, France, and Holland in 1825.[74]

Circumstances in commodity and product markets predisposed American woolen manufacturers toward innovations in the production of lower-quality woolen cloths. A shortage of skilled operatives, felt most sharply at the slubbing and cloth finishing stages, promoted a search for improvements in these operations particularly. The higher cost of capital in America, compared to England, induced innovations that cost relatively less and were suited to smaller rather than larger production units. In order to overcome bottlenecks in production, mechanization was being extended at all stages of production by the late 1820s. And at this point woolen technology was benefiting by the spinoff from cotton manufacturing technology.

13

The Movement of American Innovations in Cotton and Woolen Manufacturing Technologies to Britain

When I speak of improvements, there are two things I include in the word; the first is an improvement in the machinery itself, the other is the improvement in working this machinery. Now if we had not made any improvements in the latter, namely in working the machinery, we should have found it very difficult, up to this moment, to stand against foreign competition. . . . The other matters, which are the improvements in the machine itself, . . . have been the inventions of America, and they will come in, no doubt, to enable us to produce the goods cheaper; but they may be conveyed to foreign countries, which may, perhaps, in time, work them as well as we can do.[1]

So Kirkman Finlay, a Scottish merchant in the American trade, testified before the Select Parliamentary Committee on Manufactures, Commerce and Shipping in 1833. Compared with westward flows of industrial textile technology, the reverse movement tended to be more rapid. As the first industrial nation, Britain presented special attractions to foreign inventors. Numerous machine shops in London and the manufacturing districts—Manchester, Leeds, and Glasgow—offered an unrivaled reservoir of machine building experience and skill, resting on intense specialization of labor and on a new range of machine tools. In them, new inventions might be built in the most approved manner and further refined.[2] And Britain's manufacturing districts made up one of the world's largest markets for capital goods.[3] Large-scale manufacturers, many of whom pioneered industrial technology in Britain, remained interested in adopting new advances in powered and mechanized techniques, as the rising number of patented inventions suggested. Aware of this sort of potential, Moses Brown of Providence, Rhode Island, in summer 1792 advised Samuel Dorr to take his rotary cloth shear to England. In Brown's opinion, the machine was too complicated and expensive for American woolen manufacturers and was much better suited to circumstances in England's woolen industry.[4]

The most used channel of technology transfer from the United States to Britain was in the form of British patents of American inventions. Nearly 10 percent of the textile patents in the period 1790–1830 studied came from foreigners living abroad; by the late 1820s the proportion was over 14 percent (see table 3.3). Table 13.1 demonstrates the frequency with which Americans brought or sent their textile innovations to Britain. Every major American innovation in textile technology before 1815 was apparently patented in Britain. After 1815 the expansion of American manufacturing and the reshaping of American technology made the practice unnecessary in some cases. So Paul Moody's patents, his company's property, were built and perfected in the Boston Manufacturing Company's machine shop and used in the northern New England mills, which provided an adequate capital goods market. Conversely the Rhode Island cap and ring spindles and the Matteawan woolen power

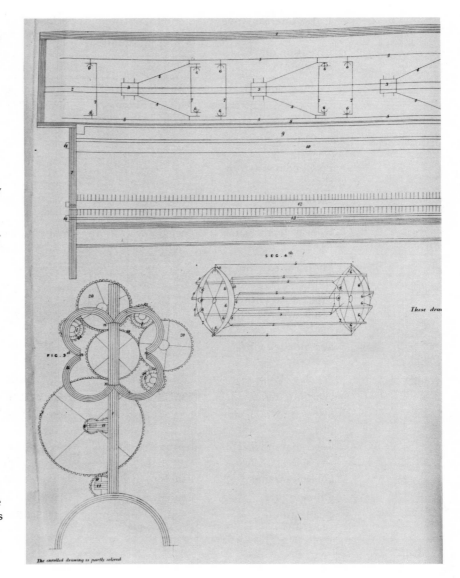

13.1 Samuel Dorr's British patent of 1793 for his cloth shearing machine. Most notable are Dorr's efforts to find new rotary movements for the shear blades. Whereas he registered a wheel of blades in his American patent of the previous year (illustration 12.1), his British design reveals a groping toward the ultimately successful form of the lawnmower (sec. 4th). (G.B. patent 1945; courtesy Boston Public Library.)

13.1. American Inventions and Innovations, 1790–1830, Patented in Britain

Invention or Innovation	American Patent		British Patent		Time Lag
	Patentee	Date	Patentee	Date	
Circular cloth shear	Samuel G. Dorr	Oct. 20, 1792	Samuel G. Dorr[a]	Apr. 9, 1793	5 mos.
Card clothing machine	Amos Whittemore	June 5, 1797	Amos Whittemore[a]	June 26, 1799	2 yrs.
Other improvements on card clothing machine	Amos Whittemore		Joseph Chesseborough Dyer[a]	Oct. 30, 1811	3–5 yrs.
			J. C. Dyer[a]	Dec. 15, 1814	
Powered hand cloth shears			Isaac Sanford[a]	Nov. 14, 1801	
Twill card clothing machine	Pliny Earle	Dec. 6, 1803			
Helical cloth shear	Beriah Swift	May 25, 1806	John Lewis[b]	July 27, 1815	10 yrs.
	David Dewey	June 27, 1809			
Multibladed vibrating cloth shear			Isaac Kellogg[a]	Aug. 21, 1809	
Cloth shear with crank-driven cutting blade			James Mallory[a]	Aug. 7, 1811	
Multibladed vibratory cloth shear			William Henry Hart[a]	Mar. 24, 1812	
Zig-zag cloth shear	Edmund Durrin	Jan. 21, 1814	John Bainbridge[c]	July 31, 1823	9½ yrs.
Warping frame	Paul Moody	Mar. 9, 1816			
Self-acting loom temple	Ira Draper	June 7, 1816			
Reed-making machine	Jeptha Avery Wilkinson	July 3, 1816	Jeptha Avery Wilkinson[a]	Aug. 23, 1817	1 yr.
Dressing frame	Paul Moody	Jan. 17, 1818			
Filling frame	Paul Moody	May 6, 1819			
Dead spindle	Used in Waltham-system mills	c. 1813–1820	Robert Montgomery[d]	Apr. 26, 1832	12 yrs.

13.1. (Continued)

Invention or Innovation	American Patent		British Patent		Time Lag
	Patentee	Date	Patentee	Date	
Differential gear for roving frame	Aza Arnold	Jan. 21, 1823	Henry Houldsworth	Jan. 16, 1826	3 yrs.
Taunton or tube speeder	George Danforth	Sept. 2, 1824	Joseph Chesseborough Dyer[a]	July 16, 1825	11 mos.
Woolen card condenser	John Goulding	Dec. 15, 1826	John Goulding[a]	May 2, 1826	6 mos.[e]
Plate speeder		Mid-1820s	Introduced in Glasgow[d]	1835	10 yrs.
Matteawan broadcloth power loom	William B. Leonard	May 23, 1827	William Collier[c]	Nov. 10, 1827	6 mos.
Cap spindle	Charles Danforth	Sept. 2, 1828	John Hutchinson	July 30, 1829	16 mos.
Ring spindle	John Thorp	Nov. 20, 1828	George William Lee	May 2, 1829	6 mos.
Eclipse speeder	Gilbert Brewster	Apr. 18, 1829	Introduced in Manchester[d]	1835	6 yrs.

a. American patentees.
b. The connection between Lewis and the U.S. inventors is not clear.
c. The correlation of the American and British patents is not wholly certain because the American patent gives only a short title.
d. See Montgomery, *Practical Detail*, pp. 61, 67.
e. Note that this was patented first in Britain.

loom, the products of smaller and more commercially isolated machine shops, were quickly taken to England in the 1820s. When an invention was already known or in use in England, foreigners could not take out a British patent. For this reason, most likely, the rotary temple received no English patent.[5]

Only a tiny minority of American inventions were double-patented in England and the United States. Most of these were technologically and commercially highly successful. Evidently most independent or semi-independent American patentees waited for at least six months to a year to see whether it was worth protecting their inventions on both sides of the Atlantic. This was not Gilbert Brewster's strategy, however. Combining his own spindle with the Rogers' trumpet flyer in upstate New York, he explained that to eliminate "any further expense in trying Experiments at home, I have at present adopted a new plan. That is to try them abroad." A little later in 1824, with his new cotton roper expected in operation in ten days' time, he urged his Connecticut employer, "No time should be lost in securing this in England & France. That is, the principle of rolling the ropen as it comes from the rolls & passing it on to spools, the spools being covered to wind up the ropen by friction or otherwise."[6] What steps Brewster took to expedite this is not known. Certainly Oliver G. Rogers, one of the patentees of the trumpet flyer, visited England soon after his American patent award (August 28, 1824) and returned to New York on July 24, 1826, being listed aboard the ship *Josephine* as an American citizen and machinist.[7]

There were exceptions to this trend. Aza Arnold totally neglected to cover his differential in England, and it was pirated by Richmond and Houldsworth.[8] John Goulding, at the other extreme, took out an English patent before protecting his condenser with an American one. The Nashua Manufacturing Company also contemplated taking out an English patent before an American one. In September 1827 its Boston treasurers instructed Ira Gay, their agent and inventor of a self-acting mule, to send a statement of work done in the two mills "and any other remarks which you think would be useful to send to England, together with the specification."[9] At this point apparently Gay had no American patent, and his improvement received no protection in England.

One other feature in table 13.1 is important: the tendency, growing in the 1820s, of trusting English agents, merchants mostly, to register patents on behalf of their foreign inventors. Joseph Chesseborough Dyer, an American resident in England, was one of the first to serve American inventors in this way.[10] Another was Timothy Wiggin, the Boston merchant who recruited Prince for the Merrimack Manufacturing Company. He evidently played some part in Charles Richmond's transactions of 1825, in which Richmond took Arnold's differential gear and the Taunton speeder from New England to Manchester. The differential was given to Henry Houldsworth, Jr., in exchange for a printing machine, and the Taunton speeder was patented in England by Dyer.[11]

The oft-heard claim that American inventions were improved after patenting in England was unquestionably true in some cases. Dorr radically modified his cloth shear from a wheel of knives to a rotating cylinder of blades in between his American and British patent awards.[12] Dyer improved Whittemore's card clothing machine and Danforth's tube

13.2 Taunton or tube roving frame patented in 1824 by
George Danforth of Taunton, Massachusetts, and pat-
ented in Britain in 1825 by Joseph Chesseborough Dyer.
The tubes that insert temporary twist are marked *e*.
(Source: Leigh, *Science*. Courtesy Boston Public Library.)

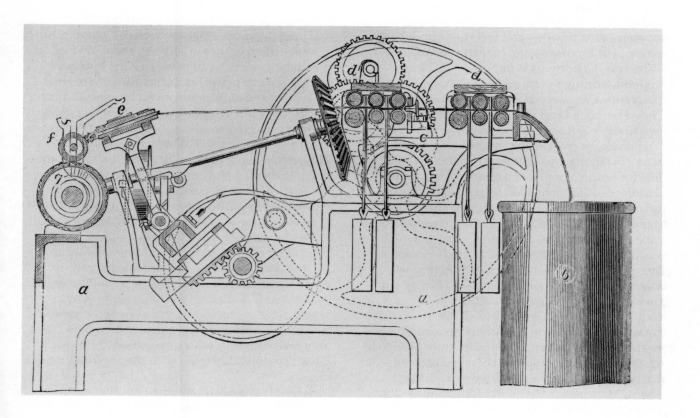

speeder with patents of his own.[13] And the helical cloth shear was the subject of numerous English improvements.

Supplementing English patents for American inventions was the listing of American patents in English journals. This belatedly commenced in the *Repertory* in October 1830. A few patent abstracts, derived from the *Franklin Journal*, appeared soon after this date; they include Thorp's cap spindle, which was patented soon after Danforth's.[14]

The patent systems of the Atlantic economy were supplemented by other channels of diffusion in the eastward transmission of textile technology. Chief was the growing number of Americans in England. Through them some of the new American innovations were established in England. Joseph Chesseborough Dyer was one prominent expatriate.[15] A Rhode Island importer with a gift for mechanics, he crossed the Atlantic several times before marrying and settling in England in 1811. Besides acting as a patent agent for American inventors, he established a card clothing manufacturing business first at Birmingham and then at Manchester. In the 1820s and 1830s, it produced great quantities of card clothing, some of which went to America under Board of Trade license. Dyer's firm began in 1812 as a partnership with another American living in England: Henry Higginson of Boston, then an American banker in London. Higginson, as Dyer acknowledged, shared in the work of improving the American card clothing machine. Jeptha Avery Wilkinson, from Otsego in upstate New York, imported his reed making machine in 1817 and also started a profitable business in Manchester.[16] Thomas Robinson Williams, a Rhode Island inventor, arrived in the mid-1820s,

set up as a London engineer, and took out a variety of patents in the following decades.[17]

American visitors to Europe might also be entrusted by mill managements with the task of promoting American inventions abroad. Thus the Nashua Manufacturing Company's treasurers wanted a Mr. M., then about to sail for Germany, to look at Ira Gay's self-actor mules because "possibly our improvement in the Mule may be of value to the Company in Saxony."[18]

Not much early American textile machinery was exported to Europe. Gilbert Brewster sent machines of unspecified nature to Prussia, and early in 1827 a broadcloth power loom made in New York was shipped to one of the Leeds factories.[19] This may have been one of the Matteawan broadcloth power looms. Only one woolen broadcloth power loom "from a foreigner residing abroad" was patented in Britain in the late 1820s. The patent, in the name of William Collier, a fustian shearer of Salford, Lancashire, and dated November 10, 1827, incorporated a warp let-off motion, a variable batten speed, a picker motion, and a cam shedding arrangement.[20]

Few Englishmen troubled to visit America to collect technical data. Henry Wansey was one at the beginning of the period, but he went before any distinctive American innovations emerged. Lawrence Greatrake, an English paper manufacturer, in 1813 asked E. I. du Pont for "a drawing of your whole place . . . including your Woolen Factory . . . [which] I purpose to send to Europe, or may probably take it myself, with some others."[21] What Greatrake intended to do with the drawings he did not say.

Before 1840 a relatively small number of American innovations were adopted in England. Card

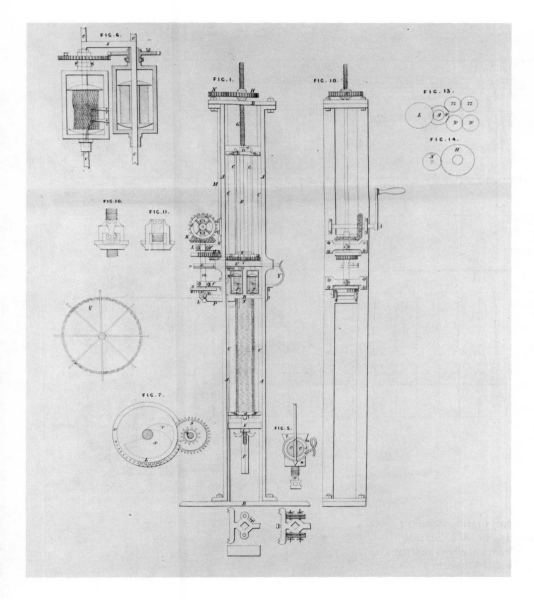

13.3 Reed-making machine patented in America in 1816 and in Britain in 1817 by Jeptha Avery Wilkinson of Otsego, New York. Figure 1 shows a front elevation of the whole machine. As the reed ribs pass through the machine, so flat steel or metal wire dents are fed between them from reel *Q*. Yarns from the two bobbins and flyers (just above *B*) bind the dents to the ribs. (G.B. patent 4162; courtesy Boston Public Library.)

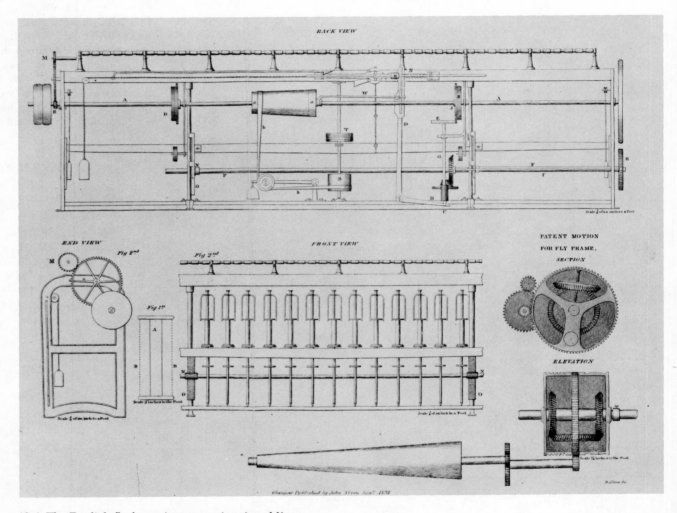

13.4 The English fly frame incorporating Arnold's
American differential; a drawing of 1832. (Source:
Montgomery, *Carding and Spinning Master's Assistant*.
Courtesy Merrimack Valley Textile Museum.)

clothing machines spread slowly and were resisted until the 1860s in Yorkshire, where hand setting with child labor was cheap. Significantly, the American card clothing machine needed modification before it could be introduced in England. Dyer recalled that he and Higginson "by our joint labors [made the machine] far more simple in several of its movements, whereby it could be worked with much greater safety and speed than in its previous state."[22] Goulding's condenser was attacked until the 1890s by men who could remember the billy yarn. Indeed the condenser was said to cause a weaker yarn, a cloth with a lighter covering of fibers for fulling and raising, and a finish "papery to handle and destitute of the usual suppleness of the best west cloth."[23] Yorkshire immigrants in America in the 1820s used the condenser only with reluctance.[24] British manufacturers preferred the piecing machines, developed in the late 1820s, which replicated the billy's movements.[25]

On the other hand, Wilkinson's reed-making machine was established in Manchester, and Danforth's cap spindle was said to be more popular in England than in America in the early 1830s.[26] Arnold's differential, now erroneously attributed to Houldsworth, spread through and beyond British cotton manufacturing districts. Within a year of Houldsworth's patent, Richard Roberts purchased an improved bobbin and fly frame (incorporating the differential) for export to Mulhouse in France. His firm's correspondent described it to their French customers:

The opinion of its value is such, that numerous orders have been given by parties, who had machines upon the former principle. We think it therefore very desirable that you should possess such a machine, the more so because we know that Messrs. Risler Freres and Dixon have made great exertions, not only to procure a machine, but also to engage a workman, who had been employed in making similar machines, in either of which attempts we do not learn that they have yet succeeded.[27]

Coarse spinners in Glasgow took up the Taunton or tube speeder in the early 1830s.[28]

These were individual inventions that could be incorporated into English cotton and woolen manufacturing production lines without too much difficulty. The whole range of innovations that constituted the new system of cotton manufacturing technology in northern New England was an entirely different proposition. Kirkman Finlay told the Select Parliamentary Committee in 1833 that most new improvements were easily adapted to other machines, "till the late ones by Americans, which would require the whole machinery to be taken out and the new machine to be put in." The reason was explained by James Kempton, a Connecticut manufacturer who also testified in 1833. Having stated that he had seen the best machinery in Manchester, he was asked, "Have you seen none equal to what you have in America?" Kempton replied, "Not for the manufacture of coarse goods." In other words, the Waltham system's cotton manufacturing technology, shaped for the efficient production of coarser yarns and fabrics, had no parallel in Lancashire, where the technology was designed for the production of fine yarns and goods. American technology, it was observed, had the advantage of economically using better cotton qualities.[29] Even so, Kempton's criticisms still remained true until the mid-nineteenth century and after.

In their *Reports* on American manufacturing of

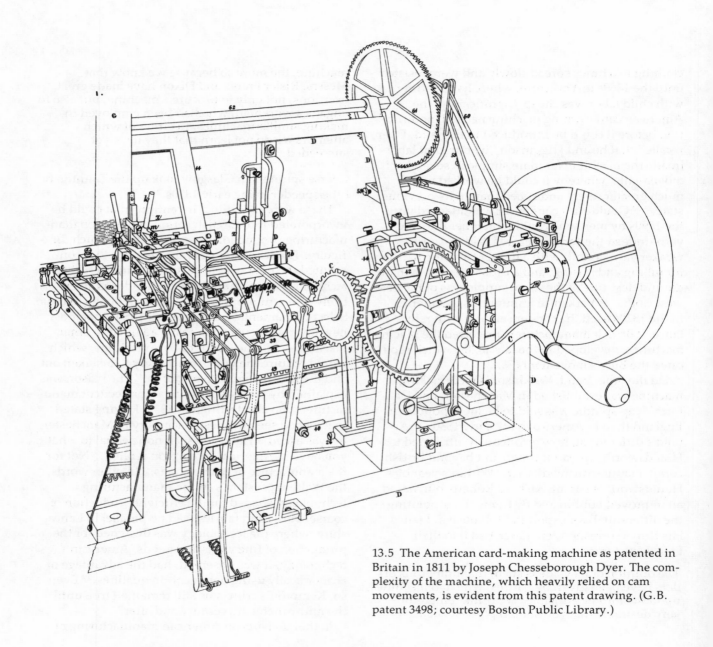

13.5 The American card-making machine as patented in Britain in 1811 by Joseph Chesseborough Dyer. The complexity of the machine, which heavily relied on cam movements, is evident from this patent drawing. (G.B. patent 3498; courtesy Boston Public Library.)

the early 1850s, Wallis and Whitworth detected in American fabrics neglect of "that sterling quality in make, which always forms so important an element in the judgment of the European consumer." They noted about American machinery: "Wherever it can be introduced as a substitute for manual labour, it is universally and willingly resorted to."[30] In short, American technology, so widely admired for its labor-saving characteristics, was inappropriate for the manufacture of the quality textile fabrics to which sophisticated European consumers were long accustomed, fabrics that could then be made only by traditional skills. Even so, Britain, the first industrial nation, continued its older habits of learning at a technical level from foreign followers and rivals.[31]

14

Conclusion: Some Perspectives on the Transatlantic Diffusion of Early Industrial Textile Technologies

The westward transatlantic diffusion of textile technologies had startling economic and social results. Few who watched the first fruits of transfer, a hand-powered carding machine and an eighty-spindle jenny, laboriously operated on a rattling carriage in Philadelphia's 1788 Independence Day parade could have imagined what these omens of a machine age portended.[1] Yet within twenty years or so, the United States boasted nearly 100,000 cotton mill spindles. Between 1810 and 1820 this figure more than tripled and then it more than tripled again in the 1820s.[2] "I suppose," a Rhode Island cousin wrote in 1826 to the machinist Aza Arnold, then at Great Falls, New Hampshire, "you are building factories & filling them with Machinery at a great rate—so that I should hardly know your village were I to land in it from a balloon."[3] By 1831 perhaps 700 firms had over 1.2 million cotton spindles and 33,500 looms (mostly power-driven) in water-driven mills on never-failing streams, chiefly east of the Alleghenies and mostly between New Hampshire in the north and Maryland in the south. In the early 1830s, fixed capital investment in the United States cotton industry, computed at nearly $45 million, was over a third of the size of the fixed capital investment in the contemporary British cotton industry.[4] This rate of growth, accomplished within forty years, was an astonishing testimony to the success of technological diffusion.

Americans were not merely imitators. Although their first experiments in factory spinning closely copied British models, the involvement of Boston's leading business interests in the research and development stage of power loom weaving led to a series of innovations in the organization and design of cotton manufacturing equipment that disquieted rival British manufacturers as early as 1826.[5] These innovations made the Waltham-type coarse-goods cotton factory at least 10 percent more efficient in throughput of materials than its British counterpart.

Modified technology permitted, and indeed was partly the product of, modified factory discipline. Before it soured in the late 1830s, the benevolent paternalism of the Lowell corporations made their name an international byword for enlightened industrialism. Between the late 1820s and the Civil War Lowell was to the New World what New Lanark was to the Old, making a visit to Lowell nearly as obligatory for foreign tourists as a sight of Niagara Falls.[6] Dickens in 1842 admired the cleanliness, vigor, dress, and deportment of the Lowell operatives, and especially their musical and literary talents, evidenced in the boardinghouses' "joint stock pianos," circulating libraries, and the girls' own periodical, the *Lowell Offering*.[7] Although the meaning of the Lowell mill and boardinghouse disciplines continues to intrigue historians, there can be no doubt that, to the beholder, the contrast between conditions in the Massachusetts mills and British cotton factories was sharp.[8] Dickens, writing as reporter rather than novelist, saw the difference as "between the Good and Evil, the living light and deepest shadow."[9]

In Rhode Island and other states to the south, the technology and organization of cotton factories remained closer to English models. As a Pennsylvania state investigating committee discovered in 1837, a fifth of factory employees were children under the age of twelve; adults and children alike worked eleven to fourteen hours a day, six days a week; and "no particular attention is paid to the education or morals of the children, by the employers."[10]

In the American woolen manufacture, the results of technology diffusion were less spectacular, not least because the woolen fiber inhibited technical change. Even so, diffusion led to a measure of technical and economic transformation. American woolen men visiting England's clothing districts in the mid-1820s disdainfully noted the persistence of billies, small jennies, and hand looms. Condenser cards, large jennies, jacks, and power looms distinguished the bigger, best-practice American woolen factories, especially those between the Hudson and Merrimack valleys, where woolen and cotton manufacturing were often contiguous or shared the same managements.[11] Glimpses of the spirit of change and improvement pervading American woolen manufacture in the 1820s occasionally surface in the record. One New Hampshire mill manager told his directors in 1826, "The adoption of the new system of loco-motive machinery throughout our buildings and the enlargement of our Dye House, now nearly equal to another Factory, will justify the expenditure of the time and money devoted to these objects."[12] Condenser cards and power looms formed part of this "loco-motive" production line. For an aggregate estimate of economic change, the absence of fixed capital investment data in the contemporary British woolen industry denies an international comparison. However, by 1820, when half of the American woolen industry was located in New York, Connecticut, and Massachusetts, the whole American woolen manufacture was just under a quarter of the size of the American cotton industry in fixed capital investment and between a tenth and a sixth in spindleage. (See appendix D.)

In attempting to understand the complex process of technology diffusion that lay behind America's early industrial transformation, this study reaches a number of conclusions about the circumstances hindering and promoting transfer and about conditions inducing technical modifications. To begin, the investigation lends definition to the roles played by immigrant artisans in transatlantic transfer. Most obviously, the artisan emerges as the preeminent technology carrier in this period. The finding hardly comes as a surprise. Requiring new skills, the new technologies appeared in published verbal forms only belatedly and then incompletely. However other grounds for emphasizing the importance of the artisan become plain when obstacles to transfer are considered. In Britain, mill-level secretiveness, resistance to the publication of technical knowledge, and a chaotic search system in the London patent repositories all elevated the knowledge and skill of key industrial workers in the diffusion process. Even the severe but ineffective laws against artisan emigration underscored the value of skilled workers to foreign rivals. The gathering pace of inventive activity, and with it innovation, the increasing complexity of machinery employed in textile processing, and the incremental nature of technical change together implied that overseas imitators would place heavy reliance on experienced British artisans.

As technology carrier, the skilled worker was not without problems. Conflicting British regional traditions in manufacturing and machine making, the British pursuit of higher-quality textile products, resistance to technical change in some trades, and widespread attitudes of secretiveness, to say nothing of union militancy in the 1820s, conspired

against a wholly sanguine view of artisans as technology carriers. For these reasons, technical publications and other rational vehicles of technology diffusion, like international exhibitions and manufacturers' conferences, would extensively supplant migrant artisans later in the nineteenth century. [13]

The contribution of artisans in diffusing the four textile technologies to America exhibited no simple repeating pattern. At the first stage, the conveyance of technical potential to the United States, immigrant workers proved indispensable in the transfer of cotton factory spinning and woolen machine carding and spinning. But with cotton power loom weaving and calico printing, American visitors to Britain largely displaced the immigrant artisan as carrier. The difference reflects partly the intensifying of American business interests after 1807 and partly the state of the technologies. Power loom weaving caught American attention before the completion of mechanization, forcing Americans to share in research and development costs, while calico printing comprised a complex group of processes combining partially understood industrial chemistry with design and mechanics, requiring more thorough investigation by American imitators. In the establishment of prototype mills and technological models, immigrant artisans predominated where the technology necessitated skills acquired only by lengthy learning periods: for several machines in woolen manufacturing (billy, jenny, gig, cloth shear), for mule spinning and mill managing in cotton yarn production, and for calico printing skills and printery superintendence in cotton finishing. With internal diffusion in the United States, immigrants seemed most conspicuous before 1807, when the process primarily rested on kin groups and trade networks. Slater's impact, for example, owed much to this situation. With increases in the scale of production, associated with power loom weaving, capital goods firms became more important vehicles of diffusion, as the Locks and Canals Company at Lowell demonstrated. Selling agents serving numbers of manufacturing firms also disseminated technical information among their clients. Nevertheless a few immigrants, possessing skills that eluded mechanization or rapid learning, figured in internal diffusion. The Scholfield family in woolen manufacturing and the dozens of Lancashire printers who moved among the New England printworks in the 1820s were cases in point.

In an assessment of the role of British textile workers as agents of technology transfer, a clear distinction must be made between individual artisans and artisans in aggregate. The initial conveyance across the Atlantic of new information and skills arose from the emigration of a relatively small number of British workers, perhaps a few dozen, many of whom shared in setting up prototype mills and to some extent in the process of internal diffusion. On the other hand, the mass of textile workers reaching the United States from Britain present a very different picture. Quantitative analyses dispel any notion that they, collectively, represented the new British textile technologies. At any time between 1770 and 1831, data show, at least 74 percent of immigrant textile operatives were weavers, almost certainly hand loom weavers. The proportion of operatives with new industrial skills rose from 11 percent of those immigrating (and staying with their trades) on the eve of the War of 1812 to 17.5 percent (of

3,632 new immigrants) in the 1820s. However, numbers of woolen manufacturers arriving annually rose appreciably between the War of 1812 and the 1820s (from 7 to nearly 24) and even more so did numbers of general machine makers (from 10 to 100). Yet a substantial level of transience was suggested by the overall 60 percent of workers in their teens and twenties and a similar proportion unaccompanied by family.

The immigrants who shared in transatlantic transfer did not conform to the pattern displayed by immigrants in aggregate. Besides having industrial experience, most of the technology carriers were much older men. John Murray, who set up an early though not very successful power loom, was in his mid-forties. So too was John D. Prince, the calico printery manager. The Thorp brothers who introduced roller calico printing in 1809 were in their thirties when they left England. And the Scholfields were in their mid-thirties when they set out for Boston. Slater, certainly, had just reached his majority when he sailed from London, and John McDonald, who set up an early power loom, was in his twenties. Not all the technology carriers were unaccompanied when they set out for America. The Scholfields went in a family group. It might be concluded that greater experience and emotional stability promoted successful transatlantic transfer. All that can certainly be said is that the successful artisan carriers of the technologies under review did not conform to the typical picture of the British textile immigrant in the United States: young, unaccompanied males possessing preindustrial skills. These conclusions underscore the importance of studying technological migrations at both the individual and aggregate levels.[14]

The implications of this aggregate immigration pattern are intriguing. Did the superfluity of hand loom weavers support the persistence of traditional technologies and thereby slow industrial growth? The question is hard to answer, even for Philadelphia, renowned for its hand loom weavers until the Civil War. All that can be said now is that Philadelphia was not an especially preferred destination for immigrant British weavers in the period 1770–1831; neither was the Philadelphia region, which in comparison with New England or New York was short of skilled industrial immigrants from Britain's textile industries. On this basis, any delay that Philadelphia experienced in industrializing cannot be attributed to immigrant weavers, though a detailed local study could well alter this conclusion.

The significance for American industrialization of the supply of skilled industrial immigrants also invites further exploration. A comparison between numbers of new immigrants linked with all textile trades and numbers added to the expanding American cotton and woolen industries in the 1820s discloses large shortfalls at operative, managerial, and machine-making levels. At the most, fresh immigration could have met about a quarter of the demand for new operatives in the two American industries during this decade. At the managerial level, the presence of manufacturers may overstate the potential contribution of immigration to the two industries' labor expansion of the 1820s. Including manufacturers, immigration could supply about half the new demand for cotton mill managers; excluding them, it could meet less than 5 percent. In the case of woolen managers, prospects were brighter for Americans. Assuming that clothiers were manufacturers and not just finishers, current immigration could have provided nearly 70 percent of new American demand. Current immigration of machine makers could have furnished perhaps 40 percent of new demand for American textile machinery makers. The relative abundance of immigrant machine makers (1,001) compared with skilled industrial operatives (under 600) may well help to explain the low wages of American machine makers, which Rosenberg hailed as "truly startling."[15] The closeness between American and British machine makers' wage levels and therefore the lower labor costs in building American capital goods presented an additional reason for Americans to prefer capital-intensive equipment.

These estimates, for the moment at least, are the best available, but they are also imperfect. Nevertheless, in view of the conservative nature of my quantitative estimates, I maintain that new immigration could at most supply 25 to 50 percent of the growing demand for textile workers in the United States during the 1820s. By this date American manufacturers had crossed the threshold into independence of Britain's technical know-how. And this seems quite consistent with my earlier finding that before 1807, individual immigrant artisans secured textile technology transfer at the first three stages of diffusion and thereafter figured primarily at the introduction and prototype stages. Their diminishing role in the internal diffusion stage simply parallels industrial growth. By the 1820s the cotton and woolen industries expanded, in part because the general reservoir of technical knowledge and skills in the United States had both spread beyond the confines of the immigrant community and risen in level. Only technologies new in

Britain (and America), or older ones embodying hard-won, nonmechanized skills, needed the presence of British immigrants in the 1820s.

Even were my figures perfect, this quantitative bird's-eye view could not be the complete picture. For example, some trades in Britain that seemed applicable to the United States proved inappropriate because of the diverging courses of British and American innovation; English mule spinners and weaving overseers made this discovery at Dover, New Hampshire, in the 1820s. The hazard of overemphasizing statistics will become evident in local studies pursuing the economic and social adaptation of immigrant workers. Charlotte Erickson's *Invisible Immigrants* demonstrates the kind of detailed career profiles and contrasting British and American local contexts that can be assembled to investigate the motivations, movements, and adaptation of immigrant Englishmen. From this work, it is clear that insecurity, loneliness, illness, and early death awaited some proportion of immigrants; skilled industrial immigrants, plagued by these misfortunes, saw their economic potential crumble. Machine maker James Standring, machine engraver Thomas Lonsdale, and mule spinner James Leard, all ruined by alcoholism, illustrate the kind of destructive effect that noneconomic influences could wreak on promising industrial immigrants. Although this study throws new light on the role of artisans in the diffusion of textile technologies to early industrial America, much local work remains to be done.

Another conclusion relating to the channels of technology transfer also confirms the familiar: industrial espionage inevitably accompanies industrial competition, and no amount of public or private protection can eliminate the borrowing or piracy. Artisans slipped through the British customs net because it lacked an infallible mechanism for identifying occupations and because an individual emigrant's intentions could never be positively established. Yet it does seem that Britain's prohibitory laws raised technology acquisition costs for foreign borrowers. Insurance on smuggled equipment added at least 30 to 45 percent to purchase prices in the 1820s, though this halved by the late 1830s. Circuitous export routes, additional warehousing costs, and disassembly and repackaging charges probably raised total smuggling costs in the 1820s nearer to 50 percent of machinery retail prices. Since the most astute industrial spies "always take the easiest and cheapest course," Americans naturally imported relatively small amounts of equipment from Britain.[16] They included key parts or even whole machines when design, quality, or price was better than that obtainable in the United States, and American visitors evinced a keen interest in Britain's best-practice technology. In official applications for machinery export orders made to the British Board of Trade from 1825 through 1843, American destinations figured in only 7 percent of the total. Although machinery values are unknown, it may be surmised that a similar fraction, or even less, of smuggled British machinery exports was bound for America.

The influence of general business conditions in the receptor economy presents another lesson to be drawn about the circumstances for international technology transfer. For the United States in the period, noneconomic assets (a measure of political stability; language and social and legal institutions deriving from the originating country; a Puritan work ethic; and a relatively high degree of social mobility) and economic ones (abundant raw materials and water power, a reservoir of skilled traditional craftsmen, and the will to make profits) looked unpromising when set against high capital and labor costs and constricted product markets. But combined with secure and expanding markets, or expectations of such, they enabled American capitalists to surmount their factor problems with the aid of technological solutions frequently imported by skilled industrial immigrants.

Prospects of greater political stability and the emergence of an economic infrastructure, together with the propagandist activities of a pro-industry pressure group, lent high expectations to new industrial ventures in the early 1790s. In this context, Arkwright cotton spinning technology was successfully transmitted to America. When European war indirectly choked such prospects and British textile imports flooded the American market, few early mills survived. Between 1807 and 1815 chances for American textile manufacturers improved again with the restraints on British imports and the demands of a wartime economy. Over these years cotton factory spindleage in the United States swiftly rose from under 50,000 to around 300,000 spindles. And in these expanding conditions, the diffusion of power loom technology occurred. Peace and the resumption of European trade in 1815 severely curbed American textile manufacturing; however, population growth, tariff protection, proliferating canal and road networks, and falling cotton prices greatly enhanced business prospects in the 1820s. New manufacturing investment in power loom weaving appeared profitable, and, later, calico

printing technology offered a likely way of maintaining unit profits. So the stage was set for the diffusion of calico printing technology.

Market conditions do not alone explain successful technological transfer. Even in the adverse business circumstances of the 1790s, immigrants like Slater and the Scholfields survived. Besides the backing of American capitalists and good measures of technical skill and business caution, they possessed the noneconomic assets of adaptable personalities, kin support, and integrity in their commercial dealings, all of which inspired confidence among the American community in them. The technologies they carried became more readily accepted. In contrast, dishonesty discouraged the employment of immigrants, delayed investment, and retarded technological transfer. Indeed a case might be made for the moral basis of the location of the American cotton industry; more scrupulous immigrants in the Philadelphia region might have secured the establishment of the cotton spinning industry in Pennsylvania and New Jersey ahead of Rhode Island. But this interpretation is taking one line too far. Reliability in the artisan was as necessary as skill or business prospects conducive to investment; technical transfer was the product of several, rather than one, economic and noneconomic predisposing conditions. At the same time market conditions unquestionably constituted a major variable in forming the contexts for technological transfer.

On the transatlantic movement of technology, this study confirms that rates of transfer could be surprisingly fast. Both spinning technologies, water frame and mule, appeared in the United States within twenty years of their emergence in Britain.

British power loom weaving technology took about a dozen years to reach America, and within a decade of 1803, when Radcliffe's dresser opened the way to mechanical weaving, the Boston Associates were developing an alternative power loom technology. At the level of individual mechanisms, some transfer rates were faster still. Horrocks's variable batten speed motion, patented in 1813, was being built into Rhode Island power looms four years later. On the other hand calico printing, the most complex of the textile technologies under review, took up to twenty-five years to reach America, if the Thorp brothers' printworks at Bristol, Pennsylvania, marks the introductory stage. If internal diffusion, as through the New England firms, signaled successful transfer, then fifteen years can be added to this rate. The timing of invention in Britain and changing business conditions in the United States largely explain these differing westward transfer rates. In the opposite direction, most major American innovations arrived in Britain between six months and twelve years of their American patent dates. Exceptional were the innovations of the Boston Manufacturing Company and its offspring, the northern New England promotions of the Boston capitalists, which designed, built, and consumed their own machinery and pursued a different course of technical innovation from that then prevailing in British cotton manufacturing districts. For other American inventors, Britain possessed the best available workshops where new inventions might be perfected and the largest market for equipment where they might be sold. One wider observation is obvious: the rapid movement of textile technology in both directions across the Atlantic strengthens the impression of close links between the two countries and adds another dimension to the relationships of the Atlantic community.

The second theme pursued in this technology-diffusion study is the question of modifications to imported technologies. Here a number of conclusions emerge. Foremost is the finding that a combination of economic and social mechanisms and influences induced the reshaping of British technologies, a finding that a company-level study of the cotton industry leader in northern New England reveals with new clarity. In commodity markets, better-quality staples allowed Americans to start power loom weaving with a technology cruder (imposing greater physical forces on textile fibers during processing) than the contemporary British technology. Cruder technology was quicker to develop and get into production than more refined mechanisms. Product markets favored innovations suited to the mass production of coarse but durable cotton fabrics, which led to the use of nonversatile manufacturing equipment for long production runs of standardized goods. When conditions in product markets encouraged large-scale operations, the Boston capitalists at Waltham met the problem of rising overhead costs. Of these, labor seemed the most easily reducible. At Waltham, therefore, when mill expansion occurred, implying increased output and falling prices and profit margins, managers sought technical innovations designed to cut wage bills and so restore profit margins. As the technology was modified to permit the substitution of cheaper, unskilled, female labor for more expensive, skilled, male labor, so social circumstances in New England prevented the gross exploitation of young people in the cotton mills, in contrast to that which disgraced England's cotton factories. Consequently the technology allowed a more humane treatment of the labor force than before. Competition in product markets led the Waltham capitalists to hedge their technology with patent protection, itself a further incentive to invent novel technical solutions. The technology selected as a result of all these pressures certainly lowered labor costs, but the higher throughput of materials that it attained surely conserved capital costs too. Rosenberg's insistence that the choice of technological alternatives conforms to the underlying mechanism of factor proportions and factor prices therefore seems to be borne out in this examination of the Waltham firm, with the proviso that one noneconomic influence also played some important part.[17]

For the labor-saving improvements contrived in the Rhode Island cotton manufacturing system and in American woolen manufacturing, commodity and product markets also diverted inventive efforts toward innovations appropriate to coarse, standardized products. In Rhode Island, however, versatility remained a technical goal, not least because of the persistent influence of British models, mediated by Slater and his disciples. And in the Rhode Island system, judging by comparative machinery costs, the smaller scale of cotton manufacturing led to efforts to save capital. Firm-level investigations should further illuminate the economic and social mechanisms behind the innovations of the Rhode Island system and of American woolen manufacturing.

This study also raises the possibility that immigration encouraged the employment of a more labor-saving technology in America. The high wages of immigrant mule spinners at Waltham drew attention to labor costs, inducing the application of capital so that cheaper labor could replace more expensive labor. Supporting the direction of this tendency was the relative abundance of machine makers among British immigrants. Relatively low American machine makers' wages add substance to the impression, as do the comments of British visitors. A Glasgow cotton spinner, James Dunlap, who toured cotton factories between Massachusetts and Maryland before returning to Britain in 1822, claimed that most American machinery was built by immigrant Englishmen and Scots, among whom he especially noted "artificers in the metals, such as iron and steel turners" (the skills needed to make the key textile machine components of rollers and spindles).[18] If immigration was helping to lower the costs of capital goods manufacturers and machinery prices were consequently falling, then we have a further explanation for the greater application of capital in American textile manufacturing.

It does, however, remain to be seen whether the labor element in capital goods manufacturing was falling in the early nineteenth century. One detailed estimate of the cost of building a warping and dressing frame for an Exeter, New Hampshire, cotton factory in 1820 shows that capital charges amounted to nearly 63 percent and labor charges to 37 percent of the total cost of $136.50.[19] A number of other detailed cases are needed to discern a trend. And then, of course, an increase in native-born mechanics might have contributed as much as or more than immigrant machine makers to any fall in costs of making capital equipment. To settle these and other questions, a study on early American machine making that is far more thorough and systematic than any available will be needed.

The subject of capital and labor costs raises the important question of regional variations. In the Philadelphia region, labor seems to have been much cheaper than in New England, for example. Adult male cotton mule spinners on the Brandywine in 1811 earned only $1.50 a day, compared to the Boston Manufacturing Company's mule spinners, who averaged $20 a week in 1817.[20] But neither the American nor the British, let alone the transatlantic, labor markets behaved perfectly. Variations between cotton mule spinners' earnings in Manchester, for comparable work and hours, differed between firms in the mid-1830s by as much as 35 percent. More local studies are required to define the picture.[21]

One last theme to emerge from this investigation concerns the impact of the new American textile technology, particularly that created at Waltham: the constant-batten-speed, cam power loom; the high-speed dead spindle; the stop-motion warper; the rapid-drying dresser fitted with warp measuring and shut-off device; the four-cone double speeder with a low piecing requirement; the filling frame with a cop winding motion; and the self-acting loom temples. From management's viewpoint, these inventions unquestionably cut labor costs. One dollar of labor in the Massachusetts cotton industry in 1831 processed 20 percent more cotton, by weight, than the same amount of labor in the Rhode Island cotton industry.

From the operatives' point of view, work in a Waltham or Lowell cotton mill, or in most other northern New England cotton mills, was free of the exhaustion, oppression, and cruelty that marked British cotton factories.[22] The very purpose of the Waltham innovations with regard to operatives was to admit educated daughters of respectable and respected New England rural families to the cotton mill. Not only would overseers treat their charges with some caution, under a puritanical moral code (which presumably applied as much to overseers as to operatives), but the technology would have to accommodate the social as well as economic characteristics of the new labor force. A high labor turnover pointed toward shorter operative learning periods. Educated and active minds, which quickly found machine minding very dull, required equipment with maximum automaticity. And younger, less physically strong and less well-coordinated operatives needed less onerous and hazardous machine-tending tasks. So the Waltham innovations disposed of mule stretching and mule spinning and their long learning periods; compensated for operative inattention with numerous stop motions; reduced piecing in the roving frame and in filling preparation with the filling frame; and lowered operative interference with moving parts, as with the self-acting temple. Only in dresser tending did the work remain onerous, because it required prolonged and concentrated attention to the sized warp yarns to check that none broke.

If the girls were made responsible for no more equipment than they could comfortably manage, the technology lowered the levels of skill, mental alertness, and physical hazard involved in machine minding. If overseers were tolerant, then boredom or fear of offending the mill's narrow moral code constituted the operatives' greatest afflictions. Waltham technology thus permitted a more humane regime at the purely physical (as opposed to the moral or regulatory) level of operative work.

Whether the introduction of Waltham technology

to Britain—which hardly occurred in the period under review—would have ameliorated working conditions seems doubtful. Almost any technology can be applied either in a humanizing or a dehumanizing way; it all depends on the human priorities of those wielding the technology, which in turn are shaped by much wider social and economic considerations. The absence of enlightened and humane values in sufficient force in early-nineteenth-century Britain suggests that American innovations would have made little impact on British working conditions. Indeed even in New England the Waltham technology was operated with new severity from the 1840s, when the labor supply situation drastically eased.[23]

Some Lancashire manufacturers tried to follow the northern New England mills, but their efforts produced no more than poor imitations of American cotton goods.[24] Waltham technology, as a manufacturing system, was not implemented in Britain, in this period at least. Presumably the cost of quality cotton staples, higher in Britain than in the United States, limited Lancashire options. It is very likely therefore that in the Atlantic economy, New England manufacturers moved into a position that complemented rather than competed with British mills.[25]

In summary, this investigation finds that for a new technology still partially understood or not yet reduced to verbal or mathematical forms, the experienced practitioner must be the most efficient agent of international diffusion. Modifications to imported technologies conform to prevailing economic and social circumstances in the receptor country—in particular, conditions in commodity and product markets, relative factor prices, and so-

cial values. But technology diffusion is a complex process. Although some generalizations can be made about it, different technologies and different regions demonstrate variations of some significance on familiar themes. Using stage analysis concepts, and trying to understand the interplay between the technical and economic dimensions of the technologies concerned, this study has sifted technology diffusion in the cotton and woolen manufacturing industries, which underpinned America's industrial revolution. Clearly, without technology diffusion, industrialization is long delayed or never takes place. For the United States, transatlantic connections and circumstances facilitated technology transfer and so created the possibility of a transatlantic industrial revolution.

Appendix A

British Patent Specifications for Cotton and Woolen Textile Inventions (Mechanical) Published in the *Repertory of Arts*, 1794–1830

	Patent Data				*Repertory of Arts* Publication Date	
No.	Patentee	Subject		Date	Date Published	Form (S-specification; D-drawing)
1982	John Harmar	Shearing frame		1794	1801	S,D
1984	Richard Cartwright	Carding device		1794	1817	S,D
2029	Thomas Connop	Batting frame		1795	1795	S,D
2100	Richard Varley	Carding and spinning		1796	1799	S,D
2122	Robert Miller	Power loom		1796	1798	S,D
2322	Amos Whittemore & Clement Sharp	Card making machine		1799	1804	S
2460	John S. Ward	Doubling frame		1800	1804	S,D
2523	Anthony Bowden	Batting frame		1801	1802	S,D
2672	Joseph Fryer	Shearing frame		1802	1802	S,D
2633	William Walmsley	Batting frame		1802	1802	S,D
2791	Jacob Buffington	Stretching frame		1804	1805	S,D
2869	Peter Marsland	Sizing device		1805	1805	S
2872	Henry Maudslay	Calico printing press		1805	1806	S,D
2896	Earl Dundonald	Spinning device		1805	1807	S
2944	William Clarke & Joseph Bugby	Mule, jenny		1806	1807	S,D
2992	Samuel Williamson	Handloom, picking		1806	1807	S,D
3024	Archibald Thomson	Spinning device		1807	1815	S
3027	Samuel Williams	Preparing device		1807	1808	S,D
3094	John Leigh Bradbury	Spinning device		1807	1810	S,D
3202	Archibald Thomson	Spinning device		1809	1815	S
3388	John Towill Rutt, John Tretton, & John Webb	Card making machine		1810	1814	S,D

Patent Data				*Repertory of Arts* Publication Date	
No.	Patentee	Subject	Date	Date Published	Form (S-specification; D-drawing)
3550	William Henry Hart	Shearing frame	1812	1812	S,D
3633	Joseph Raynor	Roving frame	1813	1813	S,D
3639	Matthew Bush	Calico printing machine	1813	1821	S,D
3728	William Horrocks	Power loom	1813	1814	S,D
3812	William Sellars	Spinning frame	1814	1815	S,D
3939	William Lewis	Cloth rack	1815	1820	S,D
3945	John Lewis	Cloth shear	1815	1820	S,D
3951	Stephen Price	Cloth shear	1815	1816	S,D
3960	George Austin & James Dutton	Fulling improvement	1815	1816	S
4013	George Lewis	Fulling improvement	1816	1820	S,D
4089	William Dean	Calico printing machine	1816	1817	S,D
4189	John & William Lewis & William Davis	Gig	1817	1820	S,D
4196	John & William Lewis & William Davis	Cloth shear	1818	1820	S,D
4216	Benjamin Taylor	Dobby for power loom	1818	1818	S,D
4230	James Collyer	Gig	1818	1821	S,D
4243	George Whitham	Spindle grinder	1818	1819	S,D
4269	Thomas Homfray	Bobbin of metal	1818	1818	S
4280	Richard Ormrod	Calico printing rolls	1818	1819	S
4307	James Hadden	Roving frame	1818	1820	S,D
4378	John & William Lewis & William Davis	Gig	1819	1821	S,D
4379	John & William Lewis & William Davis	Brushing machine	1819	1821	S,D

| | Patent Data | | | Repertory of Arts Publication Date | |
No.	Patentee	Subject	Date	Date Published	Form (S-specification; D-drawing)
4391	Joseph Clisild Daniel	Gig	1819	1823	S,D
4430	Joseph Main	Preparatory and spinning	1820	1823	S,D
4487	William Davis	Cloth shear	1820	1821	S,D
4543	Stephen Wilson	Jacquard loom	1821	1824	S,D
4661	Samuel Robinson	Cloth shear	1822	1823	S,D
4764	William & John Crighton	Card cylinders	1823	1824	S,D
4780	William Southworth	Calico drying machine	1823	1825	S,D
4810	Thomas Stansfeld, Henry Briggs, William Richard, William Barraclough	Woolen power loom	1823	1826	S,D
4875	Archibald Buchanan	Self-stripping card	1823	1830	S,D
4984	John Leigh Bradbury	Spinning stop motion	1824	1825	Abstract
4999	John Fussell	Cloth lustering machine	1824	1826	S
5015	Philip Chell	Drafting rollers	1824	1826	Abstract
5058	Pierre Jean Baptiste Victor Gossett	Loom shuttle	1824	1825	S,D
5070	William Hirst	Wool spindles	1825	1826	Abstract
5140	Maurice de Jongh	Self-actor mule	1825	1827	Abstract
5365	Francis Molineux	Spinning bobbin	1826	1827	Abstract
5464	Philip Jacob Heisch	Roving frame	1827	1828	S,D
5509	Lambert Dexter	Spinning	1827	1829	S,D
5564	Samuel Sevill	Cloth shear	1827	1830	S,D
5613	Augustus Applegarth	Block printing	1828	1829	Abstract
5627	George Scholefield	Power loom improvements	1828	1831	S,D
5640	William Marshall	Cloth shear	1828	1829	S,D
5773	George Haden	Cloth brushing machine	1829	1830	S,D
5782	Benjamin Cook	Calico printing rollers	1829	1830	S
5787	George William Lee	Ring spindle	1829	1830	S,D
5800	Charles Brooks	Throstle	1829	1830	Abstract
				1830	S,D

Patent Data				*Repertory of Arts* Publication Date	
No.	Patentee	Subject	Date	Date Published	Form (S-specification; D-drawing)
5874	William Clutterbuck	Improvement to hand & Harmar shears	1829	1831	S,D
5901	John Frederick Smith	Cloth steam spray	1830	1830	Abstract
5907	Henry Hirst	Roll boiling machine	1830	1830	Abstract
				1831	S,D
6039	Thomas Sands	Powered bobbin	1830	1831	Abstract
6042	Robert Dalglish	Calico printing machine	1830	1831	Abstract
6055	Daniel Papps	Gig	1830	1831	Abstract
6058	John Ferrabee	Gig	1830	1831	Abstract

Note: Twenty-two patents were published between 1790 and 1812, thirty-three between 1813 and 1824, and twenty between 1825 and 1830.

Appendix B

British Textile Emigrants to the United States, 1773–1831, by Trade

Trade	Number, 1773–1775[a]	Wool	Cotton	Trade	Number, 1773–1775[a]	Wool	Cotton
Operatives				Finishing			
Preparatory				Cloth dresser	3	X	
Wool comber	21	X		Woolen cloth worker	3	X	
Flax dresser	13			(and dyer)			
Hemp dresser	1			Felt maker	1	X	
Hemp and flax dresser	1			Bleacher	1		
Total	36			Calico printer	2		X
				Print cutter	1		
Spinning				Dyer	15		
Spinner	1			Linen dyer	1		
Twine spinner	1			Silk dyer	3		
Thread maker	1			*Total*	30		
Throwster/silk throwster	5						
Rope maker	14			**Managers**			
Total	22			Cloth manufacturer	1	X	
				Clothier	11	X	
Weaving				*Total*	12		
Weaver	262						
Broadcloth/cloth weaver	5	X		**General machine makers**			
Woolen weaver	4	X		Millwright	9		
Serge weaver	1	X		Tin plate worker	4		
Shag weaver	1	X		Whitesmith	25		
Worsted weaver	2	X		*Total*	38		
Linen weaver	19						
Canvas weaver	1			**Textile machine makers**			
Sailcloth weaver	5			Wire drawer	5		
Sack weaver	1			Card maker	1		
Silk weaver	15			Turner	12		
Ribbon weaver	2			Clock/watch maker	29		
Lace weaver (and dyer)	1			Shuttle maker	1		
Net/net hood weaver[b]	2			Stocking frame maker	1		
Stocking weaver	11			*Total*	49		
Framework knitter	2						
Total	334						

Trade	Number, 1809–1813[c]	Wool	Cotton	Trade	Number, 1809–1813[c]	Wool	Cotton
Operatives				**Managers**			
Preparatory				Manufacturer	28		
Scourer	1	X		Woolen and cloth	33	X	
Wool carder	1	X		manufacturer and			
Wool comber	2	X		clothier			
Cotton carder	4		X	Blanket manufacturer	1	X	
Flax dresser	1			Cotton manufacturer	12		X
Hemp dresser	1			Calico manufacturer	1		X
Total	10			Muslin manufacturer	1		X
				Twine and thread	2		
Spinning				manufacturer			
Spinner	19			Flax manufacturer	1		
Spinner and weaver	2			Total	79		
Wool spinner	7	X					
Cotton spinner	38		X	**General machine makers**			
Mule spinner	1		X	Millwright	12		
Rope maker	3			Mechanic	1		
Total	70			Engineer	1		
				Machinist	6		
Weaving				Machine maker	27		
Weaver	492			Tinplate worker	3		
Part-time weaver	9			Whitesmith	8		
Spinster and weaver	1			Total	58		
Woolen weaver	2	X					
Cloth weaver	2	X		**Textile machine makers**			
Satinet weaver	1	X		Cotton machine maker	2		X
Cotton weaver	7		X	Card maker	2		
Muslin weaver	2		X	Wire drawer	11		
Counterpane weaver	1			Spinning wheel maker	1		
Haircloth weaver	1			Turner	10		
Linen weaver	5			Spindle maker	1		
Hosier and stocking weaver	2			Loom maker	1		
Total	525			Reed maker	3		
				Shuttle maker	2		
Finishing				Sley maker	1		
Bleacher	2			Total	34		
Dyer (including joint	14						
occupations)							
Fuller	1	X					
Shearman	2	X					
Cloth dyer	1	X					
Velvet and fustian cutter	4		X				
Calico printer	5		X				
Silk dyer	1						
Total	30						

Trade	Male	Female	Total 1824–1831[d]	Wool	Cotton	Unspecified Fiber
Operatives						
Preparatory						
Carder	17	2	19			
Wool stapler	5		5	X		
Wool sorter	14	1	15	X		
Wool picker	1		1	X		
Wool dresser	2		2	X		
Wool comber	16		16	X		
Cotton carder	5		5		X	
Flax dresser	12		12			
Heckler	1		1			
Total	73	3	76			
Spinning						
Spinner	182	71	253			X
Spinstress		1	1			X
Twine/thread spinner	2		2			X
Yarn dresser	1		1			X
Wool spinner	5	1	6	X		
Worsted spinner	1		1	X		
Cotton spinner	118	6	124		X	
Cotton twister	1		1		X	
Cotton reeler		1	1		X	
Flax spinner	1		1			
Silk throwster	1		1			
Silk twister	1		1			
Silk dresser	2		2			
Ropemaker	31	1	32			
Total	346	81	427			
Weaving						
Warper	7	3	10			X
Dresser/sizer	11		11			X
Weaver	2,429	140	2,569			X
Woolen/cloth weaver	7		7	X		
Cloth maker	1		1	X		
Flannel weaver	1		1	X		

Trade	Male	Female	Total 1824–1831[d]	Wool	Cotton	Unspecified Fiber
Operatives: Weaving (cont.)						
Stuff weaver	1		1	X		
Cotton dresser	1	1	2		X	
Cotton weaver	11		11		X	
Cord weaver	1		1		X	
Velvet/plush weaver	1		1		X	
Linen weaver	6		6			
Canvas weaver	3		3			
Silk weaver	15	2	17			
Tape weaver	1		1			
Knitter/hosier/stocking weaver	27	1	28			
Carpet weaver	5	5	10			
Lace/bobbinet maker	14	1	15			
Total	2,542	153	2,695			
Finishing						
Bleacher	26	2	28			X
Dyer	105	7	112			X
Blue dyer	3		3			X
Madder dyer	1		1			X
Color mixer	2		2			X
Machine printer	1		1			X
Print cutter	8		8			X
Cloth worker	2		2	X		
Cloth finisher	3		3	X		
Cloth washer	2		2	X		
Wool dyer	1		1	X		
Fuller	20		20	X		
Cropper/cutter	3		3	X		
Cloth dresser	33	1	34	X		
Cloth lapper	2		2	X		
Cotton dyer	1		1		X	
Cotton/calico printer	188	18	206		X	
Calico bleacher	1		1		X	

Trade	Male	Female	Total 1824–1831[d]	Wool	Cotton	Unspecified Fiber
Operatives: Finishing (cont.)						
Calico dresser	1		1		X	
Cotton dresser	1		1		X	
Linen bleacher	1		1			
Silk dyer	1		1			
Total	406	28	434			
Managers						
Overseer	1		1			X
Manufacturer	148		148			X
Wool manufacturer/ clothier	188	1	189	X		
Worsted manufacturer	2		2	X		
Cotton manufacturer	10		10		X	
Muslin manufacturer	3		3		X	
Damask manufacturer	1		1		X	
Silk manufacturer	7		7			
Silk mercer	1		1			
Ribbon manufacturer	1		1			
Lace manufacturer	3		3			
Total	365	1	366			
General machine makers						
Artificer	3		3			
Mechanic	553	8	561			
Machinist	35		35			
Machine maker	31		31			
Millwright	77		77			
Engineer	105		105			
Civil engineer	3		3			
Engineer & surveyor	1		1			
Engine smith	1		1			
Engine turner	1		1			
Fitter	1		1			
Steam engineer	4		4			
Tool maker	3		3			

Trade	Male	Female	Total 1824–1831[d]	Wool	Cotton	Unspecified Fiber
General machine makers (cont.)						
Filer	5		5			
File cutter	1		1			
Pattern drawer	3		3			
Pattern maker	7		7			
Designer	3		3			
Tinplate worker	1		1			
Whitesmith	2		2			
Total	840	8	848			
Textile machine makers						
Cotton machine maker	1		1		X	
Silk machine maker	1		1			
Wire drawer	5		5			
Card maker	4		4			
Comb maker	2		2			
Turner	74		74			
Iron/steel turner	4		4			
Spindle maker	2		2			
Roller maker	3		3			
Loom maker	2		2			
Stocking frame maker	6		6			
Reed/sley maker	17	1	18			
Shuttle maker	1		1			
Block cutter	12		12			
Engraver	18		18			
Total	152	1	153			

a. PRO, T 47/9, 10, 11, 12.
b. One female in each of these two trades.
c. USNA, RG 45, 59, War of 1812 Papers, Returns of Enemy Aliens.
d. USNA, RG 36, Passenger Lists for New York, Philadelphia, and Boston, 1824–1831.

Appendix C

British Textile Workers Registered as Enemy Aliens during the War of 1812, by Trade

Trade	Number	Wool	Cotton	Unspecified Fiber
Operatives				
Preparatory				
Scourer	1			
Carder	4			
Wool carder	2	X		
Wool comber	3	X		
Cotton carder	4		X	
Flax dresser	3			
Hemp dresser	1			
Total	18			
Spinning				
Spinner	23			X
Spinner and weaver	3			X
Wool spinner	8	X		
Cotton spinner	48		X	
Mule spinner	1		X	
Rope maker	9			
Total	92			
Weaving				
Warper	1			X
Weaver	759			X
Part-time weaver	16			X
Spinster and weaver	1			X
Woolen weaver	3	X		
Cloth weaver	5	X		
Satinet weaver	1	X		
Cotton weaver	12		X	
Muslin weaver	3		X	
Counterpane weaver	1			
Coach lace weaver	2			
Fringe weaver	2			
Hair cloth weaver	1			

Trade	Number	Wool	Cotton	Unspecified Fiber
Linen weaver	9			
Hosier and stocking weaver	10			
Total	826			
Finishing				
Bleacher	4			X
Dyer (including joint occupations)	37			X
Fuller	2	X		
Shearman	3	X		
Cloth dyer	2	X		
Velvet and fustian cutter	4		X	
Calico printer	11		X	
Silk dyer	1			
Total	64			
Managers				
Manufacturer	50			
Woolen and cloth manufacturer and clothier	64	X		
Broadcloth manufacturer	1	X		
Blanket manufacturer	1	X		
Cotton manufacturer	24		X	
Calico manufacturer	1		X	
Muslin manufacturer	1		X	
Bunting manufacturer	1			
Twine and thread manufacturer	10			
Coach lace manufacturer	1			
Haircloth manufacturer	2			
Lace manufacturer	1			
Hemp manufacturer	1			

Trade	Number	Wool	Cotton	Unspecified Fiber
Flax manufacturer	1			
Oil floorcloth manufacturer	2			
Workshop and factory superintendent	3			
Total	164			
General machine makers				
Millwright	36			
Mechanic	3			
Engineer	2			
Machinist	10			
Machine maker	45			
Steam engine builder	1			
Tin plate worker	14			
Whitesmith	14			
Total	125			
Textile machine makers				
Cotton machine maker	9		X	
Card machine maker	1			
Card maker	3			
Wire drawer	15			
Comb maker	1			
Spinning wheel maker	1			
Turner	20			
Spindle maker	1			
Loom maker	1			
Reed maker	4			
Shuttle maker	2			
Sley maker	1			
Stocking loom maker	1			
Total	60			
Total	1,349			

Source: USNA, RG 45, 59, War of 1812 Papers, Returns of Enemy Aliens.

Appendix D

Fixed Capital Investment, Capacity, and Employment in U.S. Cotton and Woolen Industries, 1820.
Source: U.S. National Archives, Record Group 29, Bureau of the Census: MS Returns of the 1820 Census of Manufactures on USNA Microfilm M.279 (27 rolls, Washington, D.C., 1967). "South": Virginia, North Carolina, South Carolina, Georgia: "West": Kentucky, Indiana, Ohio, Tennessee, Illinois.

Table D.1. Fixed Capital Investment in the U.S. Cotton Industry, 1820

State	Cotton Spinning Amount	No. of Firms	Cotton Spinning & Weaving Amount	No. of Firms	Cotton & Wool Spinning & Weaving Amount	No. of Firms	Total Amount	No. of Firms	Percentage of U.S. Aggregate
Maine	$ 148,000	4	$ 22,000	1	$ 17,000	1	$ 187,000	6	1.64
Vermont	13,500	4	97,200	7			110,700	11	0.97
New Hampshire	280,012	17	320,400	14	22,950	3	623,362	34	5.46
Massachusetts	261,095	14	1,281,650	44	25,000	2	1,567,745	60	13.74
Rhode Island	325,322	24	1,340,500	45	65,000	2	1,730,822	71	15.17
Connecticut	317,000	17	936,355	24	148,000	3	1,401,355	44	12.28
New York	233,000	6	1,795,960	35	46,200	3	2,075,160	44	18.18
New Jersey	210,000	5	297,000	9			507,000	14	4.44
Pennsylvania	185,700	16	490,700	12	88,600	3	765,000	31	6.70
Delaware	227,000	6	203,000	3			430,000	9	3.77
Maryland	195,600	6	1,140,000	5	41,000	4	1,376,600	15	12.07
South	38,000	2			24,500	2	62,500	4	0.55
West	208,740	32	215,338	5	149,800	6	573,878	43	5.03
Total	2,642,969	153	8,140,103	204	628,050	29	11,411,122	386	100

Note: Although 386 firms disclosed their capitalizations, another 53 firms made returns without mentioning the level of their capital investments. If the national firm average of fixed capital investment ($29,562.5) is credited to those 53 firms, then an additional $1,566,812.5 can be added to the known figure of $11,411,122, making a maximum U.S. fixed capital investment in cotton manufacturing in 1820 of $12,977,934.5.

Table D.2. Spindleage in the U.S. Cotton Industry, 1820

State	Cotton Spinning		Cotton Spinning & Weaving		Cotton & Wool Spinning & Weaving		Total		Percentage of U.S. Aggregate	
	Low	High	Low	High	Low	High	Low	High	Low	High
Maine	2,480	2,480	890	890	576	576	3,946	3,946	1.18	1.15
Vermont	632	632	3,470	3,470	32	80	4,134	4,182	1.24	1.22
New Hampshire	9,328	9,368	8,383	8,543	710	710	18,421	18,621	5.52	5.44
Massachusetts	8,174	8,414	32,992	32,992	802	1,418	41,968	42,824	12.57	12.51
Rhode Island	17,770	17,770	49,164	49,164	3,374	3,374	70,308	70,308	21.06	20.54
Connecticut	8,477	8,477	25,440	26,212	1,500	1,500	35,417	36,189	10.61	10.57
New York	10,968	13,692	50,236	50,236	908	908	62,112	64,836	18.61	18.94
New Jersey	8,126	8,580	11,432	12,004			19,558	20,584	5.86	6.01
Pennsylvania	8,566	8,566	12,935	13,237	2,050	2,050	23,551	23,853	7.06	6.97
Delaware	6,904	6,904	7,300	7,300			14,204	14,204	4.25	4.15
Maryland	5,344	5,344	17,680	17,680	752	1,324	23,776	24,348	7.12	7.11
South	1,030	1,270			216	216	1,246	1,486	0.37	0.43
West	8,161	9,917	6,168	6,168	800	800	15,129	16,885	4.53	4.93
Total	95,960	101,414	226,090	227,896	1,720	12,956	333,770	342,266	99.98	99.97

Note: These figures are based on the returns of 421 firms. Another 18 firms omitted details of spinning equipment. Since the low and high average numbers of spindles per firm were 792.8 and 813, respectively, another 14,270 or 14,634 spindles, at least, could be added to the U.S. aggregate, making a possible maximum U.S. capacity of 356,900 cotton spindles in 1820. "Low" refers to low estimate, based on smallest machine size for machines of unknown spindle number; "high" means the opposite.

Table D.3. Multipliers Used in Calculating Spindle Capacities of Cotton Firms Listing Only Numbers of Machines, and Not Spindleage, in Census Returns

Machine	Lowest Known Spindleage and Census Source[a]		Highest Known Spindleage and Census Source[a]	
Water frame	54	(Mass., 71)	64	(N.H., 34)
Spinning frame/Frame	64	(Mass., 85)	84	(Mass., 155)
Throstle	42	(Kentucky, 132)	160	(Ohio, 690)
Double throstle	76	(Kentucky, 41)	120	(Kentucky, 98)
Jenny	20	(R.I., 2)	100	(Mass., 139)
Spinning machine	32	(Georgia, 44)	80	(N.H., 27)
Mule	144	(Mass., 31)	480	(R.I., 43)
Pair of mules	400	(N.Y., 1394)	400	(N.Y., 1394)

a. State and number of return on film M.279.

Table D.4. Power Looms in the U.S. Cotton Industry, 1820

State	Number of Firms Recording Power Looms		Number of Recorded Power Looms	
	Cotton	Cotton & Wool	Cotton	Cotton & Wool
Maine	1		12	
Vermont	3		38	
New Hampshire	5		120	
Massachusetts	13	2	321	10
Rhode Island	17	1	315	20
Connecticut	11	1	168	10
New York	16		334	
New Jersey	4		80	
Pennsylvania	3	1	23	2
Delaware	1		20	
Maryland	4		188	
South				
West	3		4	
Total	81	5	1,623	42

Note: The minimum total number of power looms in U.S. cotton mills in 1820 was 1,665 in 86 firms.

Table D.5. Employment in the U.S. Cotton Industry, 1820

	Males		Females[b]		Sex Unspecified		
State	Men	Boys	Women	Girls	Adults	Children	Adults & Children
Maine	12		35			47	
Vermont	23	5	90	17		43	
New Hampshire	104	14	233	29		211	30
Massachusetts	381	13	628		13	965	225
Rhode Island	429	35	599	36		1,313	
Connecticut	272	14	322	13		671	50
New York	667[a]	63	402	91		1,034	
New Jersey	181	23	105	34		462	155
Pennsylvania	195	26	133	24		355	50
Delaware	93	6	110	5		191	
Maryland	90	140	75	176		223	
South	16	3	5			48	
West	174	96	53	9		246	10
Total	2,637	438	2,790	434	13	5,809	520

Note: Figures include cotton and woolen spinning and weaving mills.
a. Includes 250 prisoners in New York City jail.
b. In New Hampshire, another 139 females were counted, but their ages were unspecified.

Table D.6. Fixed Capital Investment in the U.S. Woolen Industry, 1820

State	Woolen Spinning		Woolen Spinning & Weaving		Total		Percentage of U.S. Aggregate
	Amount	No. of Firms	Amount	No. of Firms	Amount	No. of Firms	
Maine	$ 24	1	$ 29,900	4	$ 29,924	5	1.15
Vermont			119,035	21	119,035	21	4.57
New Hampshire			34,000	4	34,000	4	1.30
Massachusetts			268,864	31	268,864	31	10.32
Rhode Island			67,485	10	67,485	10	2.59
Connecticut			470,400	45	470,400	45	18.05
New York	7,700	2	688,335	53	696,035	55	26.71
New Jersey			100,350	20	100,350	20	3.85
Pennsylvania			319,895	32	319,895	32	12.27
Delaware			144,000	5	144,000	5	5.53
Maryland			57,000	11	57,000	11	2.19
South	8,700	3			8,700	3	0.33
West	9,800	2	280,550	17	290,350	19	11.14
Total	26,224	8	2,579,814	253	2,606,038	261	100

Note: All together 293 woolen spinning or spinning-and-weaving firms reported in the 1820 Census. If the national firm average of fixed capital investment ($9,984.8) is credited to the 32 firms not stating their fixed capitals, then another $319,513.6 can be added, making a maximum U.S. fixed capital investment in woolen manufacturing in 1820 of $2,925,551.6.

Table D.7. Spindleage in the U.S. Woolen Industry, 1820

State	Woolen Spinning		Woolen Spinning & Weaving		Woolen Spindles in Cotton & Woolen Spinning & Weaving Mills		Total		Percentage of U.S. Aggregate	
	Low	High	Low	High	Low	High	Low	High	Low	High
Maine	32	80	319	575	75	75	426	730	1.29	1.39
Vermont			2,146	3,826			2,146	3,826	6.49	7.31
New Hampshire			732	912	210	210	942	1,122	2.85	2.14
Massachusetts	40	40	3,435	4,847			3,475	4,887	10.52	9.34
Rhode Island			1,070	1,710	484	484	1,554	2,194	4.70	4.19
Connecticut	80	80	6,226	7,532	1,020	1,020	7,326	8,632	22.17	16.49
New York	150	230	6,088	9,480	90	250	6,328	9,960	19.15	19.03
New Jersey			3,090	4,518			3,090	4,518	9.35	8.63
Pennsylvania			2,893	7,187	274	610	3,167	7,797	9.58	14.90
Delaware			1,080	3,050			1,080	3,050	3.27	5.83
Maryland			1,316	2,108	84	260	1,400	2,368	4.24	4.52
South	298	522			40	200	338	722	1.02	1.38
West	96	240	1,182	1,800	496	496	1,774	2,536	5.37	4.84
Total	696	1,192	29,577	47,545	2,773	3,605	33,046	52,342	100	99.99

Note: These figures are based on the returns of 294 firms (including 22 cotton firms with woolen spindles). Another 21 woolen spinning or spinning and weaving firms omitted details of woolen spinning equipment. Since the low and high average numbers of spindles per firm were 112.4 and 178.03 spindles, respectively, another 2,360 or 3,738.6 spindles at least could be added to the U.S. aggregate, making a possible maximum U.S. capacity of 56,080 woolen spindles in 1820.

Table D.8. Multipliers Used in Calculating Spindle Capacities of Woolen Firms Listing Only Numbers of Machines, and Not Spindleage, in Census Returns

Machine	Lowest Known Spindleage and Census Source[a]		Highest Known Spindleage and Census Source[a]	
Jenny	20	(R.I., 2)	100	(Mass., 139)
Patent jenny	100	(Conn., 102)	120	(N.J., 36)
Jack	30	(Ohio, 426)	50	(Pa., 821)
Spinning machine	32	(Georgia, 44)	80	(N.H., 27)
Brewster or water-powered wool spinning machine	144	(R.I., 84)	250	(Conn., after 216)

a. State and number of return on film M.279.

Table D.9. Power Looms in U.S. Woolen Industry, 1820

State	Number of Firms Recording Power Looms	Number of Recorded Power Looms
New Hampshire	1	4
Massachusetts	2	14
Connecticut	2	10
New York	2	15
Total	7	43

Table D.10. Employment in the U.S. Woolen Industry, 1820

| State | Males | | Females[a] | | Sex Unspecified | |
	Men	Boys	Women	Girls	Children	Adults & Children
Maine	17	4	7	1	8	
Vermont	143	40	37	15	30	
New Hampshire	24		16		9	
Massachusetts	253	39	76	6	79	
Rhode Island	57	7	19		20	
Connecticut	271	29	116	4	154	6
New York	406	51	50	11	215	
New Jersey	122	23	21	15	50	
Pennsylvania	282	59	82	12	120	
Delaware	111	21	20	4		
Maryland	72	30	18	2	23	
South	10	8				
West	124	30	33	5		200[b]
Total	1,892	341	495	75	708	206

a. In Delaware, another 46 females were counted, but their ages were unspecified.
b. In a woolen and papermaking establishment.

Glossary

Back-Off Motion On jenny or mule, a movement at the end of the draw. The spindles are reversed to release a few turns of yarn so that the faller wire can be lowered, positioning the yarns in readiness for winding.

Bat A continuous wad of cotton from the batting or lapping machine, ready for carding.

Batten In loom, the frame holding the reed and shuttle boxes. It beats up the filling between changes of shed. Also known as the beater, lathe, or lay.

Batting Machine A machine for loosening the cotton in readiness for carding.

Beam In loom, a wooden roller. The warp unwinds from the warp beam, and the cloth is wound onto the cloth beam.

Beating Up In loom, one of the five principal motions. In it the batten drives the filling into the fell of the cloth.

Bedtick Material, frequently cotton, for bedtick or covering for pillows and mattress; usually distinguished by stripes.

Beer or *Bier* A specific number of dents or splits in the reed, usually nineteen or twenty.

Beetling Machine In cotton finishing, a machine for hammering goods to impart a soft, thready appearance like that of linen.

Billy A machine used only in the woolen industry at this time for piecing together rolls from a finisher card.

Bleachery Equipment and processes designed to remove natural color from textile fibers. By the 1830s calico bleaching in England involved about twenty-five processes.

Block Cutter In calico printing, one who cuts the print pattern, usually in relief, on the surface of a wooden printing block or roller.

Sources: Catling, *Spinning Mule*; Leigh, *Science*; Montgomery, *Carding and Spinning Master's Assistant* and *Practical Detail*; Nichols and Broomhead, *Standard Cotton Cloths*; North, "American Textile Glossary"; Rees, *Cyclopaedia*; Ure, *Dictionary*.

Block Printing Calico printing by means of hand-held wooden blocks.

Blowing Machine In cotton preparation, a machine that blows apart and loosens cotton staple in readiness for carding.

Blue Dipper One skilled in dipping cotton goods in indigo dyes.

Bobbin A small wooden or metal cylinder with a flange at both ends and a hole drilled down its center so that it could sit on a spindle.

Bobbin and Fly Frame A roving frame; used for making both coarse and fine roving.

Bobbin Winder A machine for winding yarn or thread onto a bobbin.

Bobs Levers on the cloth hand shears, needed to pull the blades together.

Braying Scouring woolen cloth to remove the oil and grease applied in spinning.

Breaker Card The first card in a set of carding machines; its rollers are covered with a relatively coarse card clothing. Also known as a scribbler.

Broadcloth A very fine cloth woven in plain weave and heavily milled.

Brushing Laying the nap of a cloth.

Builder Motion In roving or spinning machines, the movement by which yarn or thread is distributed up and down the bobbin or spindle.

Bump Cotton Waste cotton used to make coarse yarns.

Burling In woolen finishing, the removal of extraneous matter, such as burrs, from a piece of cloth in order to improve its appearance.

Calender In cotton finishing, a machine with pressure rollers and a heating element that imparts a glazed finish to cotton goods.

Calico A common cotton fabric of many specifications.

Cambric A cotton fabric of plain weave made from medium-count yarns and used for printing.

Can Frame An early cotton roving machine in which twist was imparted to the slivers by the rotation of the cans into which they fell.

Cap Spindle A spindle for which the traditional flyer was replaced by an inverted metal cup; a continuous spinning device. Patented in the United States in 1828 by Charles Danforth.

Card A carding machine or a hand implement for carding.

Card Can A can about a meter high designed to receive the sliver of cotton produced by a card.

Card Clothing Sheets of leather pierced by long wire staples that are pointed and bent at an angle; clothing was secured to the working surfaces in a card.

Carding The action of brushing cotton or wool fibers to get rid of dirt, placing the fibers in a roughly parallel alignment, and then forming them into a thick rope in readiness for spinning. Cotton carding produced a continuous rope or sliver; woolen carding, before the condenser, yielded short lengths of rope called rolls.

Carriage In spinning jack or mule, the wheeled frame carrying the spindles.

Cassimere A light, durable woolen cloth made in twill weave and patented in England by Francis Yerbury in 1766.

Chambray A plain weave cotton fabric with colored warp and grey or bleached filling.

Check A fabric with regular colored bands in both warp and filling, making a pattern of squares as in a chessboard.

Clam Machine Probably some sort of mechanical device for clamping; possibly needed in connection with making or installing engraved printing rollers or shells.

Clasp In jenny, the pair of bars that grip the roving and draw it away from the spindles during spinning.

Clearer In roller card, the roller that lifts the stock off the worker roller and returns it to the main cylinder of the machine for renewed carding. Also known as the stripper.

Cloth The fabric made in woolen manufacturing; characterized by carded yarns and a milled surface.

Coatings A woolen cloth of imprecise definition used for outer garments.

Comb In the card, a toothed blade for removing fibers from the doffer roller. Not to be confused with the hand tool employed in the worsted industry.

Comb Plate The comb blade in a card. See comb.

Compound Motion In the roving frame, a system of gears designed to slow down the speed of the bobbins as their outer circumference increases during winding.

Condenser In the woolen card, a device for removing the film of fibers off the finisher doffer and turning it into a number of continuous slivers or ropings ready for spinning. First successful one was patented in the United States by John Goulding in 1826.

Cone Pulley A pulley in the shape of the frustum of a cone.

Continuous Spinning A method of spinning in which all three component actions (drafting, twisting, and winding) are carried on simultaneously.

Cop A cylindrical yarn package with conical ends from one of which the yarn can be removed axially without package rotation.

Cords A fabric with a pile cut in ribs running in the direction of the warp.

Count A relative estimate of yarn fineness expressed as the number of length units in one weight unit (inverse count system) or vice-versa (direct system). Cotton counts (or numbers) commonly gave the number of hanks of 840 yards each that would weigh one pound avoirdupois. The higher the count, the finer the yarn.

Counterpane A corruption of counterpoint or *point contre-point*, stitch against stitch, denoting something sewn on both sides alike. Synonymous with quilt.

Creel A frame or rack holding bobbins or spools.

Cropping Shearing the surface fibers of a woolen cloth.

Cylinder Printing Machine In cotton finishing, a machine for printing cloth from engraved metal cylinders to which color is transferred from the color box by a color roller.

Dandy Loom A compact hand loom with a cloth take-up motion driven by the movement of the batten. Invented by William Radcliffe of Stockport, Cheshire, around 1802.

Dash Wheel A compartmentalized wheel that revolves, partly submerged, in a cistern of water; calico placed in the compartments is washed and rinsed.

Dead Spindle In continuous spinning machines. The flyer fits loosely on the spindle shaft; the flyer, with its arms taken down to a base pulley, receives power while the spindle is stationary. The design reduced flyer wobble and allowed higher spindle speeds.

Dent In the loom reed, a cane or wire or the space between the canes or wires.

Devil See picker.

Differential In the roving frame, the most efficient compound motion, using four bevel gears. Patented by Aza Arnold in the United States in 1823; now found in the rear axle of every motor vehicle.

Dimity A cotton fabric of plain weave with a slight cord effect on its surface, an effect gained in drawing in by alternating several warp ends per heddle with one warp end per heddle.

Dobby In the loom, a variation of the jacquard mechanism, which controls the shedding sequence by means of a pattern chain, each bar of which represents a pick in the weave.

Doctor In calico printing machine, a blade to remove excess color from printing or color cylinders.

Doeskin A closely woven woolen cloth with a smooth face.

Doffer Cylinder In cards, the cylinder that lifts the cotton or wool fibers off the main cylinder before the formation of the lap, sliver, or roll.

Doffing Stripping fibers, as in the card, or removing full bobbins to make way for empty ones, as in spinning frames.

Double (Acting) Lapper A cotton preparatory machine, following and preceding or substituting the card, in which laps from the blower are continuously delivered by means of a pivoted conveyor belt shifted between two lap rollers. Patented in Britain in 1824 by Johann Georg Bodmer.

Double Loom A loom in which two warps are woven simultaneously.

Double Speeder A roving machine, so-called because two of its movements, bobbin rotation and spindle rail traverse, underwent a change of speed during winding.

Double Throstle Throstle frame with spindles on each side.

Drafting Stretching the card sliver, roving, or yarn so that the cotton or wool fibers slide over each other and form a finer end. One of the three component actions of spinning, the others being twisting and winding.

Draw In jenny, jack, or mule, the outward movement, away from the spindles, of the clasp or carriage during which the yarn is drafted by the movement of the clasp or carriage and is twisted by spindle rotation.

Drawing Frame In cotton manufacturing, a machine in which drafting rollers stretch the card sliver and then, by doubling (feeding the drafted slivers together back through the rollers), make a more even sliver.

Drawing In The manual operation of drawing the warp ends through the harnesses and heddle eyes in readiness for weaving. The process contributes to the formation of the weave in producing various surface effects.

Dresser or Dressing Machine In cotton weaving, a machine in which the warp is taken through troughs of sizing (starch solution) in order to enable it to withstand the movements of the loom during weaving. In woolen cloth finishing, a machine for brushing the cloth surface. A dresser is also a person who tends a dresser or dressing machine.

Drilling Cotton fabric made with a twill weave having a warp face.

Drop Box In the loom, a box on each side of the batten holding a number of shuttles, each carrying its own color.

Eclipse Speeder A variant of the tube speeder, using belts rather than tubes to insert temporary or false twist during drafting. Patented in the United States in 1829 by Gilbert Brewster.

End A single sliver, roving, or yarn.

Engraving Machine In cotton finishing, a machine for engraving patterns in the surface of calico printing cylinders or cylinder shells. It employed the steel roller dies invented by Jacob Perkins and reportedly was first developed in 1808 by Joseph Lockett of Manchester before Perkins left America for England.

Everlasting Cloth A worsted fabric common in the eighteenth century.

Faller Wire A guide wire in a jenny, jack, or mule running across the yarns just before they reach the spindles; used to build cops.

Fell The weft edge of the cloth when set in the loom.

Felting See fulling.

Fillet Card Clothing A narrow continuous strip of card clothing that is wound around a card cylinder and covers the cylinder surface completely with wire points, creating a maximum working surface.

Filling The American word for weft: the yarn interlaced at right angles to the warp in any woven fabric.

Filling Frame A spinning frame designed to spin filling (softer) yarn and packaged onto bobbins in the form of cops suitable for shuttles.

Finisher Card The final card in a set; covered with finer card clothing than the breaker and having a delivery mechanism for making a sliver or rolls.

Finishing Processes All processes that follow weaving (see table in introduction).

Flannel In cotton manufacture, one of the few fabrics with a nap. In the woolen manufacture, on the contrary, flannel receives minimal finishing; made from slightly twisted yarn, it is merely washed and pressed after being woven; this processing makes a soft, spongy cloth.

Flats Bars of wood with card clothing on their underside; positioned over the main cylinder in a cotton card.

Flyer Inverted U-shaped arm on top of a spindle that rotates around the bobbin and inserts twist in the yarn.

Fly Frame See bobbin and fly frame.

Fly Shuttle Any shuttle, with or without wheels, that is knocked from side to side of the loom by a picking mechanism. Patented in Britain by John Kay in 1733.

Frame Machine or, less frequently, a rack; also wooden or metal members supporting moving parts in a machine.

Frizing Machine Used in woolen finishing; the frizer, or plank with a rough underside surface of sand and glue, rubs the nap into small knots while the cloth is held down on the frizing table.

Fulling A woolen finishing process in which the cloth, immersed in warm, soapy water, is pounded by wooden hammers in order to be cleansed, shrunk, and felted. Felting describes the action by which a proportion of wool fibers move out of their respective yarns and intertwine, adding strength and giving a nap to the cloth.

Fulling Mill The water-driven mill in which cam-tripped hammers accelerate the fulling process.

Fulling Stocks Wooden hammers of a fulling mill.

Fustian Originally a cotton fabric with a linen warp and a cotton filling. By the early nineteenth century it was any stout-twilled cotton fabric with a short nap; it included corduroy, velverett, velveteen, thicksett, and other thick fabrics used for men's wearing apparel.

Fustian Cutter One who cut the flushed warp to raise the pile in a fustian fabric such as velveteen.

Gig Mill In woolen finishing, a large cylinder on a horizontal axle with teasel heads set into its surface. Rotated against the cloth, it raises the wool fibers in the nap in readiness for shearing.

Gin Contraction of engine. The cotton gin utilized rotating circular saw blades and brushes to remove seeds and dirt from the cotton staple.

Gingham A plain-weave cotton fabric distinguished by two or more filling colors; used for women and children's clothing.

Goods Collective noun for cotton fabrics.

Grist The thickness of a sliver, roving, or yarn, expressed as a relationship between length and weight. See count.

Harness In the loom, a frame holding the heddles in position and by which the shedding motion is achieved. Also known as leaf, wing, and shaft.

Head A self-contained group of drafting rollers in a drawing, roving, or spinning frame; each head of rollers feeds a group or head of spindles, usually two or four.

Headstock In mule, the assembly that controls the mule's movements.

Healds See heddle.

Heck Box In a warper or warping mill, a box holding two sets of metal heddles by which the warp is separated into two alternate sets of ends to facilitate drawing in.

Heddle In the loom, cords or wires, vertically stretched in the harness frame, with a loop in the middle of each through which the warp ends pass. Designed to control shedding. Also known as healds.

Heddle Eyes The loops in the middle of heddles.

Hollands Originally a linen fabric from Holland; copied in the cotton industry as a plain-weave standard print fabric.

Intermittent Spinning A method of spinning in which winding follows, in time, the other two spinning actions of drafting and twisting.

Jack In the woolen industry, a machine working on the intermittent spinning principle. It may be compared to a cotton mule without drafting rollers and with the bobbins replaced by spools for the roving.

Jacquard Loom In the loom, a shedding device capable of moving warp ends individually rather than in groups. Invented by Joseph Marie Jacquard of Lyons between 1801 and 1806, it utilized a pattern chain of perforated cardboard cards, each card representing a pick in the weave and the perforations in the cards representing the warp ends to be lifted by the lifter bar.

Jean A cotton fabric made with a twill weave and used for under and outer clothing.

Jenny The spinning machine patented in Britain by James Hargreaves in 1770. Working on the intermittent principle, it made coarse but soft yarn suitable for filling. Though superseded by the mule in the cotton industry, it survived much longer in the woolen trade, where its compact size and manual operation suited the domestic system.

Kersey A coarse woolen cloth.

Kerseymere See cassimere.

Knotting Repairing broken warp ends in the warper, dresser, or loom.

Lap Sheet of fibers.

Lapper In cotton preparation, a machine for forming a lap and winding it around a lap roller, usually in readiness for carding.

Lathe or Lay See batten.

Leash In weaving, either the system of crossing the warp ends in readiness for drawing in, or another word for heddle.

Ledger Blade Fixed blade in a shearing frame, against which the moving blades acted.

Let-Off Motion One of five principal motions in the loom; a mechanism for unwinding the warp from the warp beam.

Lifter Rail See traverse rail.

Loom A machine for weaving a piece of fabric from any yarn or thread.

Mandrel The metal shaft or axle onto which a calico printing machine cylinder or roller was fitted.

Mill Either a factory or work place housing one or more powered machines, or a machine of any description.

Milling Another word for fulling.

Mixed Cloths Fabrics made from two or more different textile fibers.

Mule The spinning machine invented by Samuel Crompton in the 1770s that combined the drafting rollers of Arkwright's water frame with the moving clasp or carriage of Hargreaves's spinning jenny. Working on the intermittent spinning principle, technically it was the most versatile of the spinning machines, capable of making all counts and twists of yarn and with maximum uniformity.

Mule, Hand The first mule that was powered and controlled wholly by its operative.

Mule, Self-Acting A mule in which all parts of its spinning cycle (draw, back off, and winding on) are mechanically powered and regulated. Such a mule was perfected by Richard Roberts of Manchester between 1825 and 1830.

Muslin A fine cotton fabric made from high count yarns.

Nap Either the fibrous surface of a felted woolen cloth, or the action that raises this surface.

Number See count.

Oiling Adding oil, like olive oil, to scoured wool in order to facilitate its working in carding, spinning, and weaving.

Padding Machine In cotton finishing, an apparatus of rollers for taking goods through a mordant vat.

Pick In weaving, one movement of the shuttle across the loom to insert one filling or weft yarn into the weave of the fabric being woven.

Picker In fiber preparation, the machine for loosening and breaking apart impacted or matted locks of wool or cotton stock. Also called a devil, opening machine, willow, willower, and willey.

Picking In fiber preparation, the loosening and tearing apart of masses of fibers. One of the five principal loom motions; in it the shuttle inserts the filling.

Piece The length of goods or cloth woven in one weave in a loom, often about thirty meters long.

Piecing Joining together laps, slivers, rolls, rovings, yarns, or threads to make continuous lengths or to repair breaks.

Piecing Machine In woolen preparation, the device attached to a finisher card in order to join together the short rolls to form a continuous slubbing ready for spinning.

Pirn A small bobbin holding filling yarn and fitting into a shuttle.

Plaid A fabric having a pattern formed from colored stripes and lines in both warp and filling.

Plain Weave The simplest of the three basic weaves (the others being twill and satin), it is characterized by a woven basket effect. For it, a minimum of two harnesses is needed, one harness controlling even-numbered warp ends and the other the odd-numbered ends. The position of the harnesses is then reversed between picks. Also known as tabby weave.

Plate Speeder A variant of the tube speeder, using revolving plates instead of tubes to insert false twist during drafting of the roving. Introduced in the United States in the mid-1820s.

Porter Another word for a beer.

Preparatory Processes The line between preparatory and spinning processes is unclear, variously placed, and therefore arbitrary. Because so many spinning inventions could be applied to roving, I have chosen throughout this study to regard roving and slubbing as initial spinning stages. In them the cotton sliver and wooden slubbing, respectively, first receive twist. See table in introduction.

Prints Cotton goods woven and finished suitably for printing.

Quill Another word for pirn.

Quilting A cotton fabric used for bedspreads; made like counterpane.

Railway Carding Head A system of conveyor belts that combines the slivers from several cotton cards in order to expedite the straightening and evening of the fibers within the sliver.

Raising Working the nap of a woolen cloth.

Reed In the loom, a frame held by the batten and containing a large number of regularly spaced and vertically positioned canes or wires between which the warp ends pass. Designed to keep the warp ends evenly spaced.

Reeding The process of drawing warp ends through their proper dents in the reed.

Reedstock The horizontal bars of the reed between which the cane or wire dents stand.

Reel A revolving frame upon which yarn is wound. Since the circumference of the reel is fixed, it is used to measure the length of yarn, in the form of skeins, hanks, and other packages.

Rim Band In mule, an endless driving band by which the tin roller that turns the spindles is driven from the rim or large pulley in the headstock.

Ring Doffer In the woolen finisher card, a doffer cylinder encircled with rings of card clothing. Designed to deliver a continuous roping.

Ring Spindle A spindle for which the traditional flyer was replaced by a stationary ring. Passing to the bare spindle, the yarn is twisted as it runs through a small metal loop or traveler, which is pulled around the ring by the drag effect of the spindle. A continuous spinning device, it was patented in the United States by John Thorp in 1828.

Rolag A short carded roll, usually of wool.

Roll Boiling In woolen finishing, a process for giving luster to cloth by scalding it, while tightly wound round a roller, in a tank of hot water or steam.

Roller Card A card in which the main cylinder is surmounted with rollers clad with card clothing.

Roller Printing Calico printing by means of engraved rollers, usually engraved metal ones, rather than wooden blocks.

Rollers In cotton drawing, roving, and spinning frames, several pairs of rollers, each pair rotating at a higher speed than its predecessor in the yarn path, which draft an end as it passes from one pair to the next.

Rolls In wool carding, the tube of wool fibers produced by the finisher card. Coming off the machine laterally, rolls could be no longer than the width of the card and therefore needed to be pieced together on the billy.

Roper In cotton spinning, the operative in charge of a stretcher that made roving.

Roving The process by which the carded and drafted cotton sliver or carded woolen slubbing is slightly twisted in readiness for spinning into yarn.

Roving Frame In cotton manufacturing, a machine for imparting twist and further draft to the carded and drawn sliver.

Satinet A mixed woven fabric with a cotton warp and a woolen filling.

Satin Weave The third basic weave (plain and twill being the others). Requiring a minimum of five harnesses, it is characterized by long floats of warp or filling. At each pick only one harness is raised (or lowered), and a twill line is avoided by ensuring that harnesses containing adjacent warp ends are not raised (or lowered) in succession.

Scourer An operative in charge of scouring.

Scouring In woolen preparation, the removal of dirt and animal grease from the wool.

Scribbler The English term for the breaker card.

Section Beam A warp beam holding a part of the warp; used in warping and dressing machines.

Self-Acting or Self-Actor Mule See mule, self-acting.

Self-Stripping Card A card in which the top rollers or flats are mechanically brushed, and embedded impurities are removed from the card clothing.

Serge A fabric with a worsted warp and a woolen filling.

Sett The spacing of dents in a loom reed, expressed as the number of dents in a length unit, such as the nail of two and a quarter inches.

Shearing Frame or Machine A machine for shearing the cloth nap evenly.

Shearman A skilled worker whose task was to shear the cloth nap.

Shears, Cloth A very large pair of shears, one blade of which was weighted down on the cloth, for shearing the nap evenly.

Shed In the loom, the separation or opening of warp ends to admit the passage of the shuttle; achieved by the movement of the harnesses.

Shedding In the loom, the creation of a shed; one of the five principal motions in weaving.

Sheeting Stout cotton fabric wide enough for bedsheets and made in plain or twill weave.

Shell Engraved cylindrical case that fitted onto a calico printing machine color cylinder.

Shirting A plain weave cotton fabric suitable for shirts and presumably much else.

Shoddy Woolen fibers retrieved from shredding rags and then reprocessed.

Shot A single filling yarn, or a single movement of the shuttle across the loom.

Shuttle In loom, a hollowed wooden block for holding a small bobbin loaded with filling yarn. Each time the shuttle is driven through the shed, the filling unwinds and one shot is laid, ready to be beaten up into the fell of the piece.

Sinking Machine Not identified.

Size In cotton weaving, a solution of flour, starch, or glue applied to the warp ends in order to bind together loose fibers and to impart strength to the warp yarns.

Skein A fixed length of yarn or thread of any fiber, measured and packaged on a reel.

Slay or Sley In loom, a reed. Slaying involved drawing the warp through the reed and also matching the reed sett to yarn count.

Sliver In cotton manufacturing, a soft rope of fibers lacking twist; produced by the finisher card.

Sliver Lapper A cotton preparatory machine in which conveyor belts place slivers in parallel to form a lap; a double lapper then divided the lap into ends to be spun upon stretchers. In this way the bobbin and fly frame and its winding problems would be circumvented. Patented by Johann Georg Bodmer in Britain in 1824.

Slubbing In woolen manufacturing, a slightly twisted rope of fibers produced on either the billy or a condenser.

Smallwares Narrow goods such as ribbons and tapes.

Sorting In the woolen or cotton industries, the process of grading the raw material according to its various properties.

Speeder In cotton spinning, a machine that drafted the carded slivers into roving. Some speeders inserted false or temporary twist, others a permanent twist.

Spindle In an intermittent spinning machine, the pointed metal shaft by which the yarn is twisted and then, by changing the yarn's angle of approach to the spindle, wound onto the spindle. In a continuous spinning machine, the metal shaft positioned inside a flyer, cap, or ring.

Spinning Head See head.

Spinning Processes Processes by which a carded and drafted sliver or roping is turned into a yarn. In this study, spinning is defined as commencing with the roving stage in cotton manufacturing and with the slubbing stage in woolen manufacturing. See table in introduction.

Spinning Wheel A hand-driven machine usually containing only one spindle.

Split In the loom, one of the interstices in a reed.

Spool A large bobbin, used in the woolen condenser and jack.

Squeezers Pressure rollers for discharging water from yarns and goods in the process of bleaching.

Staple The fiber of the raw material used in textile manufacturing.

Stocking Frame A knitting machine.

Stop Motion A device that automatically halts a machine whenever the material being processed is broken, impeded, or exhausted.

Stretch In the cotton mule, either the outward run of the carriage, or draw; or the length of spun yarn between the spindles and roller beam.

Stretcher In cotton spinning, either a small mule in which only the rollers (and not the carriage also) drafted or stretched the roving; or a roving frame very much like a throstle but with a beveled roller beam and a hollow flyer arm for laying the roving onto the bobbin. Both forms of stretcher were designed to reduce the thickness of the roving and to improve its uniformity.

Stripes Thick, colored bands in the pattern of a fabric resulting from two colors of filling yarn or, more frequently, two colors of warp yarn.

Stripper See clearer.

Superfine Cloth The finest quality of woolen broadcloth.

Surface Printing Machine In this calico printing machine, the printing rollers were formed from blocks of wood in the surface of which the pattern was cut in relief; so-called, probably, because the thickened color was applied to a tensioned surface of woolen cloth from which the rotating printing rollers took it up.

Swift The name sometimes given to the main cylinder in a card.

Take-Up Motion One of the five principal motions in the loom; by this the cloth is wound onto the cloth beam.

Tape Loom A loom for weaving narrow goods like tapes and ribbons.

Taunton Speeder See tube roving frame.

Tear Boy In calico printing, a color spreader.

Teazel Handle A hand-held implement in the shape of a wooden cross containing a number of teazel heads; used for manually raising the nap of a cloth.

Teeth In cards, the bent wires of card clothing.

Temple In loom, devices on each side of the loom frame for tensioning the fell of the piece.

Tenter-Drying In woolen finishing, drying a piece of cloth on a tenter or long wooden frame set on posts in the ground.

Thread Two or more yarns plied together.

Throstle A variant of Arkwright's water frame developed in Britain in the 1790s. Working on the continuous spinning principle, its name and popularity came from the long tin roller (possibly derived from the mule) from which endless bands individually drove each spindle. This roller was said to sing like a throstle or thrush; certainly it simplified the power transmission mechanism of the water frame and thereby permitted an increase in machine size.

Traveler In the ring spindle, the metal hook that, when pulled around its ring by the yarn, inserts twist in the yarn.

Traverse Rail In spinning machines working on the continuous principle, the rail that moves either bobbins or

flyers up and down to accomplish winding. Also known as the copping rail.

Trumpet Flyer A flyer in the shape of an inverted cone patented in the United States in 1824 by Oliver G. Rogers and Nathan Rogers.

Tube Roving Frame or Speeder In cotton manufacturing, a speeder in which the slivers are reduced in thickness but no twist is inserted; so-called because of its false twist tubes, which inserted twist temporarily during drafting. Patented in the United States in 1824 by George Danforth of Taunton, Massachusetts; hence its other name, the Taunton speeder.

Twill Card Clothing Card clothing in which the wires are set diagonally; used on flats in cotton carding.

Twill Weave One of the three basic weaves (the others being plain and satin). It is characterized by a diagonal line or variations of it, like chevrons, on the cloth surface. At least three harnesses are required for this weave, and at each pick the same number of harness is raised or lowered, the repeating sequence moving through the harnesses in turn.

Twisting One of the three component actions of spinning (the others being drafting and winding). Twisting a roving or yarn locks the fibers together and so lends strength to prevent breakages during fine drafting. Heavily twisted yarns are harder than lightly twisted ones.

Twisting Frame A machine in which two or more yarns are plied together.

Twisting Key In the woolen condenser, the equivalent of a small flyer turning on a horizontal axis, which inserts twist into the woolen rovings. Patented in Britain and the United States by John Goulding in 1826.

Velvet A cotton fabric with a thick short pile on one side; a type of fustian. If wholly of cotton, it was strictly called velveteen, which was patented in England by James Wolstenholme in 1776.

Warp The yarns running lengthwise in a piece of woven fabric. Warps, but not pieces, were between two hundred

Glossary

294

and three hundred yards long in the early nineteenth century.

Warp Protector Stop Motion In power loom, a stop motion designed to prevent the shuttle from being smashed through the warp shed by the batten.

Warp Stop Motion In power loom, a stop motion designed to halt the loom when a warp end broke.

Warp Tables Tables showing the kind of data that warper and weaver needed for preparing a warp for any given type of cloth; for example, yarn count, total number of warp ends (in beers), sett, reed width, sequences of drawing in, and reeding.

Warper In warping, either a machine that takes yarns from a rack of bobbins and lays them side by side in the formation of a warp, winding them up onto a beam or a section beam; or the operative who carries out this work.

Warping Bars A series of pegs, in a wooden frame or on a wall, over which the warp was formed by hand.

Warping Mill A hand-powered warping mechanism consisting of a creel filled with bobbins, a heck box, and a large reel onto which the warp was wound.

Water Frame In the cotton industry the spinning machine, operating on the continuous spinning principle, patented by Richard Arkwright in 1769. Its distinctive features were drafting rollers, vertical spindles, and flyers. Because drafting and twisting are physically separated in the machine, it produced a low-count, hard yarn suitable for warps. To drive, it needed more power than a single human operative could supply and therefore originally relied on water power, hence the term water twist for its yarn.

Water Loom A power loom.

Weave The structural pattern of a woven fabric.

Weaving The operation by which one series of parallel yarns (warp) is interlaced, at right angles, with a second series (filling or weft). See table in introduction.

Weft The British name for filling; also called the woof.

Weft Stop Motion In power loom, a stop motion designed to stop the loom when a filling end, from the shuttle, broke.

Willow or Willey Another name for the picker. See picker.

Winding One of the three component actions of spinning (the others being drafting and twisting). It consists of packaging the roving or yarn in a form suited to immediate technical and economic requirements.

Winding Block A late eighteenth-century device for winding rovings from the can frame onto bobbins, by hand or by power.

Winding Machine In weaving, a device for winding filling yarn onto bobbins or pirns to be placed in shuttles.

Winding-on Motion In the jenny, jack, or mule, the return movement of the clasp or carriage in which the yarn is wound onto the spindles in a distinctive cop shape.

Wiper Cam.

Worker In card, a roller clad with card clothing, the function of which is to move against the rotation and card teeth of the main card cylinder, thereby carding the cotton or wool staple.

Worsted Long-fibered wool that is combed, spun hard, and then woven into a fabric (collectively known as stuffs) that is only lightly or never raised. Worsted manufacture is a separate branch of the textile industry from the woolen manufacture.

Yarn A single strand of fibers spun together by drafting and twisting. In cotton, coarse yarns were nos. 20s and below, middling yarns were nos. 30s–50s, and fine yarns were nos. 60s and above.

Yarn Tables Tables showing the weights of a range of length units for a range of yarn finenesses.

Abbreviations in Notes and Bibliography

Baker MSS	Division of Manuscripts, Baker Library, Harvard Business School
BMC	Boston Manufacturing Company
BT	Board of Trade
C	Chancery
CORHAL	*Contributions of the Old Residents' Historical Association, Lowell, Massachusetts*
CSL	Connecticut State Library, Hartford, Connecticut
CRO	County Record Office
Cust.	Customs
DAB	*Dictionary of American Biography*
DNB	*Dictionary of National Biography*
EMHL	Eleutherian Mills Historical Library, Greenville, Wilmington, Delaware
FCC	First (Massachusetts) Circuit Court of the United States
FRC	Federal Record Center, United States
GB	Great Britain
MC	Manufacturing Company
MVTM	Merrimack Valley Textile Museum, North Andover, Massachusetts
NHHS	New Hampshire Historical Society, Concord, New Hampshire
Pat.	Patent
PC	Privy Council
PP	*Parliamentary Papers*
PRO	Public Record Office, London
RG	Record Group
RIHS	Rhode Island Historical Society, Providence, Rhode Island
T	Treasury (British)
TRISEDI	*Transactions of the Rhode Island Society for the Encouragement of Domestic Industry*

USFRC   United States Federal Record Center

USNA   United States National Archives,
Washington, D.C.

Notes

Introduction

1. Rogers and Shoemaker, *Communication of Innovations*, surveys over fifteen hundred books and articles on diffusion research in the areas of anthropology, sociology, rural and medical sociology, education, communications, and marketing.

2. Rosenberg, "Factors," p. 6. Economists like Feller and Temin argue this case strongly, as does Chandler in "Anthracite Coal."

3. Sawyer, "Social Basis," drew attention to non-economic factors, like value systems, family patterns, and social stratification, in explaining British rejection of American methods of manufacturing in the 1850s and the 1950s. A greater freedom of movement between professions and skills in the United States than in Britain could explain in part the latter's slowness to accept technical change, as Burrage, "Democracy and the Mystery of the Crafts," suggests. Wealth, class, and individualism presumably underlay the traditional British preference for tailor-made rather than mass-produced sporting arms which, as Fries, "British Response," showed, heavily influenced British manufacturers in the nineteenth century.

4. Arrow, "Classificatory Notes," p. 33.

5. Ames and Rosenberg, "Changing Technological Leadership."

6. Rosenberg, "Factors."

7. Pursell, *Early Stationary Steam Engines*; Brittain, "International Diffusion." See also Harris, "Attempts to Transfer English Steel Techniques."

8. Harris, "Skills, Coal"; Feller, "Draper Loom," and "Diffusion and Location"; Temin, "Steam and Waterpower"; Coleman, "An Innovation."

9. Robinson, "Transference."

10. Wilkinson, "Brandywine Borrowings"; Sloat, "Dover Manufacturing Company"; Harris, "Saint Gobain and Ravenhead."

11. Rosenberg and Vincenti, *Britannia Bridge*.

12. Scoville, "Minority Migration," and "Huguenots."

13. Rosenberg, "Economic Development"; Wilkins, "Role of Private Business"; Harris, "Skills, Coal"; Uselding, "Henry Burden"; Hindle, "Transfer of Power." For steam power and iron, Hindle thought that artisans were less important than drawings, models, specifications, or visits to the initiating country. Robinson seems to give priority to artisans in Musson and Robinson, *Science and Technology*, pp 216–230, but more recently in "Early Diffusion," doubts that eighteenth-century transfers were confined to one route.

14. "For too long the history of technology [has been] pursued as the antiquarian study of actual techniques in their own right rather than viewed in relation to the dynamics of economic change as a whole," observed Peter Mathias in 1974; reported in Mathias, "Skills and Diffusion," p. 113.

15. Layton, "Technology as Knowledge," pp. 37–38.

16. Wilkins, "Role of Private Business," pp. 168–171. Another variation of Wilkins's stage model has been applied by Tann and Brechin ("International Diffusion").

Chapter 1

1. Hopkinson, *Account*, pp. 11–12 and passim.

2. Main, *Social Structure*, pp. 67, 195.

3. Nettels, *Emergence*, pp. 109–129; see also Bruchey, *Roots*, for a probing discussion of America's varied potential for industrialization.

4. Nettels, *Emergence*, chap. 7.

5. Ibid., chap. 12.

6. Wilkinson, "Philadelphia Fever," p. 44; Jeremy, *Wansey*, p. 106; Nettels, *Emergence*, pp. 237–238; Morison, *Maritime History*, p. 91.

7. Jeremy, *Wansey*, p. 138.

8. Mann, *Cloth Industry*, p. 132; Fitton and Wadsworth, *Strutts and the Arkwrights*, pp. 93, 96–97; Owen, *Life*, 1:40.

9. *American Museum* 2 (1787): 168, 250.

10. USNA, RG241, Specifications 1, p. 2, William Pollard to the U.S. Patent Commissioners, June 29, 1790.

11. Main, *Social Structure*, pp. 70–77.

12. Warville, *New Travels*, p. 273.

13. Jefferson, *Notes*, pp. 156–158. Hamilton opposed this view, of course. See Rezneck, "Rise," pp. 790–791.

14. Jones, "Ulster Emigration," p. 52.

15. Main, *Social Structure*, pp. 77–79, 274.

16. U.S., Bureau of the Census, *Historical Statistics* 1:8–38; Main, *Social Structure*, pp. 26–34, 66–67, 271–276.

17. Nettels, *Emergence*, pp. 251–255; Jeremy, *Wansey*, pp. 63–75; Toulmin, *Western Country*, p. 28.

18. Nettels, *Emergence*, pp. 255–262.

19. Main, *Social Structure*, pp. 37–38; Jeremy, *Wansey*, pp. 101, 105.

20. Nettels, *Emergence*, pp. 184, 388. According to Rees, *Cyclopaedia*, s.v. "Manufacture of Cotton," a bale of cotton weighed between 3¼ and 3½ hundredweight.

21. Bridenbaugh, *Colonial Craftsman*, provides the best survey of manufacturing skills in Revolutionary America.

22. Jeremy, "British Textile Technology," p. 30. For other examples of woodworkers and builders engaged in making early textile machines, see Baker MSS, Worcester Manufactory Accounts, pp. 3, 6, 12; Jeremy, *Wansey*, pp. 148–149; Thompson, *Moses Brown*, p. 214.

23. Lovett, "Beverly Cotton Manufactory," p. 223n; Wallace and Jeremy, "Pollard," p. 419; Thompson, *Moses Brown*, p. 257; Rivard, "Textile Experiments," p. 38.

24. Baker MSS, Worcester Manufactory Accounts, p. 3; Hardie, *Philadelphia Directory, 1793*, passim.

25. Kittredge and Gould, *Card-Clothing Industry*, pp. 12–17.

26. PRO, BT 5/2, p. 302.

27. USNA, RG 46, Evans's petition to the Senate, January 8, 1806; Kittredge and Gould, *Card-Clothing Industry*, pp.

12–17; Wroth, *Abel Buell*, p. 3; Cole, *Correspondence of Hamilton*, pp. 16–17.

28. Kittredge and Gould, *Card-Clothing Industry*, pp. 12–42.

29. For science, see Nye, *Cultural Life* and Hindle's studies, *Pursuit of Science* and *David Rittenhouse*. For examples of mechanics, see Bathe and Bathe, *Oliver Evans*, and *Jacob Perkins*; Wroth, *Abel Buell*; and Ferguson, *Early Engineering Reminiscences*.

30. Solow, "Capacity," pp. 480–488.

31. North, *Economic Growth*, pp. 19–21.

32. Nettels, *Emergence*, pp. 109–111.

33. North, *Economic Growth*, p. 219.

34. There are numerous editions of Hamilton's *Report on Manufactures*. I used Jacob Cooke's, in which this quotation appears on p. 160.

35. Rezneck, "Rise."

36. Colquhoun's estimates of 1789 are quoted in Smelser, *Social Change*, p. 121.

37. Chapman, "Fixed Capital Formation," pp. 60–71.

38. Ibid., pp. 71, 75.

39. Jeremy, "British Textile Technology," pp. 28–29, 40. *Pennsylvania Magazine* 1 (1775): 157–158.

40. Neel, *Phineas Bond*, pp. 55–61; Jeremy, "British Textile Technology," pp. 29–30.

41. Neel, *Phineas Bond*, p. 57; PRO, T. 47/9, p. 145, where the emigrant weaver's name is given as John Hage.

42. Historical Society of Pennsylvania, Minutes of the Manufacturing Committee, I, under January 19, 22, March 12, 1788.

43. Bagnall, *Textile Industries*, 1:84–86.

44. Ibid., pp. 89–100; Lovett, "Beverly Cotton Manufactory," pp. 218–241.

45. Bagnall, *Textile Industries*, 1:150–151; White, *Slater*, p. 62; Thompson, *Moses Brown*, pp. 210–220; Lovett, "Beverly Cotton Manufactory," p. 228.

46. Bagnall, *Textile Industries*, 1:127–131; Baker MSS, Worcester Manufactory Accounts, pp. 1–12.

47. Bagnall, *Textile Industries*, 1:122–127.

48. Ibid., pp. 127, 131–134.

49. For American hand loom weaving, see Hargrove, *Weavers Draft Book*. For cotton finishers, see Montgomery, *Printed Textiles*, pp. 88–99; Neel, *Phineas Bond*, pp. 55–61.

50. Jenkins, *West Riding Wool Industry*, pp. 8–9, 118–120, 128–129, 171, 208–210.

51. Mann, *Cloth Industry*, pp. 50–62, 123–144.

52. Cooke, *Reports of Hamilton*, p. 197.

53. Jeremy, *Wansey*, p. 68n; Cole, *Correspondence of Hamilton*, pp. 7–11.

54. Ibid., p. 16.

55. Fitzpatrick, *Diaries of Washington*, 4:27–28.

56. Ibid., p. 68.

57. Wily, *Treatise*, pp. 8–9, 24; Mann, *Cloth Industry*, p. 300.

58. This analysis of labor inputs, of operatives, managers, and machine makers, is based on the accounts of early industrial cotton manufacturing given in Rees, *Cyclopaedia*, s.v. "Cotton Manufacture" and "Manufacture of Cotton," and in Montgomery, *Carding and Spinning Master's Assistant*.

59. Pollard, *Genesis*, chap. 5, contains a clear exposition of early factory discipline.

60. Chapman, "Textile Factory." See Penrose, *Theory*, pp. 68–71, for "Balance of Processes."

61. The properties of cotton are found in any modern technical textbook. I relied on an older text, Merrill, *American Cotton Handbook*. Brazilian and American staples, especially upland cotton from the United States, supplanted West Indian cottons especially after Whitney's gin spread in the mid-1790s. See Edwards, *Growth*, chap. 5.

62. Montgomery, *Carding and Spinning Master's Assistant*, p. 63.

63. Pollard, *Genesis*, chaps. 5, 6.

64. Collier, *Family Economy*, pp. 39–41, 71. For managers' salaries, see Pollard, *Genesis*, pp. 141–145.

65. PP (Commons), 1841 (201), 7:99.

66. PP (Commons), 1824 (51), 5:475.

67. Ibid., p. 381.

68. Collier, *Family Economy*, p. 41. The functions of the new machine makers were compared to that of the watchmaker in 1805: "one person only like the watchmaker can collect the parts, put them together, and set them in motion." See PRO, BT 1/26, bundle A.1, John Rae to the Board of Trade, October 9, 1805.

69. See the entry in *DNB* for an outline of Arkwright's career, a profile that will be superseded by R. S. Fitton's biography of Arkwright.

70. This account of mule spinning derives from Rees, *Cyclopaedia*, s.v. "Manufacture of Cotton"; Catling, *Spinning Mule*; and personal observation of jack spinning at the MVTM. Indicative of the mule spinner's ability to change the quality of his yarn is rule four of the mill rules at Taylor & Co.'s factory at Halliwell, Lancashire, in force in February 1804: "Any spinner altering his Geering Wires or nocking out parts or by any means striving to alter the counts of the Twist &c. without the permission of the overlooker, for every offence to be stopped 2s 6d." See Bolton Reference Library, Gray Papers, Notebook, pp. 53–55.

71. William Kelly of Lanark, Scotland, first powered the outward draw of the mule in 1792 (GB Pat. 1879). For mule spinners' wages, see Collier, *Family Economy*, p. 17. For a convenient summary of wage rates, see Ashton, *Economic History*, pp. 220, 232.

72. I am indebted to Harold Catling for this information, which he gave in conversations at the Fourth Pasold Textile History Conference, held at Manchester University,

September 1977. Dr. Catling cited the replacement of the rim band, which wore out in a year or less, as a maintenance task needing skill and knowledge. The seventy- to eighty-foot rope had to be stretched, threaded around pulleys in an intricate path, and then spliced.

Five or six months was reckoned, by the machine builder, sufficient to learn to operate Gay's self-actor mule in northern New England in 1828. See Baker MSS, Nashua MC, letterbook 1825–1835, p. 164.

73. PP (Commons), 1816 (397), 3:281.

74. This analysis of labor inputs comes from descriptions in Rees, *Cyclopaedia*, s.v. "Woollen Manufacture"; Ure, *Philosophy*, chap. 4; Crump, *Leeds Woollen Industry*; and Partridge, *Practical Treatise*.

75. Mann, *Cloth Industry*, pp. 322–329; Crump, *Leeds Woollen Industry*, pp. 77–78, 89, 92, 116, 287 for Yorkshire wage rates 1808–1815.

Chapter 2

1. For examples of secretiveness, see *Pennsylvania Magazine* 1 (1775): 157; Mantoux, *Industrial Revolution*, p. 297; Ashton, *Industrial Revolution*, p. 41; Raistrick, *Quakers*, chap. 2; Musson and Robinson, *Science and Technology*, chap. 6. For the secretiveness of inventors, see Wadsworth and Mann, *Cotton Trade*, pp. 457, 478; *Trial of a Cause*, pp. 41, 43; Radcliffe, *Origin*, pp. 20–22; French, *Crompton*, pp. 55–56.

In the setting of rapid technical change, Arkwright, and presumably others, adopted double standards: secrecy toward industrial followers and espionage against industrial leaders. See *Case of Richard Arkwright*, pp. 2–3; *Arkwright versus Nightingale*, plaintiff's evidence, p. 17, and defendant's evidence, pp. 22, 31; *Trial of a Cause*, pp. 59, 62–63; Fitton and Wadsworth, *Strutts and the Arkwrights*, p. 80n.

The principle of voidance of patent rights due to public use prior to patent enrollment was established early in the seventeenth century and was laid down by Chief Justice Coke in his *Institutes*, pt. 3, chap. 184.

2. Crump, *Leeds Woollen Industry*, pp. 264–265. For other quadrangle textile factory plans see Tann, *Development of the Factory*, pp. 17, 20, 34.

3. *Trial of a Cause*, pp. 52–57.

4. Unwin, *Oldknow*, p. 10.

5. Owen, *Life*, 1:31.

6. Bolton Reference Library, Gray Papers, Notebook, pp. 53–55.

7. Henderson, *Industrial Britain*, p. 99; see also pp. 91, 102, 112.

8. PP (Commons), 1824 (51), 5:159.

9. Ibid., pp. 22, 23, 37, 278.

10. Ibid., p. 20.

11. Henderson, *Industrial Britain*, p. 38.

12. PP (Commons), 1824 (51), 5:384.

13. RIHS, Allen Papers, "Trip, 1825," p. 146 (July 25, 1825).

14. Ibid., pp. 27–28 (April 19–20, 1825).

15. Ibid., pp. 41–51 (April 26, 1825).

16. PP (Commons), 1829 (332), 3:136. William Radcliffe, the Stockport manufacturer and inventor of the dressing frame and dandy loom, promoted his partly mechanized system of weaving by means of an employers' club, in the tradition of manufacturers' and employers' associations already formed in Lancashire. He claimed that he was motivated by patriotic duty and expectation of a parliamentary grant in recognition of his work. Radcliffe, *Origin*, pp. 4–5. See Rose, "Crompton," for employers' clubs.

17. Henderson, *Industrial Britain*, pp. 7–16, 75–111. For the ease with which letters of introduction could open factory doors in the 1820s, see Smith, "Journal."

18. RIHS, Allen Papers, "Trip, 1825," pp. 52–64 (April 27, 1825); Heaton, "Gott."

19. Dyer, "Notice"; Allen, *Practical Tourist*, 1:171–172; GB Pat. 4162 (1817), for Wilkinson's reed-making machine; RIHS, Allen Papers, "Trip, 1825," p. 41 and Davy's consular despatches from Leeds in USNA, RG 59.

20. Radcliffe, *Origin*, p. 85.

21. PP (Commons), 1824 (51), 5:384.

22. Ogden, *Description*, p. 93. I am grateful to Negley Harte for this reference.

23. Duncan, *Practical Essays*, pp. vi–xii.

24. For the publication history of Rees's monumental work, see Harte, "On Rees."

25. Butterworth, *Complete History*, pp. ii–iii.

26. Montgomery, *Carding and Spinning Master's Assistant*, pp. ii–v.

27. This section is derived from Jeremy, "Damming the Flood," a fuller treatment of the subject. Only sources to new material, quotations, and corrections are given.

28. PRO, BT 1/120, f. 93 (September 1, 1817).

29. Ibid., 5/39, p. 572 (correction of Jeremy, "Damming the Flood," n. 66).

30. Customs House Archives, Customs 37/56, "Seizures," 1825–1856, ff. 131–152 for the 1830s.

31. PRO, BT 5/41, p. 417 (September 24, 1833).

32. PP (Commons), 1841 (201), 7:31, 39–40, 55–56; ibid., 1825 (504), 5:38.

33. Quoted in Robinson, "Early Diffusion," p. 92.

34. *Trial of a Cause*, p. 114.

35. Ibid., pp. 24–27, 30, 172, 181–182; Robinson, "James Watt," pp. 119–122.

36. For proposals to change specification requirements, see Robinson, "James Watt." Efforts to introduce secret patents are alluded to in ibid., p. 131; PP (Commons), 1829 (332), 3:178; *Hansard*, 2d ser. I, col. 1052 (June 14, 1820). Directed against foreign piracy and requiring statutory approval, two secret patents were awarded during this period: for Joseph Booth's carding machinery (1794) and for James Lee's flax preparatory machinery

(1813). Both inventions proved to be commercial failures. See PP (Commons), 1829 (332), 3:175, 178.

37. PP (Commons), 1829 (332), 3:99.

38. Ibid., pp. 59–62, 82, 85, 96, 99, 100, 103, for the working of the patent office. Numbers of patents can be counted in Woodcroft, *Titles of Patents*.

39. The methodological weaknesses of this criterion are readily acknowledged. My assessment of widespread usage of inventions rests on the following contemporary descriptions of prevailing technology: Rees, *Cyclopaedia*; Montgomery, *Carding and Spinning Master's Assistant*; Ure, *Cotton Manufacture*, and *Philosophy*; Partridge, *Practical Treatise*; Tomlinson, *Cyclopaedia*; and also Julia Mann's modern accounts, "Textile Industry," and *Cloth Industry*, chap. 10, and Hills, *Power in the Industrial Revolution*. The 303 patents covering cotton and woolen preparatory, spinning, weaving, and finishing mechanical inventions, 1790–1830, were identified from Woodcroft, *Subject-Matter Index*. I then read the patents, studied them in their printed form, and listed them. Against this list I checked the issues of the *Repertory of Arts*, 1794–1830. Those patents published in the *Repertory* in complete or abridged form are found in appendix A.

40. PP (Commons), 1829 (332), 3:103–104. Radcliffe, *Origin*, pp. 25–27. The patents with which Radcliffe was concerned covered his dresser, GB Pats. 2684 (1803) and 2771 (1804), patented in the name of Thomas Johnson, Radcliffe's mechanic.

41. In 1802, for example, the time lag between patent award and publication was five months; in 1807 it varied between seven and fourteen months. None of the textile specifications of 1809–1811 was published before 1813. Again in 1820, John and William Lewis published five of their patents in the space of one year; the ages of their patents ranged from two to five years. See appendix A.

42. *Repertory of Arts*, 1st ser. 6 (1796): 359–360; 3d ser. 1 (1825): 52; 2 (1826): 51; 5 (1827): 59–64; 9 (1830): 49–50. Wyatt defended his publication of patent information in the *Repertory*, 2d ser. 1 (1802): i–iv. He argued that he was serving the interests of the law and the financial prospects of the patentee by advertising new inventions. Further he believed that he followed the great societies committed to the promotion of useful knowledge (the Royal Society, the Royal Institution, and so on) in publishing useful information. Rather than weaken Britain's leadership, he wrote, the exchange of information between countries would augment "the Brilliancy of British genius."

Chapter 3

1. Luccock, *Nature*, pp. 141–144; Mann, *Cloth Industry*, p. 264; Dorset CRO, Colfox Papers D 43/B9, draft letter of Thomas Colfox & Son, February 8, 1812 (I am indebted to Julia Mann for this reference); Rees, *Cyclopaedia*, s.v. "Wool."

2. Montgomery, *Carding and Spinning Master's Assistant*, pp. 74–75. There were grades of coarseness and fineness in the old hand cards. Stock cards, for example, were much coarser than were hand cards because the initial carding or scribbling was performed with these. The progression of coarser to finer card clothing was repeated in the breaker and finisher carding machines and their rollers. Pennington's patent for a leather pricking press or stang allowed for differences; its brass cylinder was "divided into numbers for the different sorts of cards, and stop'd regularly at every motion." The development of card clothing machines in this period contributed significantly to varieties of card clothing. The dimensions of count (number of teeth per square inch), gauge (thickness of wire), cut (length of wire), crown (distance between the two points of each staple), point (shape of teeth points), bend (angle of knee bend), and set (pattern in which staples were set, usually ribbed or twill) were gradually distinguished, mechanically produced, and marketed in various grades suited for different rolls of carding machines. See Leigh, *Science*, 1:137–144; William Pennington's GB Pat. 657 (1750).

3. Jeremy, "British and American Yarn Count Systems"; PP (Commons), 1820 (314), vol. 8.

4. PP (Commons), 1808 (177), 2:10.

5. Murphy, *Treatise*, p. 419; White, *Practical Treatise*, pp. 276–278.

6. Duncan, *Practical Essays*, pp. 23, 322; White, *Practical Treatise*, pp. 276–277.

7. Duncan, *Practical Essays*, p. 23.

8. Hobsbawm, *Labouring Men*, pp. 5–17.

9. Ibid., p. 7.

10. Turner, *Trade Union Growth*, p. 57; Aspinall, *Early English Trade Unions*, p. 214; GB Pats. 1378 (1783), 1443 (1784); Musson and Robinson, *Science and Technology*, pp. 415–416, 442.

11. Rees, *Cyclopaedia*, s.v. "Copper Plate Work in Calico Printing"; Baines, *Cotton Manufacture*, pp. 270–271.

12. Rees, *Cyclopaedia*, s.v. "Copper Plate Work in Calico Printing."

13. *Rules*, p. 6.

14. Ibid., p. 11.

15. Aspinall, *Early English Trade Unions*, pp. 166–168, 183, 186, 191; see also Turnbull, *History*, pp. 191–192.

16. Hammond and Hammond, *Skilled Labourer*, pp. 126–128, 273–285; Bythell, *Handloom Weavers*, pp. 74, 176–204; Kirby and Musson, *Voice of the People*, pp. 41–42, 68; Thompson, *Making*, p. 303.

17. Turner, *Trade Union Growth*, p. 103.

18. Kirby and Musson, *Voice of the People*, pp. 28, 31, 57.

19. Hammond and Hammond, *Skilled Labourer*, pp. 167–189, 301–334; Thompson, *Making*, pp. 521–575.

20. Thompson, *Making*, p. 550 (quoting Lipson); Bythell, *Handloom Weavers*, p. 57.

21. Thomis, *Luddites*, pp. 61, 66.

22. Schmookler, "Economic Sources." The costs of patenting in Britain were high during this early industrial period, at £70 to £100 in 1815 and at least £150 in the late 1820s. It seems likely that these costs discouraged the patenting of trivial inventions. If this was so, then arguably the course of invention would more closely approximate the course of innovation in Britain than in the United States where patents cost a mere $30 (then equivalent to about £7) under the Patent Law of 1793. See Henderson, *Industrial Britain*, p. 164; PP (Commons) 1829 (332), 3:17, 179–180, 218.

The premiums or prizes of the Royal Society of Arts (RSA) offered a poor alternative to the returns of a successful patent. Never more than £50 in the period before 1830, RSA premiums scarcely covered development costs, in the case of many inventions. And inventions awarded premiums by the RSA were not permitted to be patented. See Royal Society of Arts, *Transactions* 8 (1790): 328, 349; 48 (1831): vi.

23. These 303 patents were identified from Woodcroft, *Subject-Matter Index*. It must be emphasized that they cover only mechanical inventions for cotton and woolen spinning and weaving and cloth finishing. They therefore exclude techniques involving no new mechanical devices, all chemical technologies, other branches besides weaving (such as knitting, carpet making, smallwares, and lacemaking), other fibers (such as flax, silk, and jute), and the worsted side of the wool manufacture.

24. Chapman, "Fixed Capital Formation," p. 75.

25. PRO, Cust. 9/4–17.

26. Bagnall, *Textile Industries*, 2:2179–2180.

27. PRO, Cust. 9/4–17. The average price per yard (calculated by dividing total yardage exported by total value of exports at declared values) of printed and checked calicoes in 1816 was £0.065; in 1830 it was £0.0378.

28. Wood, *History of Wages*, pp. 130–131; Mitchell and Deane, *Abstract*, p. 491.

29. PRO, Cust. 9/4–17. The average price per yard of plain or white calicoes in 1816 was £0.0594; in 1830 it was £0.0264.

30. Edwards, *Growth*, chap. 7, for yarn markets before 1815.

31. Catling, *Spinning Mule*; Dickinson, "Richard Roberts"; Chaloner, "New Light"; Kirby and Musson, *Voice of the People*, p. 32.

32. GB Pats. 2166 (1797, Aaron Garlick); 3633 (1813, Joseph Raynor); 4807 (1823, John Green); 5316 (1826, Henry Houldsworth, derived from Aza Arnold of Rhode Island). USFRC, FCC Case Files, Paul Moody v. Jonathan Fisk et al. (October 1820–October 1822), deposition of Thomas Charlton Faulkner. Rees, *Cyclopaedia*, s.v. "Manufacture of Cotton."

33. GB Pat. 5016 (1824). Bodmer's personal misfortunes and the plagiarism of his patent are recounted in his petitions to the Privy Council: PRO, PC 2/220, p. 261 (8 June 1838); 221, pp. 614–620 (December 5, 1839); 222, pp. 259–261 (May 22, 1840). Henderson, *Industrial Revolution*, pp. 7–16 gives a biography of Bodmer. Without attribution to Bodmer, his "continuous lapping machine" (GB Pat. 5016, pp. 6–11 of printed specification) is described by Ure, *Cotton Manufacture*, 2:43–44, who found it in a number of coarse spinning mills; its advantage, according to Ure, was that it displaced the card cans. Bodmer's claims of widespread infringements of his patent rights thus seem justified.

Although it lies beyond the structural limits of this study, Bodmer's railway head for a line of cards (GB Pat. 5016, pp. 18–23) was patented in the United States by William B. Leonard of the Matteawan MC, Fishkill, New York, on September 16, 1833. Leonard's system, utilizing card cans, was not the wholly continuous process Bodmer envisaged. Bodmer's patented devices were selectively adopted in the 1830s, the continuous lapper being favored in Britain and the railway head in America, according to Montgomery, *Practical Detail*, pp. 44–47 (who also did not recognize Bodmer's prior claim to invention).

34. See Ure, *Cotton Manufacture*, 2:289–312, for technical changes in power loom weaving. After Cartwright, the major technical changes in power loom weaving were registered in the following patents: Miller's cam loom, 2122 (1796); Radcliffe and Johnson's dresser, 2684 (1803); Horrocks's crank loom, 2699 (1803); Horrocks's variable batten speed, 3728 (1813); Bowman's six-harness cam power loom, 4488 (1820); Wilson's introduction of the French jacquard head, at first in the hand loom, 4543 (1821); Roberts's six-harness cam power loom, 4726 (1822); Potter's power loom dobby head, 4951 (1824).

35. The shearing patents were those of Harmar, 1982 (1794), Douglas, 2225 (1798), and Sanford, 2558 (1801); for the popularity of Harmar's shears, see Partridge, *Practical Treatise*, pp. 83–84.

The wire gig patents were those of Jotham, 2539 (1801); Sanford and Price, 3015 (1807); La Salle, 4018 (1816); Williams, 4118 (1817); Price, 4186 (1817); Lewis and Davis, 4189 (1817); Jones, 4229 (1818); Davis, 4820 (1823); Servill, 4863 (1823) and 5564 (1827); and Daniell, 5795 (1829).

36. Examples of increased machine capacities are the mule (see Catling, *Spinning Mule*) and Robertson's patent for linked mules, 2984 (1806); Tate's double-sided roller beam for the water frame, 1938 (1793); Etchell's large water frame bobbins used with his rectangular and tubular flyer driven around a dead spindle, 1949 (1793); and Lane's roving frame builder motion, which made a large package tapered at the top and bottom for easier unwinding, 5969 (1830). In the power loom, larger shuttles came with Horrocks's batten speed patent and Tetlow's arrangement for weaving two pieces simultaneously, 5020 (1824).

37. Higher speeds were sought by the dead spindles of Etchells and of Heppenstall, 2770 (1804); Thomson's ring flyer, a form of ring spinning without a traveler, 3202 (1809); and Andrew, Tarlton, and Shepley's double spindle, 5079 (1825) which, they claimed, reached six thousand revolutions per minute. These are only examples from spinning machines.

38. For combinations of machines see Foden's combined dressing and warping frame and loom, 2353 (1799); Johnson's and Radcliffe's dresser patents, 2684 (1803) and 2771 (1804); Horrocks's avoidance of section beaming in 4817 (1823); and the woolen condensing devices of Davis, 5159 (1825), Whitaker, 5486 (1827), and Dowding, 5566 (1827). Bodmer's flow movements are in his 1824 patent. Varley attempted to integrate carding and spinning in one machine, 2100 (1796), and Stansfeld tried to make dyeing and sizing one continuous process, 4991 (1824).

39. Different positions for the flyer, to facilitate doffing, attracted several patents: those of Teschemacher, 1917 (1792); Thomson, 3024 (1807); Bradbury, 3094 (1807); and Andrew, Tarlton, and Shepley, 5079 (1825).

40. The dobby, which worked from a barrel of pegs, like a musical box, to control the shedding movements, was patented by Taylor, 4216 (1818) for the hand loom, and by Potter for the power loom, 4951 (1824). Besides introducing the jacquard head in 1821, Wilson improved it in patent 4795 (1823).

In sequence of process, the following stop motions were patented in Britain: for the roving frame, by Bradbury, 3990 (1816); for the spinning frame, by Bradbury, 4984 (1824); for twisting frames, by Ward, 2460 (1800), by Margrave, 2802 (1804), and by Bradbury, 4984 (1824); for power looms, by Miller (warp protector) 2122 (1796), by Todd (warp protector), 2698 (1803), by Johnson and Kay (warp protector), 2876 (1805), and by Stansfeld, Prichard, and Wilkinson (weft and warp stop motions), 5211 (1825); and for a fustian cutter, by Brown, 1967 (1793).

41. Iron replaced wood in machine frames before 1814. Bodmer then saw cast-iron frames in Manchester devils, cards, throstles, and mules (Henderson, *Industrial Britain*, p. 100). Henry Houldsworth, Sr., dated the transition from wood to iron to the years 1805–1810 (PP (Commons), 1824 [51], 5:383). Brass, which would not rust like iron, was mentioned as a material for gig wire teeth in Lewis and Davis's patent, 4198 (1817).

For variable batten speeds, see the patents of Marsland, 2955 (1806), Horrocks, 3728 (1813), and Buchanan, 4854 (1823).

For automatic fault detectors, or stop motions, see the patents listed in note 40.

42. For friction-reducing spinning devices see, for example, the patents of Bradbury, 3094 (1807), Wood and Wordsworth, 3987 (1816), and Simpson, 4138 (1817).

Shafting and gear train arrangements for power loom weaving were patented by Sadler, 5177 (1825), 5581 (1827), and 5951 (1830).

Minimal mechanical complexity was an important consideration, as Buchanan's variable batten speed illustrates. It employed elliptical gears and reached a loom speed of 130 picks per minute without suffering a higher rate of breakages than that incurred with Horrocks's motion, running at 80 picks per minute; Ure conjectured that its complexity was one reason why Buchanan's motion was neglected in Lancashire. See Pat. 4854 (1823), and Ure, *Cotton Manufacture*, 2:310.

43. Besides the use of metal to replace wood, metal substituted cane in reeds (through the American invention of Wilkinson in 1816–1817) and cords in headle eyes (Osbaldeston's patent, 5044, of 1824).

44. Some vertical integration accompanied power loom weaving in Lancashire in the 1820s. Even so, the state of technology and external economies induced regional specialization: spinning in south Lancashire and weaving around Blackburn. See Taylor, "Concentration," p. 119.

45. Salter, *Productivity*, p. 33. Blaug, "Survey," exposes the difficulties inherent in a theoretical approach to the concepts of capital- and labor-saving biases in technology. Despite this, I have used these concepts, loosely perhaps, to identify the kinds of costs of which contemporary inventors and manufacturers were aware and aimed to reduce.

46. GB Pat. 3728 (1813), p. 4.

47. Ibid. Horrocks's motion was not the first attempt to achieve a variable batten speed. Marsland's horsehead

motion, a more cumbersome mechanism, appeared in 1806 (GB Pat. 2955). The original idea for this came from James Watt, Jr., according to Radcliffe, *Origin*, p. 37n.

48. Baines, *Cotton Manufacture*, p. 235.

49. PP (Commons), 1824 (51), 5:251; see also pp. 300, 305, 346, 474.

50. Ibid., p. 545.

51. Musson and Robinson, *Science and Technology*, chaps. 13–15.

52. PP (Commons), 1825 (504), 5:11.

53. Rolt, *Tools*, pp. 83–91; Woodbury, *History of Lathe*, pp. 96–108.

54. Ure, *Philosophy*, p. 37, quoted in Musson and Robinson, *Science and Technology*, pp. 478–479.

55. PP (Commons), 1841 (201), 7:23.

56. Ibid., pp. 96–99.

57. See Harte, "On Rees," for authorship of Rees's *Cyclopaedia* articles; see also Jackson, *Attempt*.

58. Philadelphia edition of Rees, 1:vi–vii.

59. Ibid., p. vii.

60. Montgomery, *Cotton Spinner's Manual*, p. vi.

61. Bugbee, *Genesis*, pp. 145–147.

62. Shaw and Shoemaker, *American Bibliography*, pp. 65–66.

63. The articles "Cotton Manufacture" and "Wool" are on pp. 386–396 and 396–399.

64. Baker MSS, Wendell Collection, case 14, Jacob Wendell, receipts. For Olinthus Gregory, see *DNB*.

65. USFRC, FCC Case Files, BMC v. Jonathan Fisk et al., October 1820–May 1821, special matter filed by defendants, September 15, 1820.

66. Ogden, *Thoughts*, p. 35.

67. Ibid., passim.

68. See Sinclair, *Philadelphia's Philosopher Mechanics*, p. 32 and chap. 3 for Browne.

69. For Jones and the *Franklin Journal* see Sinclair, *Philadelphia's Philosopher Mechanics*, passim.

70. Ferguson, *Bibliography*, p. 62.

Chapter 4

1. Neel, *Phineas Bond*, pp. 55–61, 65n; Jeremy, "British Textile Technology," pp. 39–40; Bagnall, *Textile Industries*, 1:84–86, 149–152; White, *Slater*, pp. 57–63.

2. Wallace and Jeremy, "Pollard"; Cooke, *Tench Coxe*, p. 107.

3. Boyd, *Papers of Jefferson*, 14:546–548, 15:476–477; Wallace and Jeremy, "Pollard."

4. Cole, *Correspondence of Hamilton*, p. 203.

5. For Pearce and Digges, see Pursell, "Digges"; Parsons, "Mysterious Digges"; Syrett, *Papers of Hamilton*, 9:85–87, 10:345–347, 11:241–249, 505–506, 566–567, 12:78, 135–136, 140–142, 298–299, 516–517. The five recruiters were Digges, Andrew Mitchell, Abel Buell, Enoch Edwards, and Robert Campbell. The *Manchester Mercury* of July 19, 1785, reported, with exaggeration no doubt, "that great Numbers of Agents have been a long Time employed in many of the great manufacturing Towns in engaging Workmen in various branches to settle in America." I am grateful to R. S. Fitton for this reference.

6. Wroth, *Abel Buell*, pp. 24–26.

7. Syrett, *Papers of Hamilton*, 10:345–346.

8. Ibid., 13:31–32.

9. Arbuckle, *Pennsylvania Speculator*, p. 144.

10. Syrett, *Papers of Hamilton*, 8:556–557.

11. Ibid.

12. White, *Slater*, p. 37.

13. Pennsylvania State Archives, Nicholson Papers, General Correspondence, Pollard to Nicholson, May 4, 1793.

14. Syrett, *Papers of Hamilton*, 11:245.

15. Ibid., p. 242.

16. White, *Slater*, pp. 73–75; Ware, *New England Cotton Manufacture*, pp. 124–125.

17. Davis, *Essays*, 1:399, 400, 462, 471, 480, 481, 484, 496; Syrett, *Papers of Hamilton*, 12:141–142.

18. White, *Federalists*, pp. 136. For early U.S. patent law, see PP (Commons) 1829 (332), 3:217–221; Bugbee, *Genesis;* and U.S., Department of Commerce, Patent Office, *Story*, pp. 1–7. For Pollard's fees, see Wallace and Jeremy, "Pollard," p. 416.

19. PRO, BT 5/10, pp. 427–428, for memorial of James Douglas, July 27, 1797; Syrett, *Papers of Hamilton*, 11:243.

20. Bagnall, *Textile Industries*, 1:186. Pollard's role is detailed in Wallace and Jeremy, "Pollard." Slater is treated in Gras and Larson, *Casebook*, pp. 209–230; White, *Slater*, remains the best available printed collection of primary sources on Slater. Surprisingly the definitive biography of Slater waits to be written.

21. Nettels, *Emergence*, pp. 60–64.

22. Edwards, *Growth*, p. 243.

23. Hedges, *Browns*, pp. 169–179.

24. Wallace and Jeremy, "Pollard."

25. Thompson, *Moses Brown*, chap. 10; Ware, *New England Cotton Manufacture*, pp. 22, 124–125.

26. Gras and Larson, *Casebook*, pp. 209–230.

27. Wallace and Jeremy, "Pollard," p. 419.

28. For attempts to develop Arkwright spinning before Slater arrived, see Rivard, "Textile Experiments"; White, *Slater*, p. 96.

29. Arrow, "Classificatory Notes," p. 34.

30. Jeremy, *Wansey*, p. 73.

31. Davis, *Essays*, 1:471.

32. Jeremy, *Wansey*, p. 127; Davis, *Essays*, 1:462, 481; Pursell, "Digges," p. 588.

33. Arbuckle, *Pennsylvania Speculator*, pp. 148–149.

34. Davis, *Essays*, 1:483.

35. Pennsylvania State Archives, Nicholson Papers, General Correspondence, Pollard to Nicholson, April 30, 1793.

36. Ibid., Pollard to Nicholson, May 12, 1793.

37. Davis, *Essays*, 1:490; Ware, *New England Cotton Manufacture*, p. 199; Gras and Larson, *Casebook*, p. 215.

38. Bagnall, *Textile Industries*, 1:216–217.

39. Ibid., pp. 251–258.

40. Wilkinson, "Reminiscences"; Woodbury, *History of Lathe*, pp. 89–93.

41. Bagnall, *Textile Industries*, 1:159, 255, 396–403, 420–422.

42. Ibid., pp. 368–369, 373–377, 392, 460–461, 501–505.

43. Freedley, *Philadelphia*, pp. 300–301; Ware, *New England Cotton Manufacture*, p. 27n.

44. Bagnall, *Textile Industries*, 1:148, 404–406.

45. *TRISEDI* (1864), pp. 76–84.

46. Hayes, "American Textile Machinery," p. 14.

47. Bagnall, *Textile Industries*, 1:335–346.

48. Wilkinson, "Reminiscences," pp. 107–108; Baker MSS, Slater, Almy & Brown Papers, day book (1799–1814), p. 318; Bagnall, *Textile Industries*, 1:388–389. In 1812 Blackburn (then aged 35), with a wife and four children between the ages of 5 and 13, lived at Walpole in southern Massachusetts. See USNA, RG 45, War of 1812 Papers, Marshals' Returns of Enemy Aliens. USFRC, FCC, Case Files, Paul Moody v. Jonathan Fisk, October 1820, contains a note of Blackburn's expenses for appearing as a defense witness; he was still at Walpole.

49. James Beaumont (born in 1778 at Denby between Huddersfield and Sheffield, emigrated to the United States in 1799 or 1800 at the invitation of two Yorkshire acquaintances who earlier had gone to Lebanon, New Hampshire. Between 1802 and 1823 Beaumont ran a cotton factory at Canton, Massachusetts, where he publicized mule spinning. See Hurd, *History of Norfolk*

County, pp. 945–949 and Bagnall, *Textile Industries* 1: 269–276 for Beaumont.

Samuel Ogden emigrated in 1806 and set up as a machine maker. He built mules (one of 192 spindles) and throstles and between 1808 and 1810 was in partnership in a cotton factory at Sterling, Connecticut. In 1812 he was in Providence, Rhode Island, with a wife and six children. His pamphlet on cotton manufacturing, *Thoughts*, appeared in 1815. He retired from factory work in 1832 and six years later lived in the Philadelphia area where four of his daughters worked as power loom weavers. See Bagnall, *Textile Industries*, 1:536–538; Wilkinson, "Reminiscences," p. 108; RIHS, Allen Papers, Miscellaneous Papers Box, Henry B. Lyman to Allen, n.d. but early 1860s; USNA, RG 45, War of 1812 Papers, Marshals' Returns of Enemy Aliens; Pennsylvania, Senate, *Report*, pp. 28–31. See chapter 3 above for comment on Ogden's *Thoughts*.

50. Gallatin, "Report," p. 434.

51. The introduction of the picker was associated with immigrants: with a Scot named Blair in Rhode Island and with Charles Hughes at New Ipswich, New Hampshire. Hughes, a cotton spinner, arrived unaccompanied in 1811 and a year later, at the age of 24, lived in Boston. See *TRISEDI* (1861), pp. 77, 80, 98, (1864), p. 80; Bagnall, "Samuel Batchelder," pp. 189–190; USNA, RG 45, War of 1812 Papers, Marshals' Returns of Enemy Aliens.

52. USFRC, FCC, Case Files, Paul Moody v. Medford Cotton Factory, October 1821, affidavit of Samuel Stevens.

53. Clark built Beaumont's first three cards, a drawing frame, a roving frame, and the mule. Emigrating from Paisley, he went, at Beaumont's suggestion, to Paterson, New Jersey, in 1800–1801. See Bagnall, *Textile Industries*, 1:183–184, and Trumbull, *History of Industrial Paterson*, pp. 56, 73, 74.

54. Hurd, *History of Norfolk County*, p. 946.

55. Gallatin, "Report," p. 433. The 1820 census of manufactures does not contain enough information to distinguish mule from throstle spindles.

56. If James Thomson was right in 1812, in estimating 3,000 draws per day on a 240-spindle mule, each draw making 1½ yards of yarn, then one English mule spindle made 4,500 yards per day. As late as 1832 Montgomery reckoned that a British throstle spindle turned out 4½ hanks, or 3,780 yards, per day at its maximum speed of 4,000 rpm. American spindles reached higher speeds and productivities than the British throstle. In New Jersey in 1824 spindles made between 4 and 6 hanks a day, and in 1832 8½ hanks per day. Although the kind of spindle was not specified the increase was attributed to technical improvements and these are known to have been concentrated on throstle-type spindles. British throstles were expected to achieve higher productivities for yarns under number 50 than mules, spindle for spindle, in 1832. See Rees, *Cyclopaedia*, s.v. "Manufacture of Cotton"; Montgomery, *Carding and Spinning Master's Assistant*, pp. 147–151; *McLane Report*, 2:135.

57. Hills and Pacey, "Measurement of Power," p. 27, quoting Archibald Buchanan; see also Montgomery, *Practical Detail*, p. 72. Buchanan's estimate of 1,000 mule spindles may be too high. Catling, *Spinning Mule*, p. 47 quotes one horsepower to 700 mule spindles.

Chapter 5

1. Gallatin, "Report," p. 433; *Emporium of Arts and Sciences* 1 (1812): 75; Chapman, "Fixed Capital Formation," p. 75; Henderson, *Industrial Britain*, pp. 34–35.

2. Zevin, "Growth," p. 123.

3. U.S., Department of the Interior, *Eighth Census*, 3:xix; Mitchell and Deane, *Abstract*, p. 79.

4. USNA, Census of Manufactures, 1820, M-279, roll 2, Rhode Island returns, 4.

5. EMHL, Acc. 500, Duplanty, McCall & Co. Papers, Archibald McCall to Raphael Duplanty, April 13, 1818.

6. Baines, *Cotton Manufacture*, pp. 228–237.

7. Henderson, *Industrial Britain*, p. 137.

8. USNA, RG 241, Specifications II, pp. 417–420, and Drawings, No. 1391 for the Curtis loom. RIHS, Allen Papers, miscellaneous papers box, Isaac P. Hazard to Allen, July 10, 1861; and Bagnall, *Textile Industries*, 1:289–290, for Williams. Manchester, "Recollections"; and *TRISEDI* (1869), pp. 90–92 (obituary) for Manchester. See also the recollections of Samuel Greene (grandson of Orziel Wilkinson, Slater's machine builder) in RIHS, Allen Papers, miscellaneous papers box, Samuel Greene to Allen, June 14, 1861. Greene recalled power looms built by the following (in sequence) in Rhode Island: Elijah Ormsbee, Samuel Blydenburgh (U.S. patent, September 12, 1810), John Thorp and Thomas Robinson Williams, all before Gilmour's loom of 1816.

9. EMHL, Longwood MSS, group 6, box 1, John Mitchell to Raphael Duplanty, April 6, 1811; and EMHL, Acc. 500, Duplanty, McCall & Co. Papers, John McDonald to E. I. du Pont, May 11, 1814. Both men are listed in USNA, RG 45, War of 1812 Papers, Marshals' Returns, but Murray's name is given as James. He was aged 46 in 1812 and had a wife and nine children; his trade was cotton spinning. McDonald was a 27-year-old machine maker in summer 1812; he had a wife and, having arrived in America in December 1810, reached the United States five months after Murray.

10. EMHL, Acc. 500, Duplanty, McCall & Co. Papers, McCall to Duplanty, April 13, 1814. McCall mentioned other power looms belonging to Walter Jarvis and Robert Miller (identified from McCall to Duplanty, McCall & Co., May 7, 1816) in letters to Duplanty of February 17, 19, 23, 1814. Thomas Cooper noted Siddall's loom in an article he wrote in the *Emporium of Arts and Sciences*, n.s. 3 (1814): 459.

11. It may be of some significance that Dunbarton, a county on the north side of the Clyde, was the home of Robert Miller, patentee of the wiper or cam loom of 1796, which Gilmour had seen in use. See GB Pat. 2122 and n.

37. USFRC, FCC, Case Files, Paul Moody v. Jonathan Fisk et al., October 1820, affidavit of William Gilmore (or Gilmour), October 26, 1820. Bagnall, *Textile Industries*, 1:548–549.

12. Manchester, "Recollections," pp. 71–72.

13. Bagnall, *Textile Industries*, 1:550.

14. U.S. Patent Office, *List of Patents*.

15. Appleton, *Introduction*, p. 7.

16. Trumbull, *History of Industrial Paterson*, p. 111.

17. For Lowell, see *DAB*; Gibb, *Saco-Lowell Shops*, pp. 4–14, 33–40; Mailloux, "Boston Manufacturing Company," pp. 26–58 and passim.

18. Mailloux, pp. 41–43.

19. Gibb, *Saco-Lowell Shops*, p. 8.

20. Appleton, *Introduction*, p. 19.

21. Henderson, *Industrial Britain*, pp. 84, 85, for example.

22. Mailloux, "Boston Manufacturing Company," chap. 2; North, *Economic Growth*, chap. 4.

23. The following letters from McCall to Duplanty refer to power loom weaving problems of a technical nature: December 6, 1814, July 11, August 3, 5, 13, October 6, 19, 27, December 7, 1815 (EMHL, Acc. 500, Duplanty, McCall & Co. Papers). By December 7, 1815, the race to develop a workable power loom suited to fine goods was lost.

24. Taussig, *Tariff History*, p. 30.

25. Taylor, *Transportation Revolution*, pp. 334–338.

26. Zevin, "Growth," p. 134, which shows the price of cotton sheeting halving from an 1819–1822 level of 21 cents a yard to just under 10 cents a yard in 1829.

27. EMHL, Acc. 500, Duplanty, McCall & Co. Papers, McCall to Duplanty, April 19, 26, May 5, 20, 27, June 2, 20, November 10, December 6, 23, 1814. EMHL, Longwood MSS, group 6, box 2, Thomas Siddall to Duplanty, McCall & Co., November 29, 1814, August 6, 1815. I am indebted to Rosemary Troy of the Hagley Museum for pointing out the linkage between this firm's adoption of the power loom and its government contract. That Du-

planty, McCall & Co. associated the power loom with their contract of April 13, 1814 is clear from McCall's letter to Duplanty, March 31, 1814, in EMHL, Acc. 500.

28. Bagnall, "Paul Moody," pp. 61–62.

29. The one immigrant known to have been connected with Lowell and Moody at Waltham does not deserve much credit for the development of the Waltham power loom. Thomas Charlton Faulkner, who left England in September 1814 and arrived at Waltham in February 1815, specialized in spinning technology. In a legal deposition drafted in 1821, he stated, "I learned the cotton spinning business in England first under my Father, and afterwards perfected myself in the Factory of Messrs. Adam & George Murray of Manchester where I had free access to every part of the works. I have been in many other Factories in England, in M'Connell & Kennedy's at Manchester, in Lees and Harrison's and Lees & Davies at Staley Bridge, and in others." USFRC, FCC, Case Files, Ebenezer Wild v. Paul Moody, July 1821, Faulkner's deposition for the defendant, July 24, 1821.

30. H. B. Lyman to Committee on Manufactures, May 28, 1861, *TRISEDI* (1861), p. 78, and partly reprinted in Bagnall, *Textile Industries*, 1:548–549.

31. No satisfactory study of the nineteenth-century New England textile mill managers exists. I am grateful to Helena Wright of the MVTM for supplying me with enough information to see that, contrary to the impression given by Josephson, *Golden Threads*, p. 103, some proportion of the northern New England mill agents did not hail from the Boston aristocracy.

32. Ware, *New England Cotton Manufacture*, chap. 8.

33. EMHL, Acc. 500, Duplanty, McCall & Co. Papers, McCall to Duplanty, March 31, 1815, March 6, 1816. EMHL, Longwood MSS, group 6, box 2, Thomas Siddall to Duplanty, McCall & Co., August 6, 13, 1815.

34. Appleton, *Introduction*, pp. 11–12.

35. Manchester, "Recollections," pp. 67–68, 67n. A warp protector stop motion was usually seen as essential in the power loom. It was, for example, incorporated in Miller's 1796 loom, Todd's 1803 loom, and Johnson and Kay's 1805 loom. And it was hailed as one of the merits of Siddall's loom. See EMHL, Acc. 500, Duplanty, McCall & Co. Papers, McCall to Duplanty, April 19, 1814, and *Emporium of Arts and Sciences*, n.s. 3 (1814): 459. Without warp protection, the Waltham loom was unquestionably inferior to the Gilmour or Scotch loom. According to Job Manchester, not until he modified some Waltham looms in a Connecticut factory was the Horrocks variable batten speed adopted by the Lowell mills.

Although the distinction between the Waltham and Rhode Island power looms is well known, its exact nature has hitherto been missed by modern writers on the subject. See Gibb, *Saco-Lowell Shops*, pp. 33–35, and Hedges, *Browns*, p. 182, for example.

36. Batchelder, *Introduction,* pp. 66–67.

37. The Radcliffe and Johnson patent of 1804 was GB Pat. 2771. The specification but not the drawing of Moody's U.S. patent of January 17, 1818, has survived in USFRC, FCC, Case Files, BMC v. Jonathan Fisk et al., May 1820. A comparison between the Moody and English dressers was given by William Gilmour in an affidavit of October 26, 1820, in this case. He commented on the close similarity between the Waltham dresser and the one he had seen in Miller's factory, Dunbarton, Scotland, in 1813–1814. The features distinguishing the Waltham dresser are discussed in chapter 10.

Presumably the dressers derived from Britain by Lowell and Gilmour were very similar in principle and therefore like the Clark and Rogers machines at Paterson, New Jersey in 1819, which dressed five hundred to six hundred yards of warp a day. see EMHL, Acc. 500, Duplanty, McCall & Co. Papers, John Clark, Jr. to McCall, May 17, 1819, and Clark and Rogers to Duplanty, McCall & Co., October 6, 1819.

38. Speech of Nathan Appleton in the Massachusetts House of Representatives, January 17, 1828, quoted in Bagnall, *Textile Industries*, 2:2019.

39. Ibid., p. 2016; Mailloux, "Boston Manufacturing Company," pp. 140–141; McGouldrick, *New England Textiles*, pp. 81–84, 88. Ware, *New England Cotton Manufacture*, p. 153, says that the BMC's profits were 20 percent between 1817 and 1825.

40. This was Thomas Charlton Faulkner.

41. Batchelder, *Introduction*, p. 81.

42. Ibid.

43. EMHL, Acc. 500, Duplanty, McCall & Co. Papers, William B. Leonard to Raphael Duplanty, March 30, June 17, 1819.

44. USNA, Census of Manufactures, 1820, M-279, roll 10, New York returns, 1425.

45. Baker MSS, Isaac G. Pierson Papers, Machine Watchman's Time Book, 1817, pp. 1–51, 79, 84. USNA, Census of Manufactures, 1820, M-279, roll 10, New York returns, 1409.

46. The Gilmour loom was sponsored by the Lyman MC of North Providence and the Coventry MC of Coventry (both in Rhode Island), and other firms. Their mechanics built, installed, and maintained some of the first of Gilmour's looms. One of the mechanics involved was Job Manchester who recalled some of his experiences in setting up the new power looms from 1817 through 1819. See Manchester, "Recollections," pp. 69–70.

47. Ibid., p. 71.

48. Bagnall, *Textile Industries*, 1:549; Gibb, *Saco-Lowell Shops*, pp. 41–42, 47; Lincoln, "Origin of Piece Work."

49. Bagnall, *Textile Industries*, 1:549; Lincoln, "Beginnings."

50. Woodbury, *History of Lathe*, pp. 89–93; Manchester, "Recollections"; Wilkinson, "Reminiscences," p. 109; Lincoln, "Beginnings." Lozier, "Taunton and Mason," p. 29, notes that Wilkinson's shop sired that of David Fales and Alvin Jenks in Pawtucket and that of James S. Brown in Troy, New York.

51. Spalding, "Boston Mercantile Community," examines the Waltham system as a series of promotional corporations offering Boston investors opportunities to speculate in land, water power, and manufacturing. Rosenberg, "Economic Development," p. 551, draws attention to diffusion by capital-goods firms.

52. Friends of Domestic Industry, *Report*, p. 16.

Chapter 6

1. Baines, *Cotton Manufacture*, p. 266, an estimate repeated in Ure, *Dictionary*, p. 218. Another measure is given in Rees, *Cyclopaedia*, s.v. "Copper Plate Work in Calico Printing."

2. USNA, RG 59, War of 1812 Papers, Marshals' Returns of Enemy Aliens; USNA, Census of Manufactures, 1820, M-279, roll 14, eastern Pennsylvania returns, 644; Boatman, "Brandywine Cotton Industry," p. 59; Siddall, *Siddalls*, pp. 2–4; Scharf and Westcott, *History of Philadelphia*, 3:2316–2317.

3. These plants can be identified in *McLane Report*, 2:200–201, and the British immigrants in Dixon, "Frankford's Early Industrial Development"; Martin, "Samuel Martin"; Bonner, "La Grange Print Works"; Robson, *Manufactories*, passim.

4. Zevin, "Growth," pp. 129, 137, 146; Ware, *New England Cotton Manufacture*, p. 153. For the Merrimack MC, see Gibb, *Saco-Lowell Shops*, pp. 64–74.

5. Ware, *New England Cotton Manufacture*, pp. 84, 94; Zevin, "Growth," p. 134. Americans were quite aware of the increased profit margins accruing from finishing. The Dover MC agent in 1826 reckoned that calico printing the company's cotton goods would increase profits from seven to twelve cents per yard. The following year the Ware MC agent told his directors that a $350 calender machine would add one cent to the market value of the company's cotton goods, per yard, and yield a gain in length of half a yard in each piece of fabric. See NHHS, Dover MC Letterbooks, John Williams to Hon. David Sears, July 20, 1826; MVTM MSS, Ware MC, Agent's Letterbook, S. V. S. Wilder to Lewis Tappan, March 8, 1827.

In England in the period 1814–1833, profits from calico printing apparently were small. In 1823 the profit on one piece was just under 3 percent, according to Baines, *Cotton Manufacture*, p. 356.

6. Ware, *New England Cotton Manufacture*, p. 83n. Pollock returned to Boston on May 1, 1822: USNA, Boston Passenger Lists, M-277, roll 1 (1822), 40.

7. Bagnall, *Textile Industries*, 2:2163–2175.

8. Penn, "Introduction," pp. 243–245. Penn's identification of the Taunton John Thorp with the Rhode Island inventor of ring spinning, and an English immigrant at that, seems most unlikely. The Taunton Thorp was certainly from England and in September 1825 spoke of returning there; the Rhode Island Thorp has long since been traced in genealogical records as a native New Englander. NHHS, Dover MC Letterbooks, John Williams to William Shimmin (company treasurer in Boston), August 19, 1825, Williams to William Payne (company president), September 16, 1825. For the Rhode Island inventor, see Clark, "John Thorp," pp. 86–92.

9. NHHS, Dover MC Letterbooks, Arthur L. Porter to Williams, September 9, 1825.

10. For the Dover MC (incorporated in 1824 with a capitalization of $1 million) see Hurd, *History of Rockingham*, pp. 816–819; Sloat, "Dover Manufacturing Company"; Candee, "Great Factory."

11. For Porter see *Collections of Dover H.S.* 1 (1894): 225–226; Ham, *Dover Physicians*, pp. 9–10; Hurd, *History of Rockingham*, p. 847. In 1835 Porter left New Hampshire and went to Detroit, Michigan, where he died in 1845.

12. NHHS, Dover MC Letterbooks, Williams to Shimmin, March 18, 26, August 31, 1825, Matthew Bridge (assistant agent) to Crocker, Richmond & Otis of Taunton, Massachusetts, May 25, 1825, Porter to John Schaffer, July 25, 1825.

13. Ibid., Williams to John Schaffer, August 22, 1825.

14. Ibid., Williams to Shimmin, August 19, 1825.

15. Ibid.

16. Penn, "Introduction," pp. 247–250; USNA, N.Y. Passenger Lists, M-237, roll 7 (1825), 594; PP (Commons), 1824 (51), 5:384. The Taunton MC directors, in October 1827 and February 1832, again decided to send to England for key print workers: a superintendent and a printer to take charge of the printing machines, on both occasions. Baker MSS, Taunton MC Papers, Directors' Records 1823–1844, pp. 16, 32.

17. Bagnall, *Textile Industries*, 2:2171–2173; USFRC, FCC, Naturalization Records, 2:346.

18. Bagnall, *Textile Industries*, 2:2174–2175; Musson and Robinson, *Science and Technology*, pp. 315–316.

19. Bagnall, *Textile Industries*, 2:2173; Cowley, "Foreign Colonies," pp. 168–170.

20. Bagnall, *Textile Industries*, 2:2175–2176. For Peasley's defection to Dover, see NHHS, Dover MC Letterbooks, Williams to Samuel Torrey (temporary treasurer during Shimmin's illness), October 15, 1825, at which date Peasley had left Lowell. The printing machine built by Peasley at Dover was a single color one. Ibid., Williams to Shimmin, April 14, 1827.

21. For the recruiter Swan, see NHHS, Dover MC Letterbooks, Williams to Shimmin, March 31, 1825, Bridge to Shimmin, June 27, 1825, Williams to Torrey, September 24, 1825, Williams to Mr. Burgess (of 116 Ann Street, Boston, who provided temporary quarters to immigrants on their way to Dover), October 12, 1825, Williams to Shimmin, January 26, 1827, Porter to Shimmin, September 19, 1827.

22. Ibid., Williams to Shimmin, January 26, 1827. Chadburne (or Chatburne) was described as "an excellent blue dipper" and a "great rogue." Ibid., October 5, 1827.

23. Ibid., January 26, 1827.

24. Before this, Porter went to Lowell again, in spring 1826, to gain information and possibly to hire workers. On directions from Aaron Peasley, he sought drawings of an engraving machine (obtainable from Edward Jackson of East Chelmsford), a comparison between the Mer-

rimack printery and plans for a Dover one (Porter was asked to make sketches of the Lowell printery), and an interview with a madder dyer (Eliphat Brown) for information and possibly recruitment. See ibid., Williams to Porter, February 27, March 9, 1826.

25. Ibid., Williams to Shimmin, August 19, 1826.

26. Ibid., Bridge to Porter, April 4, 1826, and Williams to Porter, April 10, 1826, for some of the final instructions that Porter took with him to England. The next few paragraphs are based on Williams's letter, so sources are cited only for quotations. Other instructions, of which no record remains, were given to Porter by the company directors when he reached Boston on his way to New York. Before he left, Porter also discussed the procurement of an engraving machine, a major objective for the company because it could free them from reliance on imported printing shells. Peasley was building a White engraving machine, and it was decided at first that Porter should try to obtain critical components, a prudent course at that point when costs and British export restrictions seemed uncertain. Ibid., Williams to Shimmin, March 29, 1826. Bridge's directions related to metal rolling and nail making.

27. Ibid., Williams to Porter, April 10, 1826.

28. Ibid.

29. Ibid.

30. Ibid., Williams to Shimmin, July 26, 1826.

31. Ibid., October 14, 1826, January 12, February 13, April 14, 1827; Williams to Porter, March 30, April 30, 1827, Bridge to Shimmin, June 30, 1827.

32. Ibid., Williams to Shimmin, April 19, 1827.

33. Bogle reached Boston, from Chorley, Lancashire, on March 9, 1827. He was then aged 41 and was accompanied by a wife (aged 34) and five children, all under 18 years of age. See ibid., Williams to Porter, March 30, 1827; USNA, Boston Passenger Lists, M-277, roll 3 (1827), 19. Bogle's annual salary is inferred from that paid to his successor which, at £300, did not exceed Bogle's.

NHHS, Dover MC Letterbooks, Porter to Shimmin, September 19, 1827.

34. These are named and identified in NHHS, Dover MC Letterbooks: Williams to Moses Paul (superintendent of one of the Dover MC mills), January 2, 1827; Williams to Shimmin, March 30, August 29, December 31, 1827; Williams to Charles W. Cutter (clerk of U.S. District Court, Portsmouth), April 24, 1828.

35. Baker MSS, Merrimack MC, Directors' Minutes 1822–1843, p. 33.

36. Bagnall, *Textile Industries*, 2:2178.

37. Ibid., p. 2179.

38. All USNA, Boston Passenger Lists, M-277, roll 3 (1827), 54. Presumably Richard Worswick was related to one of the Worswick engravers listed in Pigot, *Directory of Manchester for 1830:* John of Chatham Street and John & Co., engravers to calico printing, of 41 Back George Street. I am obliged to R. S. Fitton for this reference.

For Slater see James Payne to A. Gilman, November 25, 1882, in *CORHAL* 2 (1883): 327–328.

39. Baines, *Cotton Manufacture*, p. 438; the fall in wages was possibly linked to new printing machines.

40. USNA, Boston Passenger Lists, M-277, roll 2 (1826), 38.

41. NHHS, Dover MC Letterbooks, Williams to Greenhalgh, May 13, 1826.

42. Ibid.

43. Ibid., Williams to Shimmin, May 5, 1826 (Greenhalgh's contract; also shows that Greenhalgh knew Yates, the British-born Taunton printery manager), December 4, 1826 (list of immigrant printers at Dover). Greenhalgh also partly supplied a list of workers and their specializations at the Clitheroe works of James Thomson and Richard Chippendale. See Williams to Porter, December 22, 1826. In this letter, Williams reported a contract made with Greenhalgh under which the

latter would work for the Dover MC at $15 a week (six days) for his first year, $16 a week for his second year, and so on up to $20 a week in his last and sixth year. For Greenhalgh's Methodist beliefs see Williams to Shimmin, August 18, 1827.

James Thomson's firm at Clitheroe, in operation from about 1810 or 1811 to the 1850s, produced high-quality prints and was a leader in the industry. See Turnbull, *History*, pp. 78–81.

44. Bagnall, *Textile Industries*, 2:2179–2180.

45. Montgomery, *Carding and Spinning Master's Assistant*, p. 145. This limit on the fineness of throstle-spun yarn was also recognized in the United States; see *McLane Report*, 2:796.

46. MVTM MSS, Ware MC, Agent's Letterbook, Christopher Colt (agent) to Messrs. W. & S. Lawrence & Stone, June 16, 1829.

47. Baker MSS, Merrimack MC, Directors' Minutes 1822–1843, pp. 83, 85; ibid., Hamilton MC, Proprietors and Directors' Minutes 1824–1864, p. 38.

For mule-type stretchers, see Rees, *Cyclopaedia* s.v. "Manufacture of Cotton" and Montgomery, *Carding and Spinning Master's Assistant*, pp. 132–135; for throstle-type stretchers, see Montgomery, *Practical Detail*, pp. 63–65. The Dover agent in 1825 considered the use of a hot roller, applied after weaving, to destroy the undesirable, loose fibers left on the surface of print yarns spun on a throstle filling frame. See NHHS, Dover MC Letterbooks, Williams to George Bond, October 1, 1825.

48. NHHS, Dover MC Letterbooks, Williams to Shimmin, January 26, 1827. Before Porter was sent to Europe, Williams expressed an interest in English "self-governing Mules." Success with throstle-spun yarns, or Porter's impressions of the English self-actor mules (then not widespread), evidently dispelled this interest. Ibid., March 31, 1825.

49. See Kirby and Musson, *Voice of the People*, chaps. 2, 3.

50. NHHS, Dover MC Letterbooks, Bridge to Paul, June 6, 1826, Bridge to Shimmin, June 10, 1826.

51. Sloat, "Dover Manufacturing Company," p. 64.

52. Ibid., pp. 60, 64, 66.

53. McGouldrick, *New England Textiles*, pp. 83–85.

54. Ware, *New England Cotton Manufacture*, pp. 91–95.

55. *McLane Report*, 1:162, 168, 340; Friends of Domestic Industry, *Report*, p. 18; Montgomery, *Practical Detail* (author's annotated edition), p. 197; Turnbull, *History*, p. 431. In 1834 nearly 90 percent of the goods of the Merrimack and Hamilton MCs at Lowell were printed, according to *Annual Statistics of Lowell*, p. 22. Slater's biographer estimated that American printworks printed 120 million yards of cloth in the year ending April 1, 1836. See White, *Slater*, p. 404.

56. Baker MSS, Hamilton MC, Proprietors' and Directors' Minutes 1824–1864, Samuel Batchelder to Nathan Appleton, September 25, 1824.

57. Ibid., treasurer's report, October 9, 1827 (p. 25).

58. Ibid., report of directors' subcommittee, January 26, 1831 (p. 35), and treasurer's statement, July 23, 1831 (p. 184).

59. Ibid., Hamilton MC, Waste Book 1829–1831, for June 1831.

60. Ibid.

61. Dodd, *Textile Manufactures*, p. 57.

62. NHHS, Dover MC Letterbooks, Williams to Torrey, September 19, 1825.

Chapter 7

1. Jeremy, *Wansey*, pp. 68–69, 148–149. One woolen manager recruited in England to run an American cotton and woolen manufactory in the early 1790s had a background in worsteds. Ibid., p. 73n.; *American Museum* 2 (1787): 253.

2. Mitchell and Deane, *Abstract*, p. 189; Mann, *Cloth Industry*, pp. 47, 309.

3. Cole, *Correspondence of Hamilton*, pp. 7–11.

4. North, "New England Wool Manufacture," *Bulletin* 29 (1899): 214–216.

5. CSL, Scholfield Papers, copy of letter from John Scholfield (Stonington, Connecticut) to Arthur Scholfield, Sr., his father (Stone Edge Foot, Saddleworth, Yorkshire), August 3, 1810. For Yorkshire cloth production, see Mitchell and Deane, *Abstract*, p. 189.

6. Ibid., Arthur Scholfield, Sr. (Stone Edge Foot), to John Scholfield (Montville, Connecticut), February 17, 1802, in which the author complained about the expansion of the cotton industry into the Saddleworth district.

7. Ibid., Arthur Scholfield, Sr., to Captain Delano (of the ship *Perseverance*, Liverpool, the vessel that took Arthur, Jr., and John Scholfield to Boston), August 13, 1793.

8. Ibid., notebook containing cost calculations for making various types of cloth; the 24-page booklet was obviously compiled while the Scholfields were still in Yorkshire, for there are references to places like West Houghton, Bolton, and Rochdale for contacting merchant purchasers of their cloths. In the CSL inventory, the notebook is document 337.

9. Besides the newspaper clippings and notes made by late-nineteenth-century members of the Scholfield family (and not always accurately so), there are modern accounts of the Scholfields' emigration and contributions to the American woolen industry in Cole, *Wool Manufacture*, 1:87–97, and Rogers, "Scholfield Wool-Carding Machines."

10. For John Mayall, see North, "New England Wool Manufacture," *Bulletin* 29 (1899): 214–216; CSL, Scholfield Papers, John Mayall (Boston) to Arthur and John Scholfield (Montville), March 3, 1800; USNA, Census of Manufactures, 1820, M–279, roll 1, Maine returns, 34. Mayall, at Lisbon, Maine, then had a picker, three cards, a billy of thirty spindles, two spinning machines of forty spindles each (jennies presumably), five looms, and a fulling mill.

His capital investment was $7,500, with which he processed 1,000 pounds of wool a year.

11. Jeremy, *Wansey*, p. 124.

12. Ibid., pp. 102–103.

13. Partridge, *Practical Treatise*, p. 15.

14. CSL, Scholfield Papers, John Mayall to John and Arthur Scholfield, March 3, 1800.

15. Ibid., certificate attesting to the integrity of the Scholfields signed by Jedediah Morse (Congregational minister of Charlestown), November 22, 1793.

16. For Douglas see PRO, BT 5/10, pp. 427–429 (minutes for July 27, 1797); PP (Commons), 1806 (268), 3:221, 423–431; *Repertory of Arts*, 2d ser. 3 (1803): 316–317.

17. CSL, Scholfield Papers, Account Book of Charlestown Partnership of Arthur and John Scholfield and John Shaw, 1793–1794, s.v. dates.

18. Ibid.; EMHL, Winterthur MSS, group 3, box 17, Robert Smith (Philadelphia) to (?) E. I. du Pont, February 20, 1812 (letter listing "prices of the best West of England Cloths called London superfines, Imported by Wm. Bell from Joseph Woods & Sons, from the Year 1797 to 1810").

19. Cole, *Wool Manufacture*, 1:88–89; Davis, *Essays*, 2:277–278. The capital consisted of £10,000 in real and £80,000 in personal estate.

20. Bagnall, *Textile Industries*, 1:202–212; Bathe and Bathe, *Jacob Perkins*, pp. 13–21, 104.

21. Jeremy, *Wansey*, pp. 148–149. The peculiar features of the Hartford card may have derived from the maker's previous experience in textile machine making. Zenas Whiting, a general builder, constructed a carding machine in March 1789 for the Worcester MC, one of the early cotton spinning manufactories. If this incorporated flats as well as rollers, as cotton cards might, it could have inspired Whiting to reduce the number of rollers he installed in the Hartford woolen card. Alternatively Whiting may have been advised by the immigrant Englishmen who arrived at Hartford during or soon after the Revolu-

tion and were reported as having wild ideas; by the late 1780s their notions of the technology involved would also be obsolete. Baker MSS, Worcester Manufactory Accounts, p. 3; Cole, *Correspondence of Hamilton*, p. 8.

22. The 1790s woolen carding engine design is in Samuel Godwin's patent of 1804, GB Pat. 2766, drawings, sheet 5. See illustration 7.2.

23. CSL, Scholfield Papers, account of William Delany and Messrs. Scholfields, March 20, 1798, and drawings of machinery (CSL inventory 331–335).

24. Ibid., Account Book of Charlestown Partnership of Arthur and John Scholfield and John Shaw, May 7, 1794, Arthur Scholfield (Pittsfield, Massachusetts) to John Scholfield (Montville), November 14, 1802; Thompson, *Moses Brown*, pp. 215–216.

25. USNA, RG 45, War of 1812 Papers, Marshals' Returns of Enemy Aliens; CSL, Scholfield Papers, John Taylor (no place) to "Friends & Country Men" (Scholfields presumably), May 10, 1798.

26. Ibid.

27. Ibid., Arthur Scholfield, Sr. (Stone Edge Foot), to Arthur, Jr. (Montville), March 4, 1799. The Montville site was obtained on a 14-year lease at $30 per year. Ibid. By 1800 the Scholfield brothers were widely dispersed. Arthur and John were in Connecticut. Isaac was in Massachusetts, possibly with James. Joseph was back in England, returned from Newburyport, Massachusetts, but was waiting to sail for Boston. Benjamin was in Breconshire, Wales, "looking over an Enjin." Abraham was in Germany, on a third visit there, selling superfine woolens—spending to find markets. For all these movements see ibid., Arthur Scholfield, Sr. (Liverpool), to Arthur and John (Montville), July 27, 1800.

28. Ibid., Arthur Scholfield (Pittsfield) to John Scholfield (Montville), October 19, 1805, July 10, 1808.

29. Ibid., November 14, 1802.

30. James Standring, described as "a comb plate & spindle maker & an excellent workman," became Arthur

Scholfield's partner at Pittsfield, sometime before 1816. But he took to drink, ran the partnership into debt (two thousand dollars at one store), and ruined Arthur Scholfield's economic independence. See ibid., David Hiscock (Pittsfield) to John Scholfield (Waterford, Connecticut), December 20, 1816, July 18, 1817.

A Benjamin Standring took out a U.S. patent for a carding improvement on June 28, 1803. A Richard Gookins obtained a U.S. patent for batting wool on January 24, 1806. He may well have been the Gookins from whom Ezra Worthen, first superintendent of the Merrimack MC at Lowell, learned machine building, according to Bagnall, "Ezra Worthen," pp. 35–36.

31. CSL, Scholfield Papers, Account Book of Charlestown Partnership of Arthur and John Scholfield and John Shaw, September 14, 1793; Bagnall, *Textile Industries*, 1:259–260; Tryon, *Household Manufactures*, p. 118n; Jeremy, "British and American Yarn Count Systems," p. 360.

32. CSL, Scholfield Papers, Winthrop Earle (Leicester, Massachusetts) to John Scholfield (Montville), March 29, 1803.

33. Tryon, *Household Manufactures*, pp. 251–252.

34. Cole, *Wool Manufacture*, 1:91–92; CSL, Scholfield Papers, copy of John Scholfield to Arthur Scholfield, Sr., August 3, 1810; copy of John Scholfield's will, August 14, 1819.

In 1820 a James Scholfield was reported as running a woolen mill at Providence, Delaware County, Pennsylvania. The owner of the mill, Isaac Sharpless, noted, "He makes Goods more after the fashion & to suit the Market and probably may make out; he pays 70 Dollars a Quarter Rent." See USNA, Census of Manufactures, 1820, M-279, roll 14, Pennsylvania returns, 698.

Joseph Scholfield, son of John, Sr., was making satinet power looms in the mid-1820s at Stonington, Connecticut. At least one of these power looms he sold to one of the larger New Hampshire woolen factories. See Baker MSS, Wendell Papers, Case 11, Great Falls MC, Letters

1823–1827, Joseph Scholfield to Jacob Wendell, August 12, 1826.

35. Cole, *Wool Manufacture*, 1:113–114.

36. Coxe, *Statement*, p. 4. See appendix D for a summary of 1820 census of manufactures figures.

37. CSL, Scholfield Papers, Arthur Scholfield to John Scholfield, November 14, 1802.

38. Ibid., Lemuel Sawyer (Washington, D.C.) to John Scholfield (New London, Connecticut), January 17, 1810.

39. Ibid., Sylvester Dering (New Haven) to John Scholfield (Stonington), September 14, 1810.

40. Ibid., Jesse Hedges (Sag Harbor, Long Island) to John Scholfield (Montville), March 25, 1815. Yet another Long Islander wanted Scholfield to take one of his boys as an apprentice and teach him jenny spinning. Ibid., Jeremiah Moore (Southold, Long Island) to John Scholfield, May 8, 1815.

41. For the background to the Delaware woolen industry, see Gibson, "Fullers, Carders," and "Delaware Woollen Industry"; Hartman, "du Pont Woollen Venture." Clifford claimed 22 years' experience as a Gloucester clothier, which ostensibly meant that he had worked in the industry since the age of four, since he was 27 years old when he registered as an enemy alien in 1812. See du Pont, *Life of E. I. du Pont*, 8:291–295; USNA, RG 45, War of 1812 Papers, Marshals' Returns of Enemy Aliens.

42. Clifford's contract is printed in du Pont, *Life of E. I. du Pont*, 8:293–295.

43. Ibid., 9:72.

44. EMHL, Winterthur MSS, group 3, box 17, Wm. and Wm. W. Young to du Pont, Bauduy & Co., September 22, 1813, in the postscript of which Young defended the establishment of his woolen factory because it served a product market different from those of his fellow manufacturers, and for this reason "it is necessary to instruct our artificers." Bannister was recorded as a 35-year-old enemy alien with a wife and five children in July 1813,

when he had been in the United States for a year. See USNA, RG 45, War of 1812 Papers, Marshals' Returns of Enemy Aliens. Partridge, *Practical Treatise*, p. 15, noted Bannister as one of the most able of immigrant woolen manufacturers.

45. *McLane Report*, 2:218. For the Bancroft family see Bounds, "Bancroft's Mills."

46. *McLane Report*, 2:218.

47. Ellis, *Country Dyer's Assistant*; Bemiss, *Dyer's Companion*; and Bronson, *Domestic Manufacturer's Assistant*; Haigh, *Dyer's Assistant* (published at Leeds, 1778; London, 1778; York, 1787; London, 1800; Philadelphia, 1800 and 1810; and New York, 1813); Rauch, *Receipts*.

48. Partridge, *Practical Treatise*, pp. 20–21.

49. Leavitt, *Hollingworth Letters*, p. 27; Erickson, *Invisible Immigrants*, pp. 301–303.

50. Partridge, *Practical Treatise*, pp. 46–47.

51. Ibid., p. 6.

52. Ibid., p. 20.

53. One, James Booth, an immigrant weaver and cloth worker employed at Captain Towers' factory near Philadelphia, wrote to Clifford offering his services. See EMHL, Winterthur MSS, group 3, box 29, Booth to Clifford, October 16, 1811.

54. Partridge's origins are traced by Julia Mann in the Pasold Research Fund reprint of the *Practical Treatise*, pp. viii–xi, and his early associations in America in Dickman, "Technological Innovation." The rest of the information on Partridge in this paragraph is derived from: EMHL, Winterthur MSS, group 3, box 17, Arthur W. Magill to du Pont, Bauduy & Co., February 28, March 28, 1812; Acc. 500, du Pont Bauduy & Co. Papers, vol. 37, Journal 1810–1815, p. 80; Partridge's obituary in the *Binghamton Daily Republican*, July 21, 1858, a reference kindly supplied by E. Shepherd Griffith of Binghamton, New York.

55. For Jones, see EMHL, Winterthur MSS, group 5, box

4, William Partridge to Charles du Pont, September 9, 1820; Mann, *Cloth Industry*, p. 231.

56. EMHL, Winterthur MSS, group 5, box 4, William Partridge to Charles du Pont, September 9, 1820.

57. Hartman, "du Pont Woollen Venture," p. 19.

58. EMHL, Winterthur MSS, group 3, box 18, William H. Cox (Providence, R.I.) to E. I. du Pont, August 25, 1814.

59. RIHS, Allen Papers, "Trip, 1825." This covers the period April 13 to August 4, 1825. It differs in dating and content from Allen's four-volume MS diary for 1825, which became the basis of his *Practical Tourist* (1832). A photocopy of the diary is in the RIHS. The original is with the Factory Mutual Co., Norwood, Massachusetts. For Allen see *DAB*.

60. RIHS, Allen Papers, "Trip, 1825" pp. 16, 116.

61. Ibid., p. 52 (Gott's jennies were "of old construction & not very nice workmanship"); 53 (no power looms in Gott's mill); 116 (no power looms in Bradford-on-Avon). For hand shears and opposition to steam fulling, see ibid., pp. 30 and 56, respectively.

62. Ibid., pp. 20, 27. Jones's patents were GB Pats. 4229 (1818) and 4897 (1824).

63. RIHS, Allen Papers, "Trip, 1825" p. 33 (April 23, 1825). Ibid., pp. 29–40, for the Kirkstall Abbey factory and the Leeds machine shop.

64. Ibid., pp. 41–51 (April 26, 1825), for Hirst's mill; pp. 52–64 (April 27, 1825) for Gott and his mill.

65. Even at the Stanley Mills in Kings Stanley near Stroud, where he admired the "most beautiful blue cloths" and the "fine iron water wheels . . . at least 70 feet in length of float boards," Allen noted single cards, hand loom weaving, and a common screw press instead of hot pressing equipment. See ibid., pp. 127–128 (July 26, 1825). For the Stanley Mills, see Tann, *Gloucestershire Woollen Mills*, pp. 149–151.

The "beautiful mules for spinning of 200 spindles each, tended by one man & two women for two sides of 400 spindles," were in the Frome mill of (William ?) Shep-pard, "one of the most celebrated manufacturers in the West of England." RIHS, Allen Papers, "Trip, 1825," p. 119 (July 25, 1825).

66. Ibid., p. 40, and passim, for his notes and drawings.

67. USNA, RG 59, Consular Despatches: Leeds, T-474, roll 1, John Quincy Adams to Davy, May 26, September 8, 1824; RIHS, Allen Papers, "Trip," pp. 41, 65–69.

68. Ibid., p. 51.

69. Ibid., p. 134.

70. Ibid., pp. 134–140.

71. Ibid., pp. 34–40; USNA, RG 241, Drawings, no. 5610.

72. Douglas reportedly took plans of the gig to Philadelphia in the early 1790s, though in 1813 Dr. James Mease claimed that "for the purpose of raising the nap on cloths, handcards are at present universally used in Pennsylvania." See *Archives of Useful Knowledge* 3 (1813): 343–344. Walter Burt is supposed to have registered a gig mill in his U.S. patent of June 23, 1797, "raising a nap on cloths," and the Providence Woollen Co. reportedly had a brass wire gig in operation in 1812. See Cole, *Wool Manufacture*, 1:128–129. Certainly William Young in late 1814 was setting up a gig mill in his Brandywine factory. See EMHL, Winterthur MSS, group 3, box 18, William Young to Victor du Pont, December 29, 1814.

73. Seven American woolen spinning and weaving mills recorded possession of 43 power looms in 1820. These probably used cotton warps to make satinet. See appendix D.

John Goulding, according to one of his employees, ran some of the first woolen power looms to make broadcloth, at Dedham, Massachusetts, in 1822–1823.

Transmission of the Technologies: Some Conclusions

1. PRO, BT 1/76, ff. 33–38; USNA, RG 59, War of 1812 Papers, Marshals' Returns of Enemy Aliens. Denis Manion, a cotton spinner, spent at least five years working at McConnel & Kennedy's Manchester factory before he was

arrested for trying to emigrate to the United States in 1811. Manion, and his son-in-law, David Wooding, also a cotton spinner, brought before the Manchester magistrates on March 12, 1811, were committed for trial at the next Lancaster assizes. The next record of Manion, his registration as an enemy alien in the United States, places him at the Globe Mill, Philadelphia, September 1–9, 1814. At this date he had been in the United States for three years and three months, so he arrived early in June 1811. Clearly he succeeded in evading the machinery for the enforcement of the British prohibitory laws.

2. For Bancroft's notebooks, see EMHL, Acc. 467, nos. 986, 987.

3. For instances of secretiveness toward the census, itself a good test of the prevalence of manufacturers' secretiveness, see USNA, Census of Manufactures, 1820, M-279, roll 2, Massachusetts returns, 51, 68, 123. A more thorough testing of secretive attitudes might be undertaken in the census returns of the early nineteenth century.

III: Introduction

1. Heaton, "Industrial Immigrant," p. 520.

2. Berthoff, *British Immigrants*, p. 37.

3. Thistlethwaite, *Atlantic Community*, p. 31.

4. Jones, *Destination America*, p. 106.

Chapter 8

1. PRO, T. 47/9, 10, 11, 12, Treasury Emigration Lists from England, Wales, and Scotland, 1773–1776.

2. Campbell, "English Emigration."

3. Dickson, *Ulster Emigration*, pp. 60–81; Graham, *Colonists from Scotland*, pp. 23–89.

4. USNA, RG 45, 59, War of 1812 Papers, Marshals' Returns of Enemy Aliens.

5. Heaton, "Industrial Immigrant," and Herbert Heaton's letter to the author, May 6, 1969, in which he

wrote, "I don't find much duplication, and am glad, but sad to have your report on it." The Rhode Island returns were probably also incomplete since they named aliens from only 11 of the 31 towns of the state. See Coleman and Majeske, "British Immigrants in Rhode Island," p. 68.

6. USNA, RG 36, Passenger Lists for New York (microcopy M-237, rolls 5–15), Philadelphia (M-425, rolls 34–46), and Boston (M-277, rolls 1–6). I am very grateful to Charlotte Erickson for directing my attention to this source.

7. Erickson, "Who Were the English and Scots Emigrants?" p. 355.

8. Erickson, "Who Were the English Emigrants of the 1820s?"

9. Bromwell, *History of Immigration*, pp. 37–68. Albion, *Rise of New York Port*, pp. 465, 467, notes that Bromwell failed to distinguish Americans from aliens among the immigrants, implying that Bromwell's figures overstate the volume of immigration.

10. For pre-Revolutionary traditional craftsmen in America, see Bridenbaugh, *Colonial Craftsman*, pp. 84–89, for metal trades.

11. Bythell, *Handloom Weavers*, pp. 54, 57.

12. Baines, *Cotton Manufacture*, pp. 235, 239.

13. Bythell, *Handloom Weavers*, pp. 275–276.

14. Baines, *Cotton Manufacture*, p. 438; Bythell, *Handloom Weavers*, p. 135.

15. Baines, *Cotton Manufacture*, p. 418.

16. Bythell, *Handloom Weavers*, pp. 59–60.

17. Ibid., p. 43. In America, the skills of the hand loom weaver were published as early as 1792 in Hargrove, *Weavers Draft Book*. The author was an immigrant Irish weaver.

18. PP (Commons), 1827 (550), 5:65–73.

19. *Annual Statistics of Lowell*, p. 24; Crockett, *Account*, p. 94; Montgomery, *Practical Detail*, p. 107.

20. See USNA, Boston Passenger Lists, M-277, roll 2 (1826), 98. Silas Blair was a carder, James Laird (or Leard) a mule spinner, and Edward Joseph Quinn a dresser; at least three more of these cotton manufacturers were hand loom weavers. See NHHS, Dover MC Letterbooks, Williams to Shimmin, January 26, 1827.

21. Dickson, *Ulster Emigration*, pp. 87–97.

22. Erickson, "Who Were the English Emigrants of the 1820s?" pp. 25–26. I have recalculated the percentages of age groups given in the table on page 26 to obtain sizes of cohorts of age groups above 14 years of age as proportions of all emigrants over this age. Erickson's figures, however, include dependents as well as primary wage earners.

23. Anderson, *Family Structure*, pp. 25–28.

24. Erickson, "Who Were the English Emigrants of the 1820s?" p. 31.

25. Ibid., pp. 50–51.

26. Albion, *Rise of New York Port*, pp. 1–94; Hansen, *Atlantic Migration*, p. 105.

27. These districts, well defined by the 1830s, are described in Montgomery, *Practical Detail*, pp. 156–204. Lowell was the most important one to be developed after the War of 1812 and before 1831.

Chapter 9

1. Day, "Early Development." Day's figures come from an 1823 census that recorded only companies incorporated prior to 1820. It therefore certainly understated the number of mills built between 1809 and 1813, not least because none were incorporated in Maine or Rhode Island before 1820.

Griffin, "List," traces mills from early gazetteers and finds 155 cotton factories, 49 woolen factories, and 13 cotton and woolen factories in the period 1808–1818. Though many fewer than those in Day's article, some proportion of Griffin's was not incorporated.

2. Erickson, "Who Were the English Emigrants of the 1820s?" pp. 26–27.

3. Walsh, "Census," p. 3.

4. Freedley, *Philadelphia*, pp. 252–256.

5. See appendix D. Unless otherwise stated, the data used in this section come from USNA, Census of Manufactures, 1820, M-279 (MS Returns). For the relative importance of textiles in shifting economic growth from commerce to manufacturing in the Philadelphia region see Lindstrom, *Economic Development in the Philadelphia Region*, pp. 42–44.

6. Hazard, *Register of Pennsylvania* 1 (1828): 28.

7. Most of these specialties were listed by Philadelphia handweavers in their census returns in 1820.

8. For the Morrises, see Erickson, *Invisible Immigrants*, p. 147.

9. *McLane Report*, 2:200, 202.

10. By the mid- to late 1830s, Irishmen started to preponderate among the weavers admitted to Blockley almshouse. See McLeod, "Philadelphia Artisan," p. 83.

11. For sources on these finishing plants see chapter 6, n. 3.

12. NHHS, Dover MC Letterbooks, Williams to Shimmin, January 26, 1827. Williams was correct in his analysis; see Taylor, "Concentration." Compare this appraisal with the Ware MC's president's assessment of its future agent, Ephraim Bellows: "To be sure, he has been accustomed to run only 20 Looms & to manage only 40 persons, instead of 250 Looms & 400 persons. . . . Our corporation will lose nothing by exchanging their present inefficient Agent for Mr. Bellows, whom I consider an experienced, intelligent, active and practical Manufacturer in cotton goods; and judging from his own observations, I should think he understood well the operative part of machinery." See MVTM MSS, Ware MC, Agent's Letterbook, S. V. S. Wilder to James C. Dunn, July 26, 1827.

13. NHHS, Dover MC Letterbooks, Williams to Shimmin, January 26, 1827.

14. Ibid., April 20 and 27, 1825.

15. Ibid., January 26, 1827.

16. Ibid., Porter to Williams, September 9, 1825, in which Porter reported that the Thorps "possess some considerable knowledge of their business. . . . But how to secure the benefits of it is more than I can tell. They must be bound in bands stronger than Iron and sealed in blood."

17. Ibid., Porter to Shimmin, September 19, 1827, and Porter's memo following.

18. Ibid., Williams to Shimmin, September 22, 23, 27, 1827.

19. Ibid., May 20, 1826. Secretiveness led to double standards in America as much as in Britain. While the Dover management bitterly criticized immigrants' secretiveness, the manager told his assistant, "With reference to the Printing Establishment, it is indispensable that it should be entirely closed [to] *all* visitors & *even* proprietors except on special *occasions* which may be judged of by yourself or Dr. Porter. You will see the propriety of this rule, when you notice our numerous neighbours who boast that they follow close after us in improving on our mistakes." See NHHS, Dover MC Letterbooks, Williams to Moses Paul, October 27, 1825.

20. Ibid., April 20, 1825.

21. Harrison, *Drink and the Victorians*, pp. 37–45, 66–67, 82.

22. NHHS, Dover MC Letterbooks, Williams to Shimmin, December 29, 1827.

23. Ibid., March 15, 1828.

24. Harrison, *Drink and the Victorians*, p. 40.

25. NHHS, Dover MC Letterbooks, Bridge to Torrey, October 14, 1825.

26. Ibid., Williams to Shimmin, January 3, 1827.

27. Ibid., Bridge to Shimmin, June 18, 1827, for exchange rate.

28. Ibid., Williams to Shimmin, April 5, 1827.

29. Ibid., April 6, 1827.

30. Ibid., April 14, 1827.

31. Ibid., May 12, 1827.

32. Ibid., Bridge to Shimmin, June 16, 1827.

33. Ibid. These immigrants were Benjamin Dean and Thomas Roberts (block cutters), William Roberts (designer), William Ashcroft, George Burkett, Peter Dean, John Delaney, George Holte, John Oldham, John Ramsbottom, Thomas Read, Alexander Rogers, and William Waddington (all calico printers). Two other Delaneys (James and Thomas), Philip Roberts, and Ellis Waddington also appear in the July 1827 payroll but have not been positively identified as immigrants from the Dover MC Letterbooks or the Boston Passenger Lists. The payroll is in Baker MSS, Dover MC Papers, Payroll 1826–1829.

34. NHHS, Dover MC Letterbooks, Williams to Shimmin, April 21, July 10, 1827.

35. Ibid., July 10, 1827.

36. Ibid., Williams to Whitwell, Bond & Co., July 21, 1827.

37. Ibid., Williams to Shimmin, August 8, 1827 (the letter is marked August 2, but it follows the letter of August 7).

38. Ibid., July 26, 1827.

39. Aspinall, *Early English Trade Unions*, pp. 183–186, 214, 272, 304, 372; Hammond and Hammond, *Skilled Labourer*, pp. 94–135; Kirby and Musson, *Voice of the People*, chap. 2; Baines, *Cotton Manufacture*, p. 438. Compare also Smelser, *Social Change*, pp. 322–330; Turner, *Trade Union Growth*, pp. 57, 67–75, 103.

In the Manchester spinners' strike in summer 1818, blacklegs were called "knobsticks": Kirby and Musson, *Voice of the People*, p. 20.

The fall in Manchester calico printers' wages may have been due to the spread of the surface printing machine.

IV. Introduction

1. For card clothing exports to the United States, see PRO, BT 6/151, February 1, 1826, November 30, 1829; for exports (illegal) of rollers and spindles, PP (Commons), 1841 (400), 7:66; for exports of metal reeds, NHHS, Dover MC Letterbooks, Bridge to Shimmin, June 30, 1827; for exports of printing shells, Baker MSS, Hamilton MC, Waste Book, 1829–1831, for June 1831.

2. PRO, BT 5/2, p. 302; Wadsworth and Mann, *Cotton Trade*, pp. 461–465, 517; Mann, "Cardmaking Machine"; Dyer, "Origin," pp. 138–146; Aikin, *Description*, p. 267. Unless otherwise stated, information on American card clothing technology comes from Kittredge and Gould, *Card-Clothing Industry*.

3. Kittredge and Gould, *Card-Clothing Industry*, p. 31.

4. USFRC, FCC, Case Files, Amos Whittemore v. William Cutter, May 1810–May 1815. In 1809 the Whittemores had 55 patent machines at their West Cambridge, Massachusetts, works. The 37 then in operation, tended by 40 operatives, produced $2,000 worth of card clothing weekly, in the form of 80 dozen hand cards and 200 square feet of machine cards. Gallatin, "Report," p. 436.

5. USFRC, FCC, Case Files, Whittemore v. Cutter, 1810–1815, Notification of Special Matter on behalf of the defendant, April 26, 1810.

6. Ibid., affidavit of William F. Cutter, May 22, 1813.

7. Baker MSS, Worcester Manufactory Accounts, p. 5; Earle, *Earle Family*, p. 450, quoting Almy & Brown's letter to Pliny Earle, November 4, 1789.

8. Pliny Earle described his invention: "I was undoubtedly the first man in America, and, for aught I know, the first in the world, who made cards with a machine moving the leather side-wise, until the pricker strikes six times through the leather; it then falls back, and so continues to operate, falling back once for every six strokes of the pricker until completed." Quoted in Hurd, *History of Worcester County*, 1:appendix. I am grateful to Helena Wright of the MVTM for this reference. For Earle's sale of patent rights, see Gottesman, *Arts and Crafts in New York*, pp. 422–423.

9. USFRC, FCC, Case Files, Pliny Earle v. Shoswell Sprague, Pliny Earle v. Alphaeus and James Smith, Pliny Earle v. Henry Sargent and Isaac Southgate (all October 1814), Pliny Earle v. William Sprague (May 1815), Pliny Earle v. Artemas Dryden (October 1815).

John Scholfield was approached by one of the Smiths who wanted to know whether twilled cards were used in England before Earle's patent; had they been, Smith wanted Scholfield's deposition to defend himself against Earle. See CSL, Scholfield Papers, John Scholfield (Preston, Connecticut) to John Scholfield, Sr. (Montville), September 20, 1814.

10. EMHL, Acc. 500, Duplanty, McCall & Co. Papers, bill from Woodcock, Smith & Co. enclosed in McCall to Duplanty, February 28, 1814; ibid., Longwood MSS, group 6, box 2, Hodgson brothers' contract with Duplanty, McCall & Co., March 22, 1813.

11. For the machine life of card clothing, see Montgomery, *Practical Detail*, p. 44.

Chapter 10

1. See, for example, Gibb, *Saco-Lowell Shops*, pp. 27–29, and Strassmann, *Risk*, pp. 79–88.

2. Temin, "Labour Scarcity," pp. 277–278; Sawyer, "Social Basis," p. 269. The British reports are reprinted in Rosenberg, *American System of Manufactures*.

3. Habakkuk, *American and British Technology*, pp. 11–17.

4. Temin, "Labour Scarcity," p. 295.

5. Quoted in Saul, *Technological Change*, p. 5.

6. Salter, *Productivity*, p. 21.

7. Mailloux, "Boston Manufacturing Company," pp. 52–53.

8. Stigler, "Division of Labour."

9. Bagnall, *Textile Industries*, 2:2028.

10. Specification for Moody's filling frame patent is in USFRC, FCC, Case Files, Ebenezer Wild v. Moody, July 1821.

11. The pace-setting role of the power loom in integrated spinning and weaving was explicitly recognized by the Dover MC management: "The Dressing Room & Looms regulate the work of [the] mill," reported the company agent to the treasurer. See NHHS, Dover MC Letterbooks, Williams to Shimmin, March 18, 1825.

12. Rosenberg, "Factors," pp. 21–23. This flow-production mill layout, described in Montgomery, *Practical Detail*, pp. 15–16, was reversed by 1850; see Knowlton, *Pepperell's Progress*, p. 54. Helena Wright of the MVTM tells me that the Pepperell arrangement (with looms below and cards at the top) was followed by most New England companies after 1850, judging by her scanning of textile mill insurance surveys.

13. Taylor, *Transportation Revolution*, p. 192.

14. Edwards, *Growth*, p. 253.

15. Baines, *Cotton Manufacture*, p. 314; Mailloux, "Boston Manufacturing Company," p. 72.

16. Friends of Domestic Industry, *Report*, p. 7.

17. MVTM MSS, Ware MC, Agent's Letterbook, Bellows to Dunn, March 25, 1828.

18. Ibid., Wilder to Tappan, August 26, 1826.

19. U.S., Bureau of the Census, *Historical Statistics* 1:8, 12; North, *Economic Growth*, p. 64.

20. Friends House Library, Pritchard Letters, Case 18, Henry Newman (visitor from England) to George Newman (Wanborough, Illinois), May 20, 1833. I am grateful to Charlotte Erickson for this quotation.

21. Durability was the quality that gave Almy & Brown's shirtings the edge over English and Indian imports in the western market in 1809. See Hedges, *Browns*, p. 173.

22. Bagnall, *Textile Industries*, 2:2019, quoting Appleton's speech in the Massachusetts House of Representatives, January 17, 1828.

23. Ibid.

24. A point made in McGouldrick, *New England Textiles*, p. 31.

In recently searching some papers of the Office of the Quartermaster General, I came across what, to date, is the earliest known sample of cotton goods woven on a Waltham power loom. Enclosed in a letter from Benjamin C. Ward & Co. (the BMC's selling agents) to Callender Irvine (Commissary General) of January 22, 1821, the sample measures 31¾ by 3½ to 4 inches and is marked "Chain 48 [ends] to the inch; Filling 52." See USNA, RG 92, Coxe and Irvine Papers, 1794–1842, box 32A of three boxes of samples of cloth, 1800–1833. I am grateful to Michael P. Musick of the National Archives staff for bringing these boxes of cloth samples to my attention while I was a fellow with the Regional Economic History Research Centre at the Eleutherian Mills Historical Library, August 1979.

25. The Dover factory manager told his treasurer in Boston, "Many and great advantages result in having all the cloth in one Factory of one texture and width." NHHS, Dover MC Letterbooks, Williams to Shimmin, May 27, 1825.

26. McGouldrick, *New England Textiles*, p. 31.

27. Mailloux, "Boston Manufacturing Company," p. 112; Appleton, *Introduction*, pp. 12, 16.

28. Bagnall, *Textile Industries*, 2:2020, quoting Appleton's speech of January 17, 1828, in the Massachusetts House of Representatives.

29. David, "Learning by Doing," p. 529.

30. Ibid.

31. Mailloux, "Boston Manufacturing Company," p. 80.

32. Bagnall, *Textile Industries*, 2:2026.

33. Moody's patents are as follows: warper, March 9, 1816; dresser, January 17, 1818; double speeder, April 3, 1819; filling frame, May 6, 1819; improved double speeder, December 30, 1820; another double speeder improvement, January 19, 1821; two more double speeder improvements, February 19, 1821; and another filling

frame patent, February 19, 1821. Stimpson's assignment of his temple patent to the BMC is mentioned in Mailloux, ''Boston Manufacturing Company,'' p. 80. For the identification of the warper, see Jeremy, ''Innovation,'' p. 65.

34. Gibb, *Saco-Lowell Shops*, p. 47.

35. These actions may be followed in the Circuit Court Case Files of Massachusetts and Rhode Island in the USFRC, Waltham, Massachusetts. By date of action, they are: BMC v. Jonathan Fisk and Ephraim and Jacob Stevens, October 1820–May 1821 (dresser); Moody v. Fisk et al., October 1820–October 1822 (double speeder); Moody v. Medford [or Medway] Cotton Factory, October 1821 (double speeder); Moody v. Fisk et al., October 1822; Ebenezer Wild v. Paul Moody, July–November 1821 (double speeder and filling frame); all in the Massachusetts Circuit Court records. And Moody v. Daniel Lyman, June 1822 (double speeder); Moody v. Palemon Walcott, June 1822 (double speeder); Moody v. William Harris, June 1822 (double speeder), all in the Rhode Island Circuit Court records. Further searches in the court records might unearth more BMC litigation.

36. See Ware, *New England Cotton Manufacture*, chap. 7, and Porter and Livesay, *Merchants and Manufacturers*, chap. 2.

37. Temin, ''Labour Scarcity,'' p. 291.

38. Strassmann, *Risk*, pp. 87–88; Habakkuk, *American and British Technology*, pp. 56–59.

39. Williamson, ''Optimal Replacement,'' pp. 1321–1322. McGouldrick, *New England Textiles*, pp. 223–232, calculates that the average service life of textile equipment in the period 1836–1890 for all machines was 35 years.

40. Gibb, *Saco-Lowell Shops*, pp. 42, 47.

41. Baker MSS, Wendell Collection, Case 11, Williams & Wendell letters 1813–1826, Isaac Wendell to Jacob Wendell, endorsed ''November 1821.''

42. Cole, ''Evolution of the Foreign-Exchange Market,'' pp. 406–407n.

43. Montgomery, *Practical Detail*, p. 110.

44. Patent specifications have survived in the First Circuit Court records as follows: Moody's dressing frame (January 17, 1818) and the transfer of the patent to the BMC (January 14, 1819) in BMC v. Fisk et al., May 1821. Moody's double speeder (April 3, 1819) in Moody v. Fisk et al., October 1820. Moody's filling frame (May 6, 1819) in ibid. Moody's double speeder (January 19, 1821) in Wild v. Moody, July 1821. Moody's filling frame (February 19, 1821) in ibid. For the warper, see Jeremy, ''Innovation,'' pp. 64–65. For the dead spindle, see Montgomery, *Carding and Spinning Master's Assistant*, p. 177; *Practical Detail*, pp. 67, 70–73.

45. Baker MSS, Proprietors of the Locks & Canals Co., Samuel C. Oliver's Forgings List.

46. Montgomery, *Practical Detail*, pp. 70, 85–90; Jeremy, ''Innovation,'' pp. 64–66.

47. Manchester, ''Recollections,'' p. 67n.

48. *McLane Report*, 1:80.

49. Ibid.

50. McGouldrick, *New England Textiles*, p. 114. Gibb, *Saco-Lowell Shops*, pp. 50–51, did not think that the BMC's accounting methods were in very much error through failure to include all overhead costs. On the other hand the Ware MC agent in 1826 omitted depreciation in his calculations of profit and loss. He wrote that production costs included ''all expenses and charges for labour, stock, fuel, loss in weight, oil, belts, repairing of machinery, sizing, packing, sacks, watch, dyestuffs, transportation, agencies' commissions on purchasing & sales.'' See MVTM MSS, Ware MC, Agent's Letterbook, Wilder to Tappan, March 8, 1826.

51. The BMC met no substantial product competition. See Ware, *New England Cotton Manufacture*, pp. 69–73, and McGouldrick, *New England Textiles*, pp. 30–34.

52. Patrick Tracy Jackson's affidavit, dated July 25, 1821, in Wild v. Moody case file. Jackson was appointed BMC

agent on January 1, 1816, at an annual salary of $3,000, according to Bagnall, *Textile Industries*, 2:2012.

53. Jackson's affidavit of July 25, 1821, and Thomas Charlton Faulkner's deposition of July 34, 1821, in Wild v. Moody case file.

54. Paul Moody's deposition of July 25, 1821, in Wild v. Moody case file.

55. *Boston Gazette* (*Extra*), March 20, 1815, quoted in Mailloux, "Boston Manufacturing Company," pp. 85–86.

56. Montgomery, *Carding and Spinning Master's Assistant*, p. 177, and *Practical Detail*, pp. 67, 70–73; Batchelder, *Introduction*, pp. 73, 168–169. Dead spindles were patented in Britain in 1793 and 1804.

57. Mailloux, "Boston Manufacturing Company," pp. 72, 85, 89. For later experience of the problem of balancing processes and machines see Penrose, *Theory*, pp. 68–71; Jeremy, "Innovation," p. 72.

58. Bagnall, *Textile Industries*, 2:2025.

59. Jackson's affidavit, July 25, 1821.

60. Ibid.

61. Ibid. A copy of Moody's contract, dated November 1, 1813, is appended to Jackson's affidavit.

62. Strassmann, *Risk*, p. 83. John W. Lozier in his dissertation, "Taunton and Mason," pp. 23–24, chides me for underestimating the importance of textile inventions south of Massachusetts because, he claims, only 4 percent of American patents issued for cotton and woolen manufacturing between 1810 and 1837 came from the "large mill towns of the north." But "large mill towns" is too narrow a measure to be useful; patents for the whole of northern New England should have been included because the region's mill towns owed their existence to the technology represented by a handful of Waltham patents. In any case it is misleading to draw conclusions about the success of inventions from the numbers of patents awarded. As Schmookler emphasized, patents provide some kind of measure of inventive effort but not of tech-

nical or commercial success. And the significance of American patents is still harder to gauge because of two other difficulties: most of the specifications are no longer extant, and the American patent system in the period of 1793–1836 granted patents after formal registration and the payment of a thirty dollar fee, leaving the patentee to defend his claim to invention and his monopoly in the courts. It is therefore very likely that there was more room in the United States than in Britain for technically, and commercially, less successful inventions to be patented. See PP (Commons) 1829 (332), 3:217–221; Bugbee, *Genesis*, passim; White, *Federalists*, pp. 136–138.

63. Montgomery, *Practical Detail*, pp. 85–90, plate 6; Jeremy, "Innovation," pp. 63–66.

64. Moody's dresser patent, January 17, 1818, specification.

65. MVTM MSS, Ware MC Agent's Letterbook, Wilder to Tappan, June 6, 1827.

66. Moody's deposition, July 25, 1821.

67. Moody's double speeder patent of January 19, 1821.

68. Jackson's affidavit, July 25, 1821.

69. For mule-type stretchers, see Montgomery, *Carding and Spinning Master's Assistant*, pp. 132–135. They were used to produce a finer and more uniform yarn.

70. Appleton, *Introduction*, p. 10; Gibb, *Saco-Lowell Shops*, p. 37.

71. Moody's deposition, July 25, 1821.

72. Moody's filling frame patent, May 6, 1819, specification.

73. Mailloux, "Boston Manufacturing Company," p. 80.

74. USNA, Census of Manufactures, 1820, M-279, roll 16, Maryland returns, 22. Ware, *New England Cotton Manufacture*, p. 149 says that a 12-hour day was common; this basis was used to reduce the Patapsco's daily production per loom of 20 yards to an hourly figure.

75. Gibb, *Saco-Lowell Shops*, p. 739, n. 26; Mailloux, "Boston Manufacturing Company," p. 81. Draper's pat-

ents for loom temples were dated June 7, 1816, and April 1, 1829. The former is in USNA, RG 241, Specifications 3:525–533, Drawings, No. 2608; the latter is in ibid., Specifications 8:365–366, and Drawings, No. 5419. Montgomery, *Practical Detail*, pp. 102, 107.

76. Moody's filling frame patent, February 19, 1821, specification. For early American stop motions, especially those in drawing and twisting frames, see Jeremy, "Innovation," pp. 63–72.

77. USNA, Census of Manufactures, 1820, M-279, roll 2, Massachusetts returns, 33.

78. Mailloux, "Boston Manufacturing Company," p. 102.

79. Montgomery, *Carding and Spinning Master's Assistant*, p. 146.

80. Baines, *Cotton Manufacture*, p. 438.

81. PP (Commons), 1833 (690), 6:319.

82. Baker MSS, BMC Papers, vol. 80 (Wages, 1817–1818), f. 34; James Cowan's deposition, May 19, 1821, in Moody v. Medford Cotton Factory case file.

83. Baker MSS, BMC Papers, vol. 80, f. 117; USNA, RG 45, War of 1812 Papers, Marshals' Returns of Enemy Aliens.

84. Baker MSS, BMC Papers, vol. 80, f. 53; USNA, RG 45, War of 1812 Papers, Marshals' Returns of Enemy Aliens.

85. Gitelman, "Waltham System," p. 233. See Layer, *Earnings*, for New England cotton mill earnings after 1825.

86. Gitelman, "Waltham System," p. 235.

87. Baker MSS, BMC Papers, vol. 80, ff. 53, 101, vol. 82 (Wages, 1820), pp. 285, 314.

88. Dublin, "Women at Work," pp. 68–71.

89. NHHS, Dover MC Letterbooks, Williams to Shimmin, May 5, 1827; Williams made a similar complaint in ibid., Williams to George Bond, October 1, 1825, quoted in Jeremy, "Innovation," p. 47.

90. Dublin, "Women at Work," p. 73.

91. Ibid., p. 72.

92. Jeremy, "Innovation," p. 47.

93. Gersuny, "Devil in Petticoats," p. 142; cf. Pollard, *Genesis*, pp. 181–197.

94. Ware, *New England Cotton Manufacture*, chap. 9; Gersuny, "Devil in Petticoats."

95. Ware, *New England Cotton Manufacture*, p. 254.

96. Gersuny, "Devil in Petticoats," pp. 138–140.

97. *New Statistical Account of Scotland*, 6:143.

Chapter 11

1. USNA, Census of Manufactures, 1820, M-279, roll 2, Rhode Island returns, 7, 16, 17, 19, 20, 24, 27, 34, 49, 90, 91, 94, 95, 96, 104, 108, 109.

2. Stigler, "Division of Labour," pp. 187–191.

3. USNA, Census of Manufactures, 1820, M-279, roll 2, Rhode Island returns, 7, 16, 17, 19, 20, 24, 27, 34, 49, 90, 91, 94, 95, 96, 104, 108, 109.

4. Ware, *New England Cotton Manufacture*, p. 74. See also the glossary and Nichols and Broomhead, *Standard Cotton Cloths*, for technical specifications and samples.

5. Appleton, *Introduction*, p. 13.

6. For the tariff, see Taussig, *Tariff History*, pp. 16–36, and Potter, "Atlantic Economy," pp. 253–265, a study that uses quinquennial average figures but, in the case of cotton, does not distinguish fabric types.

7. *McLane Report* 1:941.

8. EMHL, Acc. 500, Duplanty, McCall & Co. Papers, McCall (Philadelphia) to Duplanty (on the Brandywine), January 27, 1819.

9. Baker MSS, Taunton MC Papers, Directors' Records 1823–1844, p. 16.

10. *DAB*, s.v. "Aza Arnold" (1788–1865). At the time he invented his differential, Arnold appears to have been working in the Pawtucket machine shop of Pitcher and Hovey.

11. RIHS, Allen Papers, Miscellaneous Papers Box, Aza Arnold to Allen, September 4, 1861. Arnold's differential

gear patent survives in two repositories. The drawing is in ibid.; the specification is in USNA, RG 241, Patent Reissues 2:157–159.

12. Manchester, "Recollections," pp. 72–75. Clark & Rogers at Paterson, New Jersey, were also building semi-twill power looms in 1819. See EMHL, Duplanty, McCall & Co. Papers, Clark & Rogers to Duplanty, McCall & Co., October 6, 1819. Samuel Batchelder introduced twill power looms to the Waltham system when he moved from New Ipswich, New Hampshire, to the Hamilton MC at Lowell, Massachusetts, in 1825. See Baker MSS, Hamilton MC Papers, Proprietors' and Directors' Minutes, 1824–1864, Batchelder to Nathan Appleton, September 25, 1824, inside front of volume.

13. USNA, Census of Manufactures, 1820, M-279, roll 2, Rhode Island and Massachusetts returns; Friends of Domestic Industry, *Report*, p. 16.

14. *McLane Report*, 1:300, 340, 934, 939, 970; Ware, *New England Cotton Manufacture*, pp. 301–302.

15. USNA, Census of Manufactures, 1820, M-279, roll 2, Rhode Island and Massachusetts returns; Friends of Domestic Industry, *Report*, p. 16.

16. Sale of stock provided Massachusetts- or Waltham-type mills with almost all their initial finance in the 1820s; by 1829 the most important group of stockholders was made up of merchants, who held over 40 percent of the equity in six such companies. See Davis, "Stock Ownership," pp. 206–207.

17. Ware, *New England Cotton Manufacture*, p. 139.

18. RIHS, Arnold Papers, box 1, Correspondence 1820–1855, I. P. Hazard (South Kingston, Rhode Island) to Arnold (Somersworth, New Hampshire), July 3, 1827.

19. For George Danforth, see Lozier, "Taunton and Mason," pp. 161–162. For an account of his speeder's invention, see RIHS, Allen Papers, Miscellaneous Papers Box, Silas Shepard to Allen, April 26, 1862.

20. Plate speeders, made by Godwin, Rogers & Co., were purchased by the Ramapo factory (in Rockland County, New York) operated by Isaac G. Pierson. See Baker MSS, Isaac G. Pierson Papers, Accounts 1817–1834 and Cotton Mill Purchases 1823–1828, passim.

21. CSL, Sanseer MC Papers, Gilbert Brewster (New Hartford, New York) to John R. Watkinson (Middletown, Connecticut), November 3, 8, 1824. Brewster's Eclipse speeder, patented April 18, 1829, is described in the *Franklin Journal*, 2d ser. 4 (1829): 58. Whereas in his speeder Danforth replaced the flyer with twist tubes and, for the same purpose, the Paterson shop employed friction plates, Brewster substituted belts for the flyer, to insert twist in the roving. By 1840 the Eclipse speeder was preferred above the other two because of its greater productivity and its simplicity of construction, according to Montgomery, *Practical Detail*, pp. 60–62.

22. Montgomery estimated that in a week, one can of the roving can frame could prepare roving for seventy jenny spindles, and one spindle of the fly frame could prepare roving for one hundred jenny spindles, but that one tube of the Taunton tube frame could make roving for five hundred jenny spindles. *Carding and Spinning Master's Assistant*, pp. 131–132.

23. Friends of Domestic Industry, *Report*, p. 16.

24. Ware, *New England Cotton Manufacture*, p. 199.

25. MVTM MSS, Ware MC, Agent's Letterbook, Wilder to Tappan, June 6, 1827.

26. Ibid.

27. Ibid., Anthony Olney to John Elliot, July 14, 1826.

28. Ware, *New England Cotton Manufacture*, p. 243. See EMHL, Acc. 500, Duplanty, McCall & Co. Papers, McCall to Duplanty, December 31, 1814, a letter that cites rates as follows: "At the Globe Mill [Philadelphia] they have heretofore paid 70 cents pr. hundred hanks, upon small single Mules of 160 spindles—this was the price for No. 8 to 30, but the numbers generally worked were 8 to 10—the spinners were not generally well supplied with rovings. In a few weeks they will have a new set of mules of 228 to 268 spindles, when they intend to reduce the price

to 50 cents for the single & 35 or 40 cents for the double mules, which, i.e. 50 & 35, is the same as is paid at Baltimore."

29. Baker MSS, Nashua MC Papers, Treasurer's Letterbook 1825–1828, directors' report, June 2, 1824; and Thomas Searle to Gay, March 23, April 28, 1827, June 9, 1828. Ibid., Letterbook 1825–1835, Asher Benjamin to George Searle, February 20, 1827.

30. RIHS, Allen Papers, Miscellaneous Papers Box, Thomas J. Hill to Allen, June 3, 1876 and James S. Brown (Pitcher's partner after Gay left Pawtucket, Rhode Island, for Nashua, New Hampshire) to Allen, June 2, 1876. Baker MSS, Nashua MC Papers, Letterbook 1825–1835, Gay to Oliver and Nathan Rodgers, March 10, 25, 1828: $4 a spindle was the rate for mules up to 200 spindles in size; over 200 spindles $2.25 per spindle was charged for building, including the patent fee.

31. Ibid., Gay to Crocker & Richmond & Co., December 29, 1830. This letterbook contains numerous other letters referring to Ira Gay's self-actor mule. By 1840 the Nashua self-actor had the reputation of being "imperfect and complex" and therefore seemed unlikely to be adopted, according to Montgomery, *Practical Detail,* pp. 81–82.

32. Navin, *Whitin Machine Works,* p. 35. See also Lozier, "Taunton and Mason," pp. 271–281, 352–372.

33. EMHL, Acc. 500, Duplanty, McCall & Co. Papers, William B. Leonard to Duplanty, March 8, 30, June 17, 1819. Leonard took out two spinning patents on January 9, 1819, but neither specification has survived.

34. CSL, Sanseer MC Papers, Gilbert Brewster to John R. Watkinson, November 3, 8, 1824. The Rogers' patent of August 28, 1824, is in USNA, RG 241, Specifications 5: 181, and Drawings, No. 3931.

35. For Charles Danforth, see *DAB;* for his family see Lozier, "Taunton and Mason," pp. 161–162.

36. Baker MSS, Isaac G. Pierson Papers, Work Ledger 1828–1833, pp. 4–16. Danforth's cap spindle patent of

September 2, 1828, is in USNA, RG 241, Specifications 7:231–232, and Drawings, No. 5214.

37. Baker MSS, Isaac G. Pierson Papers, Case 7 (Miscellaneous Papers), inventory of December 13, 1833.

38. Montgomery, *Practical Detail,* p. 70.

39. Clark, "John Thorp," and *DAB* for John Thorp (1784–1848).

40. For Thorp's ring spindle patent of November 20, 1828, see USNA, RG 241, Specifications 7:87–89, and Drawings, No. 5279. Thomson's British patent is No. 3202 (February 7, 1809).

41. RIHS, Arnold Papers, box 1, Correspondence 1820–1855, Samuel Greene (Providence, R.I.) to Arnold (Fall River, Mass.), December 15, 1830.

42. MVTM MSS, Ware MC, Agent's Letterbook, Wilder to Tappan, November 4, 1826.

43. USFRC, FCC, Case Files, Aza Arnold v. Proprietors of the Locks and Canals, May 1836; RIHS, Allen Papers, Miscellaneous Papers Box, Arnold to Allen, September 4, 1861.

44. USFRC, FCC, Case Files, Taunton MC v. Thomas Hurd, September 1827. For the adoption of the Taunton tube speeder by northern New England cotton companies, see Baker MSS, Newmarket MC Papers, Treasurers' Letterbook 1827–1830, letters to Hacker, Brown & Co., January 4, 1828, to S. A. Chase, January 28, 1828, and to John Smith & Co. (local makers of the Taunton speeder) of Salem (Massachusetts or N.H.?), May 9, 1828.

45. Baker MSS, Nashua MC Papers, Treasurer's Letterbook 1825–1828, directors' report, June 2, 1824 and Thomas Searle to Ira Gay, March 23 and April 28, 1827.

46. Baker MSS, Hamilton MC Papers, Proprietors' and Directors' Minutes, 1824–1864, Batchelder to Nathan Appleton, September 25, 1824, inside front of volume.

47. NHHS, Dover MC Letterbooks, Williams to David Wilkinson & Co., November 2, 1825.

48. For Gay see Baker MSS, Nashua MC Papers, Treasurer's Letterbook 1825–1828 and Letterbook 1825–1835. For Arnold's work at the Great Falls MC, Somersworth, New Hampshire, in the late 1820s, see RIHS, Arnold Papers, boxes 1, 2, and Arnold's Notebook, 1828–1836.

49. NHHS, Dover MC Letterbooks, Williams to Shimmin, May 12, 1826.

50. *McLane Report* 1:367, 402. Possessing an iron furnace, where the machinists of numbers of manufacturing companies necessarily left their machinery patterns for days or weeks, Leach was in a good position to borrow and share technology, either openly or covertly. Certainly he was suspected of doing something of the sort. In 1826 the Dover MC agent advised his treasurer that he had just learned that Leach was a director in the Saco MC and therefore "they may get castings cheaper than to give $10,000 for use of patterns, i.e. our patterns should be looked after." See NHHS, Dover MC Letterbooks, Williams to Shimmin, July 22, 1826.

51. Zevin, "Growth," p. 146.

52. Chandler, *Visible Hand,* p. 241.

53. Montgomery, *Practical Detail,* pp. 124–125.

Chapter 12

1. Cole, *Wool Manufacture*, 1:179–181. Unless otherwise stated, the rest of this section on household woolen manufacturing comes from Cole.

2. See appendix D for statistics of the 1820 census of manufactures. For the subsequent movement of the woolen industry into the Midwest, see Crockett, *Woollen Industry*.

3. Cole, *Wool Manufacture*, 1:194.

4. Allen A. Fannin (Brooklyn, New York) to the author, January 30, 1971.

5. Tryon, *Household Manufactures*, p. 118n. Tryon's source must be mistaken in defining a knot as 20 threads; see Jeremy, "British and American Yarn Count Systems," p. 360. My calculation is based on a 40-thread knot.

6. Crump, *Leeds Woollen Industry*, p. 276.

7. Oliver Barrett, *Domestic Roving and Spinning Machine,* April 17, 1812.

8. Allison's patents were dated April 27, 1812, March 3, 1813, and June 28, 1814.

9. *Emporium of Arts and Sciences*, n.s. 1 (1813): 461–463.

10. See U.S., Patent Office, *List of Patents, 1790–1836*, for a chronological listing of inventions.

11. *Archives of Useful Knowledge* 2 (1812): 105–107.

12. Ibid.

13. A full treatment of American innovations in the household spinning wheel is Esposito, "Development of the Wool Spinning Wheel."

14. The patentees were as follows. New York: Samuel G. Dorr (1792)**; Beriah Swift (1806)**; Russell Dorr (1807)**; Stimson Stewart, Ebenezer Hovey & James Henderson (1808); Beriah Swift (1810); Jesse Molleneaux (1811)*; Eleazer Hovey (1811); Lemuel Dickerman (1812); and George Book (1812). Vermont: Samuel Kellogg (1795); David Dewey (1809)**. Massachusetts: Liberty Stanley (1803); Friend B. Kellogg (1805); Ebenezer Stowell (1808); Ezra Willmarth (1811); Benjamin Cummings (1811). Connecticut: Ebenezer Sprague (1810); George C. Kellogg (1811); Stephen Treadwell (1812); Gershom Bostwick (1812). Rhode Island: William Stillman (1812). New Jersey: Enoch Burt (1807). Kentucky: Walter Kennedy (1812). **Specifications and drawings in USNA, RG 241. *Specifications only in ibid.

15. USNA, Census of Manufactures, 1820, M-279, roll 4, Connecticut returns, 40 (Kellogg), 83 (Stillman). Ibid., 101 for a Hovey shear. Ibid., roll 8, New York returns, 1055, 1148, for Parsons shears. Ibid., roll 20, Indiana returns, 11, for a Stanley & Dewey shear.

16. Cole, *Wool Manufacture*, 1:101–107. A ring doffer was patented in Britain by William Davis, engineer of Bourne, Gloucestershire, and Leeds, on May 7, 1825. Since at least six weeks elapsed between patent application and en-

rollment in England, Davis must have been working on the device concurrently with, and presumably independent of, Ezekiel Hale. See GB Pat. 5159.

17. Among these was the Indian Head Woolen Factory at Nashua, New Hampshire, whose agent reported, on December 26, 1825, "It is my intention to make the necessary experiment of our new *roving condenser* with one sett of Cards bought of Hovey, Stimpson's new approved Loom and a common Jenny now making in our Machine Shop." See Baker MSS, Textron, Inc. Papers, Indian Head Factory, Directors' Minutes 1825–1830, p. 28.

When Zachariah Allen, in the summer of 1825, told a woolen manufacturer at Louviers in France about "the new plan of making roving from the doffer," it "produced an effect quite astonishing and Mr. Duperier observed that the Americans were already becoming their teachers in making cloth." RIHS, Allen Papers, "Trip, 1825" p. 105 (June 15, 1825).

18. For Brewster's spinning machine, see Partridge, *Practical Treatise*, p. 46; Montgomery, *Practical Detail*, p. 81; Bagnall, *Textile Industries*, 2:1391, 1515. For glimpses of Brewster's activities, see CSL, Sanseer MC Papers, Gilbert Brewster (New Hartford, New York) to John R. Watkinson (Middletown, Connecticut), November 3, 8, 1824.

19. Besides Cole, *Wool Manufacture*, 1:chap. 6, see Calvert, "Technology," and Riznik, "New England Wool-Carding and Finishing Mills."

20. Cole, *Wool Manufacture*, 1:79–82.

21. Partridge, *Practical Treatise*, pp. 25–26.

22. Ibid., p. 26.

23. George Adey, GB Pat. 4158 (1817), printed specification, p. 2.

24. Partridge, *Practical Treatise*, p. 83.

25. Ibid., p. 84. A broadside of 1810, advertising a new shear marketed by Hall & Weld of Union Street, Boston, acknowledged the quality problems of American shears. See Hall & Weld broadside, October 3, 1810, EMHL, Winterthur MSS, group 3, box 17.

26. Vickerman, *Woollen Spinning*, pp. 241, 244.

27. USNA, Census of Manufactures, 1820, M-279, roll 17, New Jersey returns, 94.

28. Cole, *Wool Manufacture*, 1:204.

29. Mitchell and Deane, *Abstract*, p. 495.

30. Temin, *Causal Factors*, pp. 14–15, quoting Paul David's estimates which are based on rising productivity in agriculture and the movement of workers from agriculture into other occupations, especially commercial ones, before 1840.

31. Taussig, *Tariff History*, pp. 40–41, 91–92, 100; Potter, "Atlantic Economy," pp. 265–269.

32. *McLane Report*, 2:70, 82.

33. Ibid., 1:78.

34. Cole, *Wool Manufacture*, 1:198–199, 204–206.

35. USNA, Census of Manufactures, 1820, M-279, roll 3, Vermont returns, 242; roll 5, New York returns, 283; roll 10, New York returns, 1379.

36. *McLane Report*, 1:692–693, 2:70, 82.

37. James Beaumont of Canton, Massachusetts, and Abraham Marland of Andover, Massachusetts, were two of these immigrants, both of whom engaged in cotton manufacturing before switching to woolen manufacturing. For Marland see Bagnall, *Textile Industries*, 1:335–343. Ibid, p. 603, suggests that satinet was being imported to the United States in 1811.

38. Cole, *Wool Manufacture*, 1:199–201, for satinets.

39. USNA, Census of Manufactures, 1820, M-279, roll 1, New Hampshire returns, 29 (Hillsboro: 4 power looms); roll 2, Massachusetts returns, 63 (Walcott MC, Southbridge: 10 power looms), 69 (Millbury: 4 power looms); roll 4, Connecticut returns, 101 (John G. W. Trumbull and John Breed of Jewett City: 2 power looms), 216 (Pameacha MC, Middletown: 8 power looms); roll 8, New York returns, 1019 (Oriskany MC, Whitestown, Oneida Co: 13

power looms); roll 10, New York returns, 1317 (John Barrow & Sons, Ninth Ward, New York City: 2 power looms driven, like their two 240-spindle mules, by a 10-horsepower steam engine).

40. Porter and Livesay, *Merchants and Manufacturers*, pp. 26–27.

41. CSL, Scholfield Papers, Rensselaer Havens (New York) to John Scholfield (Stonington, Connecticut), October 7, November 13, 1812, May 21, 1813.

42. Ibid., November 27, 1812.

43. Ibid., November 13, 1812. See also Havens's letter of February 1, 1813, in ibid.

44. Baker MSS, Hamilton Woolen Co., Southbridge, Massachusetts, Papers, letters of selling agents to company, June 11, 1828. See also Tiffany & Sayles (Boston) to Samuel Hitchcock (Southbridge, agent), May 3, 9, 1830.

45. EMHL, Acc. 500, vol. 37, Woolen Journal 1810–1815 (du Pont firms at the Louviers mill on the Brandywine), entry for March 31, 1812.

46. Ibid., entries for April 15, 1812, November 30, 1813, March 25, 1814. For construction of handshears see Rees, *Cyclopaedia*, s.v. "Woollen Manufacture."

47. USNA, Census of Manufactures, 1820.

48. Gibson, "Delaware Woollen Industry," p. 97.

49. USNA, Census of Manufactures, 1820, M-279, roll 4, Connecticut returns, 101, 216.

50. Partridge, *Practical Treatise*, p. 46. On the other hand at least one manufacturer was very satisfied with his Brewster machine, which spun about 50 pounds of wool a day into 5½ run warp, or 112 pounds a day into 3 run weft—equivalent to 440,000 yards and 537,600 yards a day, respectively. Equal to about 2,000 yards per spindle on a 250-spindle machine, this was 30 to 40 times the amount that a billy could turn out in slubbing (not the best comparison because the billy made a much coarser end, but enough to demonstrate the efficiency of the Brewster spinner).

Perez B. Wolcott of Somersworth, New Hampshire, who had four Brewster machines running at this rate of production, signed a testimonial commending the machine, because, in his eight years' use of Brewsters, it not only spun "as good quality as Jennies but for ¼ of the expense." However the Sanseer MC, which was trying to promote Brewster's machine, was obliged to send a mechanic (Elisha J. Abel) to New Hampshire to diagnose faults appearing in Wolcott's models. See CSL, Sanseer MC Papers, Elisha J. Abel (Amesbury, Mass.) to Richard Hubbard (Middletown, Connecticut), Secretary of the Sanseer MC, November 10, 1824. Wolcott's testimonial was copied into this letter.

51. Riznik, "New-England Wool-Carding and Finishing Mills," pp. 48–49, quoting Tryon, *Household Manufactures*, p. 170.

52. MVTM MSS, Ware MC, Agent's Letterbook, Wilder to Tappan, June 8, 1826.

53. Ibid., Olney to John Elliot, July 14, 1826; Wilder to Tappan, August 26, November 4, 1826, for Olney's inventions of a cotton dresser, a double cylinder carding machine, and a self-acting temple. At this point Olney's machine-making experience included the invention of a musket barrel lathe (joint patent with Daniel Dana, August 24, 1818), which was derived from a Wilkinson lathe and which was used in a modified form for a decade at the Harpers Ferry Federal Armory. Less successful must have been Olney's animal-propelled boat, patented December 11, 1817. See Smith, *Harpers Ferry Armory*, pp. 122–124; U.S., Patent Office, *List of Patents, 1790–1836*.

54. MVTM MSS, Ware MC, Agent's Letterbook, S. H. Hewes, for Christopher Colt (agent) to James C. Dunn (Boston), April 18, 1829.

55. Ibid., Olney to Elliot, July 14, 1826.

56. Ibid., Hewes, for Colt, to Clark Knight, March 10, 1829.

57. Graham, *Statistics of the Woollen Manufactories in the United States*.

58. Wilkinson, "Brandywine Borrowings," p. 3; Jeremy, "British Textile Technology," pp. 41–42.

59. USNA, Census of Manufactures, 1820, M-279, roll 23, Ohio returns, 891.

60. Partridge, *Practical Treatise*, p. 83.

61. Cole, *Wool Manufacture*, 1:101.

62. EMHL, Acc. 500, Ledger 63, Victor & Charles du Pont, Woolen Factory Petit Ledger 1817–1823, ff. 13, 39, 68, 98, 109, 129.

63. Mann, *Cloth Industry*, pp. 318–321.

64. Cole, *Wool Manufacture*, 1:83–84.

65. Partridge, *Practical Treatise*, p. 83.

66. EMHL, Acc. 500, Ledger 63, Victor & Charles du Pont, Woolen Factory Petit Ledger 1817–1823, ff. 49, 79, 100, 126, 169. For Powell's immigration see USNA, RG 45, War of 1812 Papers, Marshals' Returns of Enemy Aliens.

67. Rosenberg, "Factors," pp. 21–23.

68. GB Pat. 5355, Drawings, figure 7 in printed version; U.S. patent dated December 15, 1826, in USNA, RG 241, Drawings, No. 4605, figure 4; the specification has not survived in the National Archives, but the American patent is assumed to have been close to the British one since the American drawings are identical to the British ones in every particular except that the mule is shown in section in the British patent and as an overall perspective view in the American patent.

69. USFRC, FCC, Case Files, John Goulding v. Benjamin Bussey, October 1828. A number of other actions, taken by Goulding against those alleged to have infringed his patents are in these files. The patents in question were those dated August 24, 1827, as well as the 1826 patent.

70. Cole, *Wool Manufacture*, 1:104–105.

71. Daniel Bonney's recollections in *Bulletin of the National Association of Wool Manufacturers* 28 (1898): 42.

72. Baker MSS, Hamilton Woolen Co., Southbridge, Massachusetts, Papers, Abraham H. Schenck to James Walcott, Jr., June 4, 1828. For other details see ibid., Peter H. Schenck to James Walcott, Jr., August 23, 1828.

73. U.S. Congress, House, *Report on Petition Relative to Duties on Imports*, p. 113. I am obliged to Helena Wright of the MTVM for this reference.

The Matteawan broadcloth power loom gained first prize in the machinery category in the 1829 American Institute Fair, New York, and broadcloth woven on the loom won first prize in the woolen cloth section of the fair in 1829 and 1830. The excellence suggested by these prizes may have owed something to the involvement of Peter H. Schenck and William B. Leonard on the management of the American Institute. Even so, the Matteawan broadcloth power loom does seem to have been one of the more successful ones developed in the American woolen industry in the 1820s. See *Niles Weekly Register* 37 (1829): 140–142, 154–157; ibid. 39 (1830): 162. For these references I am grateful to John W. Lozier. See Cole, *Wool Manufacture*, 1:124–125, for American woolen power looms.

74. On part, if not all, of this tour, Schenck accompanied Zachariah Allen. See RIHS, Allen Papers, "Trip, 1825" pp. 63, 69, 100; *Niles' Weekly Register* 40 (1831): 254.

Chapter 13

1. PP (Commons), 1833 (690), 6:39.

2. Ibid., 1841 (201), 7:72, 105, 188; Musson and Robinson, *Science and Technology*, p. 64.

3. PP (Commons), 1841 (201), 7:118, 205.

4. Thompson, *Moses Brown*, pp. 256–257.

5. Although rotary self-acting temples were little used in England until the 1840s (according to Ure, *Cotton Manufacture*, 2:318, and Montgomery, *Practical Detail*, p. 102), a forerunner of Draper's American rotary temple of 1816 was patented in Britain in 1805. Awarded to Thomas Johnson and James Kay of Preston, the patent showed two discs, one above the other, gripping the cloth selvage where their edges met at one point; around the circum-

ference of the lower disc were sunken spikes, also to grip the cloth. Neglect of the rotary temple in England may have been related to the timing of its invention—well before the existence of a technically and commercially successful power loom—and the cheapness of labor in Britain, but these are only surmises. See GB Pat. 2876 (1805).

6. CSL, Sanseer MC Papers, Gilbert Brewster (New Hartford, New York) to John R. Watkinson (Middletown, Connecticut), November 3, 8, 1824.

7. USNA, New York Passenger Lists, M-237, roll 8 (1826), 428.

8. Penn, "Introduction," pp. 247–250; USNA, N.Y. Passenger Lists, M-237, roll 7 (1825), 594.

9. Baker MSS, Nashua MC Papers, Treasurer's Letterbook 1825–1828, G. & T. Searle to Ira Gay, September 10, 1827. Ira and Asa Gay patented an "improved self-operating mule" in the United States on April 10, 1829. See U.S. Patent Office, *List of Patents, 1790–1836*.

10. For Dyer see *DNB*; Dyer, "Notice"; Musson and Robinson, *Science and Technology*, p. 62.

11. The Taunton MC stockholders, Richmond among them, on October 5, 1825, passed "the Contract entered into with Mr. Timothy Wiggin of Manchester by Charles Richmond on the part of the company" to the company directors, who promptly accepted it. Baker MSS, Taunton MC Papers, Directors' Records 1823–1844, pp. 38–39. See also Penn, "Introduction," pp. 247–250.

12. Dorr's American patent of October 20, 1792, showed twelve knives set like spokes in a wheel. Underneath this wheel was a second, on the same axle, carrying four "tangent knives." As the edges of the two sets of blades passed, they acted "in the manner of shears." The wheel with four blades sat on the cloth while the wheel of twelve blades revolved inside it. However Dorr's British patent of April 9, 1793, revealed important improvements. Now the blades were set on the circumference of a cylinder, rather like a lawnmower but without any spirality in the blades.

13. Dyer and a Boston banker, Henry Higginson, improved Whittemore's machine in England when they took out a British patent on Whittemore's behalf. See Dyer, "Origin," pp. 143–145; GB Pats. 3498 (1811) and 5909 (1830).

14. *Repertory of Arts*, 3d ser. 10 (1830): 241–248; 12 (1831): 54–58.

15. Dyer, "Origin," pp. 143–145.

16. Little appears to be known about Wilkinson. His American patent of July 3, 1816, is in USNA, RG 241, Specifications 3:509–511; his English patent (including the drawings) is Pat. 4162 (1817). The latter described him as an engineer living at Covent Garden. His work in Manchester was noted in Allen, *Practical Tourist*, 1:171–172. According to Wilkinson, "Reminiscences," p. 106, Jeptha was a nephew of Jeremiah Wilkinson of Cumberland, Rhode Island.

17. Williams's career is recorded in Bagnall, *Textile Industries*, 1:280–295; in RIHS, Allen Papers, Miscellaneous Papers Box, Isaac P. Hazard to Allen, July 10, 1861; and in Williams's British patents.

18. Baker MSS, Nashua MC Papers, Treasurer's Letterbook 1825–1828, G. and T. Searle to Asher Benjamin, April 24, 1827.

19. *New England Farmer* 5 (1827), no. 43, for May 18, 1827.
The Matteawan Company was reported in November 1827 to have "some time ago" sent its broadcloth power loom to England and at this date it was "put-up and in operation in Leeds, England." See *Niles' Weekly Register* 33 (November 24, 1827): 195, quoted in Leavitt, *Hollingworth Letters*, p. xx. I am grateful to John W. Lozier for bringing this and other references to this loom to my attention.

20. GB Pat. 5561 (1827).

21. EMHL, Acc. 146, Box 7, file 81, Lawrence Greatrake to E. I. du Pont, December 5, 1813.

22. Dyer, "Origin," p. 144.

23. Vickerman, *Woollen Spinning*, p. 242. See also Roth, *Hand Card Making*, p. 2.

24. Leavitt, *Hollingworth Letters*, p. 27.

25. See Iredale, "Last Two Piecing Machines."

26. PP (Commons), 1833 (690), 6:171, for the cap spindle's use in England. According to Montgomery, *Practical Detail*, p. 70, its greater success in Britain derived from the British improvements that it received.

Reed-making machinery, belonging to Wilkinson, was noted at Manchester by Allen, *Practical Tourist*, 1:171–172. Reeds made in England were imported by the Dover MC, New Hampshire, in 1827 and were enthusiastically described as "beautiful"; they were made of metal and were expected to save on the cane reedmaker's bill. Ironically since they were metallic, these reeds were almost certainly made on one of Wilkinson's patent machines. See NHHS, Dover MC Letterbooks, Bridge to Shimmin, June 30, 1827.

27. Sharp, Roberts & Co. (Manchester) to André Koechlin et Cie (Mulhouse, Haut Rhin, France), January 22, 1827, quoted in Chaloner, "New Light," p. 35.

For the ascription of the differential to Houldsworth and the spread of the differential in Britain see Ure, *Cotton Manufacture*, 2:61–115 and passim.

28. PP (Commons), 1833 (690), 6:310.

29. Ibid., 42, 55, 171.

30. Rosenberg, *American System of Manufacturers*, pp. 304, 387.

31. Musson, "Continental Influences" demonstrates Britain's indebtedness to Europe prior and during its period of world technological leadership.

Chapter 14

1. Hopkinson, *Account*.

2. Gallatin, "Report," p. 427 gives a maximum of 100,000 spindles in 1810–1811. For a U.S. cotton spindleage of 356,900 in 1820, see appendix D, an estimate 60 percent greater than the previously accepted figure of 220,000 spindles, cited by Rogers, *Transportation Revolution*, p. 339. Friends of Domestic Industry, *Report*, p. 16, gives a spindleage of 1,246,503 for 1831.

3. RIHS, Arnold Papers, box 1, Correspondence 1820–1855, W. A. Greene (Providence, Rhode Island) to Arnold (Great Falls, Somersworth, New Hampshire), October 22, 1826.

4. Friends of Domestic Industry, *Report*, p. 16. For the estimate of £20–£25 million for British fixed capital investment in cotton manufacturing in 1834, see Blaug's figures quoted in Chapman, *Cotton Industry*, p. 31.

5. Chaloner, "New Light," p. 41, quoting Arthur Redford et al., *Manchester Merchants and Foreign Trade, 1794–1858* (Manchester: Manchester University Press, 1934), p. 132. In November 1826 the Manchester merchants sent a memorial to the Treasury in which they opposed the free export of machinery for fear of competition in foreign markets, adding, "Such competition is no chimera; it has been felt in various markets, from the manufacturers of France, Switzerland Saxony and the United States of America. The race is begun, and we would not wantonly throw away any advantages." This American competition was publicly acknowledged in 1833. See PP (Commons), 1833 (690), 6:39, 171.

6. Josephson, *Golden Threads*, pp. 178–185.

7. Charles Dickens, *American Notes*, chap. 4.

8. For some recent discussion, see Gersuny, "Devil in Petticoats," and Horwitz, "Architecture and Culture." A good introduction to the Lowell girls is Wright, "Uncommon Mill Girls."

9. Dickens, *American Notes*, chap. 4.

10. Pennsylvania Senate, *Report*, p. 4 and passim. Wallace, *Rockdale*, pp. 73–239, has a vivid portrayal of the development of an eastern Pennsylvania textile manufacturing community within what was known as the Rhode Island system.

11. Examples were the Indian Head Factory at Nashua, New Hampshire, the Ware MC in Massachusetts, and the

Matteawan MC in New York. Others can be traced in USNA, Census of Manufactures, 1820.

12. Baker MSS, Textron Inc. Papers, Indian Head Factory, Directors' Minutes 1825–1830, p. 53 (May 31, 1826).

13. Industrial exhibitions were held in Continental Europe and the United States well before 1850. Over two hundred European exhibitions before 1850 are listed in Carpenter, "European Industrial Exhibitions." Noticeably, British participation tarried until after the repeal of the laws controlling machinery exports and the acceptance of free trade. For late-nineteenth-century international exhibitions see Curti, "America at the World Fairs."

The transition from less random to more rational methods of technology diffusion deserves further study, as Hughes, "Comment," observed.

14. This, as one recent investigator readily acknowledges, is a major weakness in a purely case-study approach to technology transfer. See Stapleton, "Transfer of Technology," p. 298.

15. Rosenberg, "Anglo-American Wage Differences," p. 224. His surprise seems to be warranted by subsequent work. Over a range of occupations it has been shown that the skilled-unskilled wage rate differential was greater in America than in Britain during the 1820s. See Adams, "Some Evidence."

16. Hamilton, *Espionage*, p. 75.

17. Rosenberg, *Technology*, p. 61.

18. PP (Commons), 1824 (51), 5:473.

19. Baker MSS, Nathaniel B. Gordon Papers, vol. 2, Memorandum Book 1816–1820, p. 8 (November 21, 1820).

20. EMHL, Acc. 500, Duplanty, McCall & Co. Papers, General File, "Actual Expenses at Hagley Cotton Factory for One Day" (February 18, 1811); compare this with the figures in table 10.3 of this book. One Hagley spinner receiving these wages was David Our, a British immigrant who described himself as a cotton manufacturer in 1812.

See USNA, RG 45, War of 1812 Papers, Marshal's Returns of Enemy Aliens.

21. Ure, *Cotton Manufacture*, 2:444–445.

22. Gitelman's opinion that "the equipment adopted at Waltham was of such design that children could not be employed in its use" ("Waltham System," p. 231) is not substantiated by a close study of Waltham innovations. Quite the reverse: a greater use of inanimate power, the integration of mechanical processes, increased mechanical work, and automatic shut-off devices permitted a more extensive application of child labor. In Britain one stop motion was invented in order to compensate for the "inattention" of "the person who attends the frame (who is for the most part a child)," but this did not positively exclude child operatives; it simply reduced their number. See Bradbury's GB Pat. 3990 (1816), printed specification, p. 7. The diminishing employment of children in the Waltham system surely reflected the influence of social as much as or more than technical or economic factors.

23. Ware, *New England Cotton Manufacture*, chap. 10. In the 1840s, wages in teaching left earnings in textiles well behind, due to the influx of unskilled Irish workers. See Gitelman, "Waltham System," p. 238.

24. Bagnall, *Textile Industries*, 2:2020, quoting Appleton's January 17, 1828, speech in the Massachusetts House of Representatives.

25. Potter, "Atlantic Economy," p. 279, raised this possibility.

Sources of Illustrations

Frontispiece   Ure, *Philosophy*, frontispiece and p. 351.
Title Page   Logbook of Captain Joseph Pattinson, ca. 1814–1824, in the possession of J. H. Pattinson of Newhaven, Sussex, England.

1.1   *Pennsylvania Magazine* 1 (April 1775): 157–158.

1.2   Rees, *Cyclopaedia*, s.v. "Manufacture of Cotton," plate IV.*

1.3   Rees, *Cyclopaedia*, s.v. "Manufacture of Cotton," plate V.

1.4   Rees, *Cyclopaedia*, s.v. "Manufacture of Cotton," plate VI.

1.5   Rees, *Cyclopaedia*, s.v. "Manufacture of Cotton," plate IX.

1.6   North Western Museum of Science and Industry, Manchester, England.

1.7   Rees, *Cyclopaedia*, s.v. "Manufacture of Cotton," plate XI.

1.8   Rees, *Cyclopaedia*, s.v. "Woollen Manufacture," plate IV.

1.9   Merrimack Valley Textile Museum, North Andover, Massachusetts.

1.10   Rees, *Cyclopaedia*, s.v. "Woollen Manufacture," plate I.

1.11   Rees, *Cyclopaedia*, s.v. "Woollen Manufacture," plate II.

1.12   Rees, *Cyclopaedia*, s.v. "Woollen Manufacture," plate V.

3.1   British patent 2166 (February 7, 1797).

3.2   Rees, *Cyclopaedia*, s.v. "Manufacture of Cotton," plate VII.

3.3   British patent 3633 (January 1, 1813).

3.4   Rees, *Cyclopaedia*, s.v. "Manufacture of Cotton," plate X.

3.5   British patent 2771 (June 2, 1804).

3.6   British patent 2122 (June 28, 1796).

3.7   British patent 2955 (August 1, 1806).

*In volume 2 of the plates of Rees's *Cyclopaedia*, these plates are alphabetized in "C," s.v. "Cotton Manufacture."

3.8 British patent 3728 (July 31, 1813).

3.9 Rees, *Cyclopaedia*, s.v. "Woollen Manufacture," plate III.

4.1 U.S. National Archives, Record Group 241, "Name and Date" patents, drawings, December 30, 1791.

4.2 National Museum of History and Technology, Smithsonian Institution.

4.3 National Museum of History and Technology, Smithsonian Institution.

4.4 Rhode Island Historical Society.

5.1 U.S. National Archives, Record Group 241, "Name and Date" patents, drawing no. 1391 (November 17, 1810).

5.2 Detail from anonymous oil painting of Waltham, Massachusetts, ca. 1830, at Old Sturbridge Village but owned by Lowell Historical Society.

5.3 Broadside Collection, Division of Manuscripts, Baker Library, Harvard University.

5.4 Detail from "View of Lowell, Mass. Taken from the house of Elisha Fuller Esq: In Dracutt," by E. A. Farrar (Boston: Pendleton's lithography, 1834). Copy in Merrimack Valley Textile Museum.

6.1 Thomas Edward's view of the Lower Bridge and Factories, Dover, New Hampshire. Boston: Senefelder Lithograph Co. (c. 1830). Copy in Boston Athenaeum.

6.2 White, *Slater*, p. 395.

7.1 British patent 2766 (May 30, 1804), sheet 5.

7.2 National Museum of History and Technology, Smithsonian Institution.

7.3 Scholfield Papers, Connecticut State Library.

7.4 British patent 4897 (January 27, 1824).

7.5 Zachariah Allen, "Journal of European Trip, 1825," Rhode Island Historical Society.

7.6 Zachariah Allen, "Journal of European Trip, 1825," Rhode Island Historical Society.

7.7 U.S. National Archives, Record Group 241, "Name and Date" patents, drawing no. 5610 (August 10, 1829).

10.1 Montgomery, *Practical Detail*, plate VI.

10.2 Montgomery, *Practical Detail*, plate VII.

10.3 Slater Mill Historic Site, Pawtucket, Rhode Island.

10.4 Slater Mill Historic Site, Pawtucket, Rhode Island.

10.5 U.S. National Archives, Record Group 241, "Name and Date" patents, drawing no. 2608 (June 7, 1816).

11.2 Aza Arnold patent drawing, January 21, 1823, in Zachariah Allen Papers, Rhode Island Historical Society.

11.3 U.S. National Archives, Record Group 241, "Name and Date" patents, drawing no. 3931 (August 28, 1824).

11.4 U.S. National Archives, Record Group 241, "Name and Date" patents, drawing no. 5214 (September 2, 1828).

11.5 U.S. National Archives, Record Group 241, "Name and Date" patents, drawing no. 5280 (November 20, 1828).

12.1 U.S. National Archives, Record Group 241, "Name and Date" patents, drawing no. 46 (October 20, 1792).

12.2 Merrimack Valley Textile Museum.

12.4 U.S. National Archives, Record Group 241, "Name and Date" patents, drawing no. 4605 (December 15, 1826).

13.1 British patent 1945 (April 9, 1793).

13.2 Leigh, *Science*, 2:193.

13.3 British patent 4162 (August 23, 1817).

13.4 Montgomery, *Carding and Spinning Master's Assistant*, pp. 108–109, plate VII.

13.5 British patent 3498 (October 30, 1811).

Bibliography

Primary Sources
Manuscript Materials

Baker MSS. Division of Manuscripts. Baker Library. Harvard Business School. Harvard University, Cambridge, Massachusetts.

Boston MC Papers. Vols. 80 (Wages, 1817–1818), 82 (Wages, 1820).

Dover MC Papers. Payroll, 1826–1829.

Gordon, Nathaniel B. Papers. Vol. 11 (Memorandum Book, 1816–1820).

Hamilton MC Papers. Minutes of Proprietors and Directors, 1824–1864; Waste Book, 1829–1831.

Hamilton Woollen Company. Southbridge, Massachusetts, Papers, letters received and letters from selling agents, 1828–1831.

Merrimack MC Papers. Directors' Minutes, 1822–1843.

Nashua MC Papers. Treasurer's Letterbook, 1825–1828; Letterbook, 1825–1835.

Newmarket MC Papers. Treasurer's Letterbook, 1827–1830.

Pierson, Isaac G., Papers. Machine Watchmen's Time Book, 1817; Cotton Mill Purchases, 1823–1828; Cotton Manufacturing Accounts, 1817–1834; Work Ledger, 1828–1833.

Proprietors of the Locks and Canals Papers. Samuel C. Oliver's Forgings List.

Slater, Almy & Brown Papers. Vol. 3 (Day Book, 1799–1814).

Slater Collection. Providence Iron Foundry Papers. Minutes, 1819–1832.

Taunton MC Papers. Directors' Records, 1823–1844.

Textron, Inc. Papers. Indian Head Factory, Nashua, New Hampshire, Directors' Minutes, 1825–1830.

Wendell Collection. Case 11, Williams & Wendell letters, 1813–1826; Great Falls MC letters, 1823–1827. Case

14, Jacob Wendell, Miscellaneous Business Papers, 1803–1854.

Worcester Cotton and Woollen Manufactory Papers. Accounts, 1789–1791.

Bolton Reference Library. Bolton, Lancashire, England.

William Gray & Son Papers. Carder's and Spinner's Notebook, 1803–1830.

Connecticut State Library. Hartford, Connecticut.

Sanseer MC Papers. In letters, November 3–December 15, 1824.

Scholfield Family Papers. About five hundred letters and accounts and eleven account books, 1793–1820. Microfilm copies in the Library of Congress and the National Museum of History and Technology, Smithsonian Institution, Washington, D.C.

Customs House Archives. London, England.

Customs 37/56, "Seizures, 1825–1856."

Dorset County Record Office. Dorchester, Dorset, England.

Colfox Papers. Draft of letter of Thomas Colfox & Son, February 8, 1812.

Eleutherian Mills Historical Library. Greenville, Wilmington, Delaware.

Accession 146, Box 7.

Accession 467, Nos. 986 and 987. Joseph Bancroft's Notebooks, 1827–1828, 1828–1830.

Accession 500, Ledger 37, du Pont Woolen Journal, 1810–1815; Ledger 62, Victor & Charles du Pont Woolen Ledger, 1814–1820; Ledger 63, Victor & Charles du Pont Petit Ledger, 1817–1823; C. I. du Pont & Co., Box 26, Bills, 1816; Duplanty, McCall & Co. Papers, 10 boxes, 1811–1821.

Accession 1001. Joseph Chesseborough Dyer to R. Whiting, August 22, 1811.

Longwood MSS. Group 6, Boxes 1 and 2.

Winterthur MSS. Group 3, Boxes 17, 18, 29; Group 5, Box 4.

Friends House Library. London, England.

Pritchard Letters. Henry Newman to George Newman, May 20, 1833.

Historical Society of Pennsylvania. Philadelphia, Pennsylvania.

Minutes of the Manufacturing Committee of the Pennsylvania Society for the Encouragement of Manufactures and the Useful Arts. 2 vols., 1787–1789.

Kress Library. Harvard Business School. Harvard University. Cambridge, Massachusetts.

Interleaved edition of James Montgomery, *A Practical Detail of the Cotton Manufacture of the United States* (Glasgow, 1840) with annotations by the author, presumably in readiness for a second edition; the annotations date no later than August 1843 (pp. 76, 166).

Manchester University Library. Manchester, England.

McConnel & Kennedy Papers. Volume of stock inventories, 1812–1827.

Merrimack Valley Textile Museum. North Andover, Massachusetts.

Ware, Massachusetts, MC, Agent's Letterbook, 1826–1831.

New Hampshire Historical Society. Concord, New Hampshire.

Dover MC, Letterbooks: vol. 1, January 1825–January 1826; vol. 2, February 1826–June 1827; vol. 3, June 1827–September 1828.

Pennsylvania State Archives. Harrisburg, Pennsylvania.

John Nicholson Papers. MS Group 96, General Correspondence, 1772–1819

Public Record Office. London, England.

BT 1, In Letters to the Board of Trade, 1791–1830s.

BT 5, Board of Trade Minutes, vols. 1–45, 1784–1843.

BT 6, Vols. 151 and 152, Registers of applications to the Board of Trade to export machinery under license, 1825–1843.

C 73, C 210, Enrollments of specifications of patents of invention, 1709–1848. Printed versions of these patents have been used.

CUST. 9, Registers of Exports, vols. 1–18, 1812–1831.

PC 2, Minutes of the Privy Council, vols. 128–226, 1782–1844.

T47, Treasury Emigration Lists, vols. 9–12, 1773–1776.

Rhode Island Historical Society. Providence, Rhode Island.

Zachariah Allen Papers. Miscellaneous Papers Box (chiefly letters to Allen from retired Rhode Island manufacturers, 1861–1876, but also the drawing of Arnold's differential, patented 1823); Box K. A432Z, Allen's "Journal of European Trip, 1825."
Aza Arnold Papers. Correspondence, Boxes 1 and 2, and Arnold's Notebook, 1828–1836.

U.S. Federal Record Center. Waltham, Massachusetts.

First (Massachusetts) Circuit Court. Case Files and Final Records, 1790–1846.

Circuit Court of Rhode Island. Final Records, 1790–1860.

Circuit Court of Connecticut. Final Records, 1790–1844.

Circuit Court of New Hampshire. Final Records, 1790–1840.

Massachusetts District Court. Naturalization Records, vol. 1:1790–1827; vol. 2:1827–1846.

U.S. National Archives. Washington, D.C.

RG 29. Department of Commerce, Bureau of the Census, MS returns of the 1820 Census of Manufactures, on NA microfilm M-279, 27 rolls, Washington, D.C., 1967.

RG 36. Department of the Treasury, Bureau of Customs, Passenger Lists, 1824–1831 for New York (NA microfilm M-237, rolls 5–15), Philadelphia (M-425, rolls 34–46), and Boston (M-277, rolls 1–6).

RG 45. Department of the Navy, War of 1812 Papers, on 3 rolls of microfilm in the library of the University of Delaware, Newark, Delaware.

RG 46. Congress, Petitions to Congress, House and Senate Petitions, 1789–1824.

RG 59. Department of State, Consular Despatches: Leeds (NA microfilm T-474, roll 1).

RG 59. Department of State, War of 1812 Papers (NA microfilm M-588, rolls 1–7).

RG 92. Office of the Quartermaster General. Philadelphia Supply Agencies, Coxe and Irvine Papers, 1794–1842, cloth samples, 1800–1833, box 32a.

RG 241. Patent Office, Specifications of Restored or Reconstructed Patents, vols. 1–29, 1790–1836; Patent Reissues, vol. 2; Drawings of "Name-and-Date" patents, 1790–1836.

Published Materials (Excluding Journals)

Aikin, John. *A Description of the Country from Thirty to Forty Miles Round Manchester.* London: John Stockdale, 1795.

Allen, Zachariah. *The Science of Mechanics, as Applied to the Present Improvements in the Useful Arts in Europe, and in the United States of America.* Providence, R.I.: Hutchens & Cory, 1829.

———. *The Practical Tourist, or Sketches of the State of the Useful Arts and of Society, Scenery, &c. &c. in Great Britain, France and Holland.* 2 vols. Providence, R.I.: A. S. Beckwith, 1832.

Annual Statistics of Manufactures in Lowell and Vicinity, 1835 and 1882. Lowell, Mass.: Vox Populi Press, 1882.

Annual Statistics of Lowell Manufactures. Lowell, Mass.: n.p., 1836, 1837.

Appleton, Nathan. *The Introduction of the Power Loom and*

Origin of Lowell. Lowell, Mass.: B. H. Penhallow, 1858.

Arkwright, Richard. *The Case of Mr. Richard Arkwright & Co. in Relation to Mr. Arkwright's Invention of an Engine for Spinning Cotton &c. into Yarn; Stating His Reasons for Applying to Parliament for an Act to Secure His Right to Such Invention, or for Such Other Relief as to the Legislature Shall Seem Meet*. N.p., 1782.

————. *Richard Arkwright, Esquire: versus Peter Nightingale, Esquire . . . Court of Commons Pleas . . . 17 February 1785*. N.p., n.d.

Aspinall, Arthur. *The Early English Trade Unions: Documents from the Home Office Papers in the Public Record Office*. London: Batchworth Press, 1949.

Baines, Edward, Jr. *History of the Cotton Manufacture in Great Britain*. 1835. Reprint. London: Frank Cass, 1966.

Barrett, Oliver. *Domestic Roving and Spinning machine*. Troy, N.Y.: O. Lyon, April 17, 1812.

Batchelder, Samuel. *Introduction and Early Progress of the Cotton Manufacture in the United States*. Boston: Little, Brown, 1863.

Beckinsale, Robert P., ed. *The Trowbridge Woollen Industry as Illustrated in the Stock books of John and Thomas Clark, 1804–1824*. Devizes, Wiltshire: Wiltshire Archaeological and Natural History Society, Records Branch, 1951.

Bemiss, Elijah. *The Dyer's Companion*. New York: Evert Duyckinck, 1815.

Bigelow, Jacob. *Elements of Technology*. Boston: Hilliard, Gray, Little and Wilkins, 1829.

Boyd, Julian P., et al., eds. *The Papers of Thomas Jefferson*. Princeton: Princeton University Press, 1950–.

Brissot. See Warville.

Bronson, J., and Bronson, R. *The Domestic Manufacturer's Assistant, and Family Directory in the Arts of Weaving and Dyeing*. 1817. Reprint. New York: Dover Publications, 1977.

Butterworth, James. *A Complete History of the Cotton Trade, with Remarks on Their Progress in Bolton, Bury,* Stockport, Blackburn, and Wigan. Manchester: C. W. Leake, 1823.

Cole, Arthur H., ed. *Industrial and Commercial Correspondence of Alexander Hamilton Anticipating His Report on Manufactures*. Chicago: A. W. Shaw Co., 1928.

Cooke, Jacob, ed. *The Reports of Alexander Hamilton*. New York: Harper and Row, 1964.

Coxe, Tench. *A Statement of the Arts and Manufactures of the United States of America for the Year 1810*. Philadelphia: A. Cornman, Jr., 1814.

Crockett, David. *An Account of Col. Crockett's Tour of the North and Down East in the Year of Our Lord One Thousand Eight Hundred and Thirty-Four*. Philadelphia: E. L. Carey and A. Hart, 1835.

Crump, William B., ed. *The Leeds Woollen Industry*. Leeds: Thoresby Society, 1931.

Dickens, Charles. *American Notes for General Circulation*. 2 vols. London: Chapman & Hall, 1842.

Dodd, George. *The Textile Manufactures of Great Britain*. London: Charles Knight & Co., 1844.

Duncan, John. *Practical and Descriptive Essays on the Art of Weaving*. Glasgow: James and Andrew Duncan, 1807.

du Pont, Bessie Gardner. *Life of Eleuthére Irénée du Pont*. 12 vols. Newark, Del.: University of Delaware, 1923–1926.

Dyer, Joseph Chesseborough. "On the Origin of Several Mechanical Inventions." *Memoirs of the Literary and Philosophical Society of Manchester*, 3d ser. 3 (1868): 134–149.

Ellis, Asa, Jr. *The Country Dyer's Assistant*. Brookfield, Mass.: E. Merriam & Co., 1798.

Erickson, Charlotte J. *Invisible Immigrants: The Adaptation of English and Scottish Immigrants in Nineteenth Century America*. Coral Gables: University of Miami Press, 1972.

Ferguson, Eugene S., ed. *Early Engineering Reminiscences (1815–1840) of George Escol Sellers*. Washington, D.C.: Smithsonian Institution, 1965.

Fitzpatrick, John C., ed. *The Diaries of George Washington, 1748–1799*. 4 vols. Boston: Houghton Mifflin, 1925.

Friends of Domestic Industry. New York Convention. *Report on the Production and Manufacture of Cotton*. Boston: J. T. and E. Buckingham, 1832.

Gallatin, Albert. "Report on American Manufactures . . . 17 April 1810." *American State Papers*. Vol. II: *Finance*, pp. 425–439. Washington, D.C.: Gales and Seaton, 1832.

Gottesman, Rita S., ed. *The Arts and Crafts in New York, 1800–1804*. New York: New York Historical Society, 1964.

Graham, William H. *Statistics of the Woollen Manufactories in the United States by the Proprietor of the Condensing Cards*. New York: William H. Graham, 1845.

Great Britain. *Parliamentary Papers (Commons)* 1806. Paper 268. 3. Minutes of Evidence Taken before the Committee Appointed to Consider the State of the Woolen Manufacture.

————. 1808. Paper 177. 2. Report from the Committee on Petitions of Several Cotton Manufacturers and Journeymen Cotton Weavers.

————. 1816. Paper 397. 3. Report of the Select Committee on the State of the Children Employed in the Manufactories of the United Kingdom.

————. 1820. Paper 314. 7. Second Report of the Commissioners Appointed by His Majesty to Consider the Subject of Weights and Measures.

————. 1824. Paper 51. 5. Six Reports from the Select Committee on Artizans and Machinery.

————. 1825. Paper 504. 5. Report from the Select Committee on the Laws Relating to the Export of Tools and Machinery.

————. 1827. Paper 550. 5. Third Report on Emigration from the United Kingdom.

————. 1829. Paper 332. 3. Report from the Select Committee on the Law Relative to Patents for Inventions.

————. 1833. Paper 690. 6. Report from the Select Committee on the Present State of Manufactures, Commerce, and Shipping in the United Kingdom.

————. 1841. Paper 201. 7. First Report from Select Committee Appointed to Inquire into the Operation of the Existing Laws Affecting the Exportation of Machinery.

————. 1841. Paper 400. 7. Second Report from Select Committee Appointed to Inquire into the Operation of the Existing Laws Affecting the Export of Machinery.

Great Britain. Commissioners of Patents. *Specifications of Patents, Old Series, 1617–1852*. About nine hundred vols. of 14,359 patent specifications and drawings, London: Eyre and Spottiswoode, 1853–1858. These are the unabridged versions of patents of invention, printed from the patent rolls now in the Public Record Office. Comparisons between original and printed versions for a handful of patents show that the latter were carefully copied from the originals.

Guest, Richard. *Compendious History of the Cotton Manufacture*. Manchester: Joseph Pratt, 1823.

Haigh, James. *The Dyer's Assistant in the Art of Dyeing Wool and Woolen Goods*. Philadelphia: J. Humphreys, 1800.

Hansard, Thomas C., ed. *Hansard's Parliamentary Debates*.

Hargrove, John. *The Weavers Draft Book and Clothiers Assistant*. 1792. Reprint. Worcester, Mass.: American Antiquarian Society, 1979.

Hazard, Samuel, ed. *The Register of Pennsylvania*. Vols. 1–4. 1828–1829.

Henderson, W. O., ed. *Industrial Britain under the Regency: The Diaries of Escher, Bodmer, May and de Gallois, 1814–1818*. London: Frank Cass, 1968.

Hopkinson, Francis. *Account of the Grand Federal Procession, Philadelphia, 1788*. Edited by Whitfield J. Bell. Boston: Old South Association, 1962.

Jefferson, Thomas. *Notes on the State of Virginia*. New York: Harper and Row, 1964.

Jeremy, David J., ed. *Henry Wansey and His American Journal, 1794.* Philadelphia: American Philosophical Society, 1970.

Leavitt, Thomas W., ed. *The Hollingworth Letters: Technical Change in the Textile Industry, 1826–1837.* Cambridge, Mass.: M.I.T. Press, 1969.

Luccock, John. *The Nature and Properties of Wool.* Leeds: E. Baines, 1805.

Manchester, Job. "Recollections." *Transactions of the Rhode Island Society for the Encouragement of Domestic Industry.* 1864.

Montgomery, James. *The Carding and Spinning Master's Assistant: or the Theory and Practice of Cotton Spinning.* Glasgow: John Niven, Jr., 1832.

————. *The Cotton Spinner's Manual: or a Compendium of the Principles of Cotton Spinning.* Glasgow: John Niven, Jr., 1835.

————. *A Practical Detail of the Cotton Manufacture of the United States of America and the State of the Cotton Manufacture of That Country Contrasted and Compared with That of Great Britain.* Glasgow: John Niven, Jr., 1840.

Murphy, John. *A Treatise on the Art of Weaving.* Glasgow: 1824.

McLane Report. See U.S. Congress.

New Statistical Account of Scotland. 15 vols. London and Edinburgh: William Blackwood & Sons, 1845.

Nicholson, John. *The Operative Mechanic and British Machinist.* 2 vols. Philadelphia: H. C. Carey and I. Lea, 1826.

Ogden, James. *A Description of Manchester.* Manchester: M. Faulkner, 1783.

Ogden, Samuel. *Thoughts, What Probable Effect the Peace with Great-Britain Will Have on the Cotton Manufactures of This Country: Interspersed with Remarks on Our Bad Management in the Business; and the Way to Improvement, so as to Meet Imported Goods in Cheapness, at Our Home Market, Pointed Out.* Providence, R.I.: Goddard and Mann, 1815.

Owen, Robert. *The Life of Robert Owen, Written by Himself.* 2 vols. London: Effingham Wilson, 1857.

Partridge, William. *A Practical Treatise on Dying of Woollen, Cotton and Skein Silk with the Manufacture of Broadcloth and Cassimere Including the Most Improved Methods in the West of England.* 1823. Reprint. Edington, Wiltshire: Pasold Research Fund Ltd., 1973.

Pennsylvania. Senate. *Report of the Select Committee Appointed to Visit the Manufacturing Districts of the Commonwealth, for the Purpose of Investigating the Subject of the Employment of Children in Manufactories.* Harrisburg, Pa.: Thompson and Clark, 1838.

Philadelphia City Directories. Philadelphia, 1793, 1801, 1810, 1816, 1821, 1830.

Pigot & Sons. *General Directory of Manchester, Salford, &c.* Manchester: Pigot & Sons, 1830.

Radcliffe, William. *Origin of the New System of Manufacture Commonly Called Power-Weaving.* Stockport: James Lomax, 1828.

Rauch, John. *Receipts on Dyeing.* New York: Joseph I. Badger & Co., 1815.

Rees, Abraham, ed. *The Cyclopaedia: or Universal Dictionary of Arts, Sciences and Literature.* 39 vols. of text and 6 vols. of plates. London: Longman et al., 1802–1820. (American edition: Philadelphia: Samuel F. Bradford et al., 1810–1824.)

Rosenberg, Nathan, ed. *The American System of Manufactures: The Report of the Committee on the Machinery of the United States, 1855, and the Special Reports of George Wallis and Joseph Whitworth, 1854.* Edinburgh: Edinburgh University Press, 1969.

Rules for the Conducting of the Union Society, of Printers, Cutters, and Drawers in Lancashire, Cheshire, Derbyshire, &c. Manchester: J. Aston and J. Seddon, 1813.

Shaw, Ralph R., and Shoemaker, Richard H. *American Bibliography. A Preliminary Checklist for 1810.* New York: Scarecrow Press, 1961.

Smith, John. "Journal of John Smith (1796–1886) from 28 November 1826 to March 1827." *Essex Institute Historical Collections* 106 (1970): 88–107.

Stopford, Joseph, and Gerrard, Nehemiah. *The Cotton Manufacturer's Useful Assistant*. Philadelphia: J. Metcalfe & Co., 1832.

Syrett, Harold C., et al., eds. *The Papers of Alexander Hamilton*. New York: Columbia University Press, 1961–.

Tomlinson, Charles. *Cyclopaedia of Useful Arts*. 2 vols. London: George Virtue & Co., 1854.

Toulmin, Harry. *The Western Country in 1793: Reports on Kentucky and Virginia*. Edited by Marion Tinling and Godfrey Davies. San Marino, Calif.: Henry E. Huntington Library, 1948.

The Trial of a Cause Instituted by Richard Pepper Arden, Esq.: His Majesty's Attorney General, by Writ of Scire Facias, to Repeal a Patent Granted on the Sixteenth of December 1775, to Mr. Richard Arkwright . . . 25th of June 1785. London: Hughes and Walsh, 1785.

United States. Congress. House. *Report on Petition Relative to Duties on Imports*. House Report No. 115, 1st Session, 20th Congress. Washington, D.C., 1828.

————. *Documents Relative to the Manufactures of the United States*. Executive Document No. 308, 1st Session, 22d Congress. Washington, D.C., Duff Green, 1833. [This was known as the *McLane Report*, after the secretary of the treasury responsible for its compilation.]

U.S. Patent Office. *A List of Patents Granted by the United States from April 10, 1790, to December 31, 1836*. Washington, D.C.: Commissioner of Patents, 1872.

Ure, Andrew. *The Philosophy of Manufactures, or an Exposition of the Scientific, Moral, and Commercial Economy of the Factory System of Great Britain*. London: Charles Knight, 1835.

————. *The Cotton Manufacture of Great Britain*. 2 vols. London: Charles Knight, 1836.

————. *Dictionary of Arts, Manufactures, and Mines: Containing a Clear Exposition of Their Principles and Practice.*

London: Longman, Orme, Brown, Green & Longmans, 1839.

Warville, Jacques Pierre Brissot de. *New Travels in the United States of America, 1788*. Translated by Mara Soceanu Vamos and Durand Echeverria and edited by Durand Echeverria. Cambridge, Mass.: Belknap Press of Harvard University Press, 1964.

White, George. *A Practical Treatise on Weaving, by Hand and Power Looms*. London: Simpkin, Marshall & Co., 1851.

White, George S. *Memoir of Samuel Slater, the Father of American Manufactures*. Philadelphia, 1836.

Wilkinson, David. "Reminiscences." *Transactions of the Rhode Island Society for the Encouragement of Domestic Industry*. 1861.

Wily, John. *A Treatise on the Propagation of Sheep, the Manufacture of Wool, and the Cultivation of Flax, with Directions for Making Several Utensils for the Business*. Williamsburg, Va.: J. Royle, 1765.

Woodcroft, Bennet. *Titles of Patents of Invention Chronologically Arranged, 1617–1852*. 2 vols. London: Queen's Printing Office, by George Edward Eyre and William Spottiswoode, 1854.

————. *Subject-Matter Index (Made from Titles only) of Patents of Invention, 1617–1852*. 2 vols. London: Great Seal Patent Office, 1857.

————. *Alphabetical Index of Patentees of Inventions, 1617–1852*. 1854. Reprint. London: Evelyn, Adams & Mackay, 1969.

Journals

The American Museum 1–12 (1787–1792).

Archives of Useful Knowledge 1–3 (1811–1813).

Bulletin of the National Association of Wool Manufacturers 1–33 (1869–1903).

Collections of the Dover, New Hampshire, Historical Society 1 (1894).

Contributions of the Old Residents' Historical Association, Lowell, Massachusetts 1–6 (1879–1904).

Emporium of Arts and Sciences 1–5 (1812–1814).

Franklin Journal 1–4 (1826–1827); thereafter *Journal of the Franklin Institute*, new series, 1–18 (1826–1836).

New England Farmer 1–10 (1822–1832).

Niles' Weekly Register 1–40 (1811–1831).

The Pennsylvania Magazine; or American Monthly Museum 1–2 (1775–1776).

Proceedings of the National Association of Cotton Manufacturers 1–45 (1866–1888).

Repertory of Arts, Manufactures and Agriculture, 1st series, 1–16 (1794–1802); 2nd series, 1–46 (1802–1825); 3rd series, 1–14 (1825–1832).

Royal Society of Arts, *Transactions* 1–48 (1783–1831).

Transactions of the Rhode Island Society for the Encouragement of Domestic Industry (1861, 1864, 1866, 1869, 1871).

Secondary Sources

Books and Published Microcopies

Albion, Robert G. *The Rise of New York Port (1815–1860).* New York: Charles Scribner's Sons, 1939.

Anderson, Michael. *Family Structure in Nineteenth Century Lancashire.* Cambridge: At the Press, 1971.

Arbuckle, Robert D. *Pennsylvania Speculator and Patriot: The Entrepreneurial John Nicholson, 1757–1800.* Philadelphia: Pennsylvania State University Press, 1975.

Ashton, Thomas S. *The Industrial Revolution, 1760–1830.* London: Oxford University Press, 1948.

———. *An Economic History of England: The Eighteenth Century.* London: Methuen, 1955.

Bagnall, William R. *The Textile Industries of the United States. Vol. 1: 1639–1810.* 1893. Reprint. New York: Augustus M. Kelley, 1971.

———. "Sketches of Manufacturing Establishments in New York City, and of Textile Establishments in the Eastern States." Edited by Victor S. Clark. 4 vols. 1908. Microfiche. North Andover, Mass.: Merrimack Valley Textile Museum, 1977. [Since the four volumes have a sequential pagination, they have been collectively cited as *Textile Industries,* 2, in the notes.]

Bathe, Greville, and Bathe, Dorothy. *Oliver Evans: A Chronicle of Early American Engineering.* Philadelphia: Historical Society of Pennsylvania, 1935.

———. *Jacob Perkins: His Inventions, His Times and His Contemporaries.* Philadelphia: Historical Society of Pennsylvania, 1943.

Berthoff, Rowland T. *British Immigrants in Industrial America, 1790–1950.* Cambridge, Mass.: Harvard University Press, 1953.

Bridenbaugh, Carl. *The Colonial Craftsman.* Chicago: University of Chicago Press, 1961.

Bromwell, William J. *History of Immigration to the United States Exhibiting the Number, Sex, Age, Occupation and Country of Birth of Passengers Arriving in the United States . . . 1819–1855.* New York: Redfield, 1856.

Bruchey, Stuart. *The Roots of American Economic Growth, 1607–1861.* London: Hutchinson, 1965.

Bugbee, Bruce W. *Genesis of American Patent and Copyright Law.* Washington, D.C.: Public Affairs Press, 1967.

Bythell, Duncan. *The Handloom Weavers. A Study in the English Cotton Industry during the Industrial Revolution.* Cambridge: At the Press, 1969.

Catling, Harold. *The Spinning Mule.* Newton Abbot, Devon: David and Charles, 1970.

Chandler, Alfred D., Jr. *The Visible Hand: The Managerial Revolution in American Business.* Cambridge, Mass.: Belknap Press of Harvard University Press, 1977.

Chapman, Stanley D. *The Cotton Industry in the Industrial Revolution.* London: Macmillan, 1972.

Cole, Arthur H. *The American Wool Manufacture.* 2 vols. Cambridge, Mass.: Harvard University Press, 1926.

Collier, Frances. *The Family Economy of the Working Classes in the Cotton Industry, 1784–1833.* Manchester: Manchester University Press, 1964.

Cooke, Jacob. *Tench Coxe and the Early Republic.* Chapel Hill, N.C.: University of North Carolina Press, 1978.

Crockett, Norman L. *The Woolen Industry of the Midwest.* Lexington, Kentucky: University Press of Kentucky, 1970.

Daly, John, and Weinberg, Allen. *Genealogy of Philadelphia County Subdivisions.* 2d ed. Philadelphia: City of Philadelphia, Department of Records, 1966.

Davis, Joseph Stancliffe. *Essays in the Earlier History of American Corporations.* 2 vols. Cambridge, Mass.: Harvard University Press, 1917.

Dickson, R. J. *Ulster Emigration to Colonial America, 1718–1775.* London: Routledge and Kegan Paul, 1966.

Dictionary of American Biography. Edited by Allen Johnson and Dumas Malone. 22 vols. New York: Charles Scribner's Sons, 1928–1944.

Dictionary of National Biography. Edited by Leslie Stephen and Sidney Lee. 63 vols. London: Oxford University Press, 1885–1933.

Earle, Pliny, comp. *The Earle Family: Ralph Earle and His Descendants.* Worcester, Mass.: C. Hamilton, 1888.

Edwards, Michael M. *The Growth of the British Cotton Trade, 1780–1815.* Manchester: Manchester University Press, 1967.

Ferguson, Eugene S. *Bibliography of the History of Technology.* Cambridge, Mass.: M.I.T. Press, 1968.

Fitton, Robert S., and Wadsworth, Alfred P. *The Strutts and the Arkwrights, 1758–1830: A Study of the Early Factory System.* Manchester: Manchester University Press, 1958.

Freedley, Edwin T. *Philadelphia and Its Manufactures.* Philadelphia: Edward Young, 1859.

French, Gilbert J. *Life and Times of Samuel Crompton.* London: Simpkin, Marshall & Co., 1860.

Gibb, George S. *The Saco-Lowell Shops: Textile Machinery Building in New England, 1813–1949.* Cambridge, Mass.: Harvard University Press, 1950.

Graham, Ian C. G. *Colonists from Scotland: Emigration to North America, 1707–1783.* Ithaca, N.Y.: Cornell University Press, 1956.

Gras, Norman S. B., and Larson, Henrietta. *Casebook in American Business History.* New York: Appleton Crofts, 1939.

Habakkuk, H. John. *American and British Technology in the Nineteenth Century: The Search for Labour Saving Inventions.* Cambridge: At the Press, 1962.

Ham, John Randolph. *The Dover (New Hampshire) Physicians.* Concord, N.H.: New Hampshire Medical Society, 1879.

Hamilton, Peter. *Espionage and Subversion in an Industrial Society: An Examination and Philosophy of Defence for Management.* London: Hutchinson, 1967.

Hammond, John L., and Hammond, Barbara. *The Skilled Labourer, 1760–1832.* London: Longmans, Green & Co., 1919.

Hansen, Marcus L. *The Atlantic Migration, 1607–1860: A History of the Continuing Settlement of the United States.* New York: Harper and Row, 1961.

Harrison, Brian. *Drink and the Victorians: The Temperance Question in England, 1815–1872.* London: Faber, 1971.

Hedges, James B. *The Browns of Providence Plantation: The Nineteenth Century.* Providence, R.I.: Brown University Press, 1968.

Henderson, W. O. *Britain and Industrial Europe, 1750–1870: Studies in British Influence on the Industrial Revolution in Western Europe.* Leicester: Leicester University Press, 1972.

Hills, Richard L. *Power in the Industrial Revolution.* Manchester: Manchester University Press, 1970.

Hindle, Brooke. *The Pursuit of Science in Revolutionary America.* Chapel Hill, N.C.: University of North Carolina Press, 1965.

————. *David Rittenhouse*. Princeton: Princeton University Press, 1964.

Hobsbawm, Eric J. *Labouring Men: Studies in the History of Labour*. London: Weidenfeld and Nicolson, 1964.

Hurd, Duane Hamilton. *History of Rockingham and Strafford Counties, New Hampshire*. Philadelphia: J. W. Lewis & Co., 1882.

————. *History of Norfolk County, Massachusetts*. Philadelphia: J. W. Lewis & Co., 1884.

————. *History of Worcester County, Massachusetts*. 2 vols. Philadelphia: J. W. Lewis & Co., 1889.

Jackson, Benjamin Dayton. *An Attempt to Ascertain the Actual Dates of Publication of the Various Parts of Rees's Cyclopaedia*. London: Pewtress & Co., 1895.

Jenkins, David T. *The West Riding Wool Textile Industry, 1770–1835: A Study of Fixed Capital Formation*. Edington, Wiltshire: Pasold Research Fund Ltd., 1975.

Jones, Maldwyn A. *Destination America*. London: Weidenfeld and Nicolson, 1976.

Josephson, Hannah. *The Golden Threads: New England's Mill Girls and Magnates*. New York: Duell, Sloan and Pearce, 1949.

Kirby, R. G., and Musson, A. E. *The Voice of the People: John Doherty, 1798–1854, Trade Unionist, Radical and Factory Reformer*. Manchester: Manchester University Press, 1975.

Kittredge, Henry G., and Gould, A. C. *History of the American Card-Clothing Industry*. Worcester, Mass.: T. K. Earle Manufacturing Co., 1886.

Knowlton, Evelyn H. *Pepperell's Progress: History of a Cotton Textile Company, 1844–1945*. Cambridge, Mass.: Harvard University Press, 1948.

Layer, Robert G. *Earnings of Cotton Mill Operatives, 1825–1914*. Cambridge, Mass.: Harvard University Press, 1955.

Leigh, Evan. *The Science of Modern Cotton Spinning*. 2 vols. Manchester: Palmer & Howe, 1873.

Lindstrom, Diane. *Economic Development in the Philadelphia Region, 1810–1850*. New York: Columbia University Press, 1978.

Main, Jackson Turner. *The Social Structure of Revolutionary America*. Princeton: Princeton University Press, 1965.

Mann, Julia de Lacy. *The Cloth Industry in the West of England from 1640 to 1880*. Oxford: Clarendon Press, 1971.

Mantoux, Paul. *The Industrial Revolution in the Eighteenth Century*. Rev. ed. London: Jonathan Cape, 1961.

McGouldrick, Paul F. *New England Textiles in the Nineteenth Century: Profits and Investment*. Cambridge, Mass.: Harvard University Press, 1968.

Merrill, Gilbert R. et al. *American Cotton Handbook*. New York: Textile Book Publishers, 1949.

Mitchell, Brian R., and Deane, Phyllis. *Abstract of British Historical Statistics*. Cambridge: At the Press, 1962.

Montgomery, Florence M. *Printed Textiles. English and American Cottons and Linens, 1700–1850*. New York: Viking Press, 1970.

Morison, Samuel E. *The Maritime History of Massachusetts, 1783–1860*. Boston: Houghton Mifflin, 1941.

Musson, Albert E., and Robinson, Eric. *Science and Technology in the Industrial Revolution*. Manchester: Manchester University Press, 1969.

Navin, Thomas R. *The Whitin Machine Works since 1831: A Textile Machinery Company in an Industrial Village*. Cambridge, Mass.: Harvard University Press, 1950.

Neel, Joanne L. *Phineas Bond: A Study in Anglo-American Relations, 1786–1812*. Philadelphia: University of Pennsylvania Press, 1968.

Nettels, Curtis P. *The Emergence of a National Economy, 1775–1815*. New York: Holt, Rinehart and Winston, 1962.

Nichols, Henry W., and Broomhead, William H. *Standard Cotton Cloths and Their Construction*. Fall River, Mass.: Dover Press, 1927.

North, Douglass C. *The Economic Growth of the United States, 1790–1860*. Englewood Cliffs, N.J.: Prentice-Hall, 1961.

Nye, Russel B. *The Cultural Life of the New Nation, 1776–1830*. New York: Harper and Row, 1960.

Penrose, Edith T. *The Theory of the Growth of the Firm*. 2nd ed. Oxford: Blackwell, 1980.

Pollard, Sidney. *The Genesis of Modern Management: A Study of the Industrial Revolution in Great Britain*. Cambridge, Mass.: Harvard University Press, 1965.

Porter, Glenn, and Livesay, Harold C. *Merchants and Manufacturers: Studies in the Changing Structure of Nineteenth Century Marketing*. Baltimore: Johns Hopkins Press, 1971.

Pursell, Carroll W., Jr. *Early Stationary Steam Engines in America: A Study in the Migration of a Technology*. Washington, D.C.: Smithsonian Institution, 1969.

Raistrick, Arthur. *Quakers in Science and Industry, Being an Account of the Quaker Contributions to Science and Industry during the Seventeenth and Eighteenth Centuries*. New ed. Newton Abbot, Devon: David and Charles, 1968.

Ratcliffe, Barrie M., ed. *Great Britain and Her World, 1750–1914: Essays in Honour of W. O. Henderson*. Manchester: Manchester University Press, 1975.

Robson, Charles, ed. *The Manufactories and Manufacturers of Pennsylvania in the Nineteenth Century*. Philadelphia: Galaxy Publishing Co., 1875.

Rogers, Everett M., with Shoemaker, F. Floyd. *Communication of Innovations: A Cross-Cultural Approach*. 2d ed. New York: Free Press, 1971.

Rolt, Lionel T. C. *Tools for the Job: A Short History of Machine Tools*. London: Batsford, 1965.

Rosenberg, Nathan, ed. *The Economics of Technological Change: Selected Readings*. Harmondsworth, Middlesex: Penguin Books, 1971.

———. *Technology and American Economic Growth*. New York: Harper and Row, 1971.

Rosenberg, Nathan, and Vincenti, Walter. *The Britannia Bridge: The Generation and Diffusion of Technological Knowledge*. Cambridge, Mass.: M.I.T. Press, 1978.

Roth, H. Ling. *Hand Card Making*. Halifax: Bankfield Museum, n.d.

Salter, W. E. G. *Productivity and Technical Change*. Cambridge: At the Press, 1966.

Saul, S. B., ed. *Technological Change: The United States and Britain in the Nineteenth Century*. London: Methuen, 1970.

Scharf, John Thomas, and Westcott, Thompson. *History of Philadelphia*. 3 vols. Philadelphia: L. H. Everts, 1884.

Siddall, Teckla Anne. *Siddalls in America (Phineas Line)*. Privately printed, 1962. [Copy in EMHL.]

Sinclair, Bruce. *Philadelphia's Philosopher Mechanics. A History of the Franklin Institute, 1824–1865*. Baltimore: Johns Hopkins University Press, 1974.

Singer, Charles et al., eds. *A History of Technology*. 5 vols. Oxford: Clarendon Press, 1954–1958.

Smelser, Neil J. *Social Change in the Industrial Revolution: An Application of Theory to the Lancashire Cotton Industry, 1770–1840*. London: Routledge and Kegan Paul, 1959.

Smith, Merritt Roe. *Harpers Ferry Armory and the New Technology: The Challenge of Change*. Ithaca, N.Y.: Cornell University Press, 1977.

Strassmann, W. Paul. *Risk and Technological Innovation: American Manufacturing Methods during the Nineteenth Century*. Ithaca, N.Y.: Cornell University Press, 1959.

Tann, Jennifer. *Gloucestershire Woollen Mills*. Newton Abbot, Devon: David and Charles, 1967.

———. *The Development of the Factory*. London: Cornmarket, 1970.

Taussig, F. W. *The Tariff History of the United States*. 8th ed. New York: G. P. Putnam's Sons, 1931.

Taylor, George Rogers. *The Transportation Revolution, 1815–1860*. New York: Holt, Rinehart and Winston, 1951.

Temin, Peter. *Causal Factors in American Economic Growth in the Nineteenth Century.* London: Macmillan, 1975.

Thistlethwaite, Frank. *America and the Atlantic Community: Anglo-American Aspects, 1790–1850.* New York: Harper and Row, 1963.

Thomis, Malcolm I. *The Luddites: Machine-Breaking in Regency England.* Newton Abbot, Devon: David and Charles, 1970.

Thompson, Edward P. *The Making of the English Working Class.* London: Victor Gollancz, 1965.

Thompson, Mack. *Moses Brown, Reluctant Reformer.* Chapel Hill, N.C.: University of North Carolina Press, 1962.

Trumbull, Levi R. *A History of Industrial Paterson.* Paterson, N.J.: C. M. Herrick, 1882.

Tryon, Rolla M. *Household Manufactures in the United States, 1640–1860.* Chicago: Chicago University Press, 1917.

Turnbull, Geoffrey. *A History of the Calico Printing Industry of Great Britain.* Altrincham, Cheshire: John Sherratt and Son, 1951.

Turner, H. A. *Trade Union Growth, Structure and Policy. A Comparative Study of the Cotton Unions.* London: Allen and Unwin, 1962.

United States. Department of Commerce. Bureau of the Census. *Historical Statistics of the United States, Colonial Times to 1970.* 2 vols. Washington, D.C.: U.S. Government Printing Office, 1975.

————. Department of Commerce. Patent Office. *The Story of the United States Patent Office.* 4th ed. Washington, D.C.: U.S. Government Printing Office, 1965.

————. Department of the Interior. Census Office. *Eighth Census, 1860.* Vol. 3: *Manufactures of the United States.* Washington, D.C.: U.S. Government Printing Office, 1865.

Unwin, George, ed. *Samuel Oldknow and the Arkwrights: The Industrial Revolution in Marple and Stockport.* Manchester: Manchester University Press, 1924.

Vickerman, Charles. *Woollen Spinning: A Textbook for Students in Technical Schools and Colleges, and for Skilful Practical Men in Woollen Mills.* London: Macmillan, 1894.

Wadsworth, Alfred P., and Mann, Julia de Lacy. *The Cotton Trade and Industrial Lancashire, 1600–1780.* Manchester: Manchester University Press, 1931.

Wallace, Anthony F. C. *Rockdale: The Growth of an American Village in the Early Industrial Revolution.* New York: Alfred A. Knopf, 1978.

Ware, Caroline F. *The Early New England Cotton Manufacture: A Study in Industrial Beginnings.* Boston: Houghton Mifflin, 1931.

White, Leonard D. *The Federalists: A Study in Administrative History.* New York: Macmillan, 1948.

Wood, George Henry. *The History of Wages in the Cotton Trade during the Past Hundred Years.* London: Sheratt and Hughes, 1910.

Woodbury, Robert S. *History of the Lathe to 1850.* Cambridge, Mass.: M.I.T. Press, 1961.

Wroth, Lawrence C. *Abel Buell of Connecticut: Silversmith, Type Founder and Engraver.* Middletown, Conn.: Wesleyan University Press, 1958.

Articles

Adams, Donald. "Some Evidence on English and American Wage Rates, 1790–1830." *Journal of Economic History* 30 (1970): 449–519.

Ames, Edward, and Rosenberg, Nathan. "Changing Technological Leadership and Industrial Growth." *Economic Journal* 73 (1963): 13–31.

Arrow, Kenneth J. "Classificatory Notes on the Production and Transmission of Technological Knowledge." *American Economic Review* 59 (1969): 29–35.

Bagnall, William R. "Sketch of the Life of Ezra Worthen."

Contributions of the Old Residents' Historical Association, Lowell, Massachusetts 3 (1884): 33–41.

———. "Paul Moody." *Contributions of the Old Residents' Historical Association, Lowell, Massachusetts* 3 (1884): 57–72.

———. "Samuel Batchelder." *Contributions of the Old Residents' Historical Association, Lowell, Massachusetts* 3 (1885): 187–211.

Blaug, Mark. "A Survey of the Theory of Process-Innovations." In Nathan Rosenberg, ed. *Economics of Technological Change*, pp. 86–113. Harmondsworth, Middlesex: Penguin Books, 1971.

Bonner, Joseph A. "La Grange Print Works." *Papers of the Frankford Historical Society* 2 (1922): 346–349.

Brittain, James E. "The International Diffusion of Electrical Power Technology, 1870–1920." *Journal of Economic History* 24 (1974): 108–121.

Burrage, Michael. "Democracy and the Mystery of the Crafts: Observations on Work Relationships in America and Britain." *Daedalus* 101 (1972): 141–162.

Campbell, Mildred. "English Emigration on the Eve of the American Revolution." *American Historical Review* 61 (1955): 1–20.

Candee, Richard. "The 'Great Factory' at Dover, New Hampshire: The Dover Manufacturing Company Print Works, 1826." *Old Time New England* 65 (1975): 39–51.

Carpenter, Kenneth E. "European Industrial Exhibitions before 1851 and Their Publications." *Technology and Culture* 13 (1972): 465–486.

Chaloner, William H. "New Light on Richard Roberts, Textile Engineer." *Newcomen Society Transactions* 41 (1968–1969): 27–44.

Chandler, Alfred D., Jr. "Anthracite Coal and the Beginnings of the Industrial Revolution in the United States." *Business History Review* 46 (1972): 141–181.

Chapman, Stanley D. "Fixed Capital Formation in the British Cotton Manufacturing Industry." In J. P. P. Higgins and Sidney Pollard, eds. *Aspects of Capital Investment in Great Britain, 1750–1850*, pp. 57–119. London: Methuen, 1971.

———. "The Textile Factory before Arkwright: A Typology of Factory Development." *Business History Review* 48 (1974): 451–478.

Clark, Charles H. "John Thorp—Inventor of Ring Spinning." *Transactions of the National Association of Cotton Manufacturers* 124–125 (1928): 72–95.

Cole, Arthur H. "Evolution of the Foreign-Exchange Market of the United States." *Journal of Economic and Business History* 1 (1928–1929): 384–421.

Coleman, Donald C. "An Innovation and Its Diffusion: The 'New Draperies.'" *Economic History Review*, 2d ser. 22 (1969): 417–429.

Coleman, Peter J., and Majeske, Penelope K. "British Immigrants in Rhode Island during the War of 1812." *Rhode Island History* 24 (1975): 66–75.

Cowley, Charles. "The Foreign Colonies of Lowell." *Contributions of the Old Residents' Historical Association, Lowell, Massachusetts* 2 (1883): 165–179.

Curti, Merle. "America at the World Fairs, 1851–1893." *American Historical Review* 55 (1949–1950): 833–856.

David, Paul. "Learning by Doing and Tariff Protection: A Reconsideration of the Case of the Ante-Bellum United States Cotton Textile Industry." *Journal of Economic History* 30 (1970): 552–601.

Davis, Lance E. "Stock Ownership in the Early New England Textile Industry." *Business History Review* 32 (1958): 204–222.

Day, Clive. "The Early Development of the American Cotton Manufacture." *Quarterly Journal of Economics* 39 (1925): 450–468.

Dickinson, Henry W. "Richard Roberts, His Life and Inventions." *Newcomen Society Transactions* 25 (1945–1947): 123–137.

Dickman, Howard. "Technological Innovation in the Woollen Industry: The Middletown Manufacturing Company." *Connecticut Historical Society Bulletin* 37 (1972): 52–58.

Dixon, William B. "Frankford's Early Industrial Development: A New Departure for Frankford." *Papers of the Frankford Historical Society* 2 (1912): 50–59.

Dyer, F. N. "Notice of the Life and Labours of J. C. Dyer." *Memoirs of the Literary and Philosophical Society of Manchester*, 3d ser. (1883): 311–325.

Erickson, Charlotte J. "Who Were the English and Scots Emigrants to the United States in the Late Nineteenth Century?" In David V. Glass and Roger Revelle, eds. *Population and Social Change.* pp. 347–381. London: Edward Arnold, 1972.

Feller, Irwin. "The Draper Loom in New England Textiles, 1894–1914: A Study of the Diffusion of an Innovation." *Journal of Economic History* 26 (1966): 320–347.

———. "The Diffusion and Location of Technological Change in the American Cotton-Textile Industry, 1890–1970." *Technology and Culture* 15 (1974): 569–593.

Fries, Russell I. "British Response to the American System: The Case of the Small-Arms Industry after 1850." *Technology and Culture* 16 (1975): 377–403.

Gersuny, Carl. " 'A Devil in Petticoats' and Just Cause: Patterns of Punishment in Two New England Textile Factories." *Business History Review* 50 (1976): 133–152.

Gibson, George H. "Fullers, Carders, and Manufacturers of Woolen Goods in Delaware." *Delaware History* 12 (1966): 25–53.

———. "The Delaware Woolen Industry." *Delaware History* (1966): 83–120.

Gitelman, H. M. "The Waltham System and the Coming of the Irish." *Labor History* 8 (1967): 227–253.

Griffin, Richard W. "A List of Locations of American Textile Mills, 1808–1818." *Textile History Review* 4 (1963): 46–51.

Harris, John R. "Saint-Gobain and Ravenhead." In Barrie M. Ratcliffe, ed., *Great Britain and Her World, 1750–1914: Essays in Honour of W. O. Henderson*, pp. 27–70. Manchester: Manchester University Press, 1975.

———. "Skills, Coal and British Industry in the Eighteenth Century." *History* 61 (1976): 167–182.

———. "Attempts to Transfer English Steel Techniques to France in the Eighteenth Century." In Sheila Marriner, ed. *Business and Businessmen: Studies in Business, Economic and Accounting History*, pp. 199–233. Liverpool: Liverpool University Press, 1978.

Harte, Negley B. "On Rees's *Cyclopaedia* as a Source for the History of the Textile Industries in the Early Nineteenth Century." *Textile History* 5 (1974): 119–127.

Hayes, John L. "American Textile Machinery." *Bulletin of the National Association of Wool Manufacturers* 9 (1879): 1–68.

Heaton, Herbert. "Benjamin Gott and the Anglo-American Cloth Trade." *Journal of Economic and Business History* 2 (1929–1930): 146–162.

———. "The Industrial Immigrant in the United States, 1783–1812." *Proceedings of the American Philosophical Society* 95 (1951): 519–527.

Hills, Richard L., and Pacey, A. J. "The Measurement of Power in Early Steam-driven Textile Mills," *Technology and Culture* 13 (1972), 25–43.

Hindle, Brooke. "The Transfer of Power and Metallurgical Technologies to the United States, 1800–1880: Processes of Transfer, with Special Reference to the Role of the Mechanics." In Centre Nationale de la Recherche Scientifique. *L'Acquisition des Techniques par les Pays Non-Initiateurs*, pp. 407–428. Paris: Centre Nationale de la Recherche Scientifique, 1973.

Horwitz, Richard P. "Architecture and Culture: The Meaning of the Lowell Boarding House." *American Quarterly* 25 (1973): 64–82.

Hughes, Thomas Parke. "Comment on Papers by Brittain and Robinson." *Journal of Economic History* 34 (1974): 126–130.

Iredale, John A. "The Last Two Piecing Machines." *Industrial Archaeology* 4 (1967): 51–56.

Jeremy, David J. "British and American Yarn Count Systems: An Historical Analysis." *Business History Review* 45 (1971): 336–368.

———. "British Textile Technology Transmission to the United States: The Philadelphia Region Experience, 1770–1820." *Business History Review* 47 (1973): 24–52.

———. "Innovation in American Textile Technology during the Early Nineteenth Century." *Technology and Culture* 14 (1973): 40–76.

———. "Damming the Flood: British Government Efforts to Check the Outflow of Technicians and Machinery, 1780–1843." *Business History Review* 51 (1977): 1–34.

Jones, Maldwyn A. "Ulster Emigration, 1783–1815." In *Essays in Scotch-Irish History*, ed. E. R. R. Green, pp. 46–68. London: Routledge and Kegan Paul, 1969.

Layton, Edwin T., Jr. "Technology as Knowledge." *Technology and Culture* 15 (1974): 31–41.

Lincoln, Jonathan Thayer. "Origin of Piece Work Revealed in Early Loom Building." *Textile World* (1932): 176–177.

———. "New Light on the Development of Early American Power Looms." *Textile World* 83 (1933): 46–47.

———. "The Beginnings of the Machine Age in New England: David Wilkinson of Pawtucket." *New England Quarterly* 6 (1933): 716–732.

Lovett, Robert W. "The Beverly Cotton Manufactory: or Some New Light on an Early Cotton Mill." *Bulletin of the Business Historical Society* 26 (1952): 218–242.

Mann, Julia de Lacy. "The Textile Industry: Machinery for Cotton, Flax, Wool, 1760–1850." In Charles Singer et al., eds. *A History of Technology*, 4:277–307. Oxford: Clarendon Press, 1954–1958.

———. "The Cardmaking Machine." *Textile History* 7 (1976): 186–189.

Martin, George C. "Samuel Martin, Proprietor of the First Textile Mill in Frankford." *Papers of the Frankford Historical Society* 2 (1916): 243–244.

Mathias, Peter. "Skills and the Diffusion of Innovations from Britain in the Eighteenth Century." *Transactions of the Royal Historical Society*, 5th ser. 25 (1975): 93–113.

Musson, Albert E. "Continental Influences on the Industrial Revolution in Great Britain." In Barrie M. Ratcliffe, ed. *Great Britain and Her World, 1750–1914: Essays in Honour of W. O. Henderson*, pp. 71–85. Manchester: Manchester University Press, 1975.

North, Simon N. D. "An American Textile Glossary." *Bulletin of the National Association of Wool Manufacturers* 23–26 (1893–1896).

———. "The New England Wool Manufacture." *Bulletin of the National Association of Wool Manufacturers* 29–33 (1899–1903).

Parsons, Lynn H. "The Mysterious Mr. Digges." *William and Mary Quarterly*, 3d ser. 22 (1965): 486–492.

Penn, Theodore Z. "The Introduction of Calico Cylinder Printing in America: A Case Study in the Transmission of Technology." In *Technological Innovation and the Decorative Arts*, edited by Ian M. G. Quimby and Polly Anne Earl, pp. 235–255. Charlottesville, Va.: University of Virginia Press, 1974.

Potter, Jim. "Atlantic Economy, 1815–1860: The U.S.A. and the Industrial Revolution in Britain." In *Studies in the Industrial Revolution: Essays Presented to T. S. Ashton*, edited by L. S. Pressnell, pp. 236–280. London: Athlone Press, 1960.

Pursell, Carroll W., Jr. "Thomas Digges and William Pearce: An Example of the Transit of Technology." *William and Mary Quarterly*, 3d ser. 21 (1964): 551–560.

Rezneck, Samuel. "The Rise and Early Development of Industrial Consciousness in the United States, 1760–

1830." *Journal of Economic and Business History* 4 (1932): 784–811.

Rivard, Paul E. "Textile Experiments in Rhode Island, 1788–1789." *Rhode Island History* 33 (1974): 35–45.

Robinson, Eric. "James Watt and the Law of Patents." *Technology and Culture* 13 (1972): 115–139.

———. "The Early Diffusion of Steam Power." *Journal of Economic History* 24 (1974): 91–107.

———. "The Transference of British Technology to Russia, 1760–1820." In Barrie M. Ratcliffe, ed. *Great Britain and Her World, 1750–1914: Essays in Honour of W. O. Henderson,* pp. 1–26. Manchester: Manchester University Press, 1975.

Rogers, Grace L. "The Scholfield Wool-Carding Machines." In U.S. National Museum Bulletin 218, *Contributions from the Museum of History and Technology,* pp. 1–14. Washington, D.C.: Smithsonian Institution, 1959.

Rose, Michael E. "Samuel Crompton (1753–1827), Inventor of the Spinning Mule: A Reconsideration." *Transactions of the Lancashire and Cheshire Antiquarian Society* 75–76 (1965–1966): 11–32.

Rosenberg, Nathan. "Anglo-American Wage Differences in the 1820s." *Journal of Economic History* 27 (1967): 221–229.

———. "Economic Development and the Transfer of Technology: Some Historical Perspectives." *Technology and Culture* 11 (1970): 550–575.

———. "Factors Affecting the Diffusion of Technology." *Explorations in Economic History* 10 (1972–1973): 3–33.

Sawyer, John E. "The Social Basis of the American System of Manufacturing." *Journal of Economic History* 14 (1954): 361–379.

Schmookler, Jacob. "Economic Sources of Inventive Activity." In Nathan Rosenberg, ed. *The Economics of Technological Change: Selected Readings,* pp. 117–136. Harmondsworth, Middlesex: Penguin Books, 1971.

Scoville, Warren C. "Minority Migration and the Diffusion of Technology." *Journal of Economic History* 11 (1951): 347–360.

———. "The Huguenots and the Diffusion of Technology." *Journal of Political Economy* 55 (1952): 294–311, 392–411.

Sloat, Caroline. "The Dover Manufacturing Company and the Integration of English and American Calico Printing Techniques, 1825–1829." *Winterthur Portfolio* 10 (1975): 51–68.

Solow, Robert A. "The Capacity to Assimilate an Advanced Technology." In Nathan Rosenberg, ed. *The Economics of Technological Change: Selected Readings,* pp. 480–488. Harmondsworth, Middlesex: Penguin Books, 1971.

Stigler, George J. "The Division of Labour Is Limited by the Extent of the Market." *Journal of Political Economy* 59 (1951): 185–193.

Tann, Jennifer, and Brechin, M.J. "The International Diffusion of the Watt Engine, 1775–1825." *Economic History Review*, 2d ser. 31 (1978): 541–564.

Taylor, Arthur J. "Concentration and Specialization in the Lancashire Cotton Industry, 1825–1850." *Economic History Review*, 2d ser. 1 (1949): 114–122.

Temin, Peter. "Steam and Waterpower in the Early Nineteenth Century." *Journal of Economic History* 26 (1966): 187–205.

———. "Labour Scarcity and the Problem of American Industrial Efficiency in the 1850s." *Journal of Economic History* 26 (1966): 277–295.

Uselding, Paul J. "Henry Burden and the Question of Anglo-American Technological Transfer in the Nineteenth Century." *Journal of Economic History* 30 (1970): 312–337.

———. "Studies of Technology in Economic History." In Robert E. Gallman, ed., *Research in Economic History.* Supplement 1, 1977: *Recent Developments in the Study of*

Business and Economic History: Essays in Memory of Herman E. Krooss, pp. 159–219. Greenwich, Conn.: Jai Press, 1977.

Wallace, Anthony F. C., and Jeremy, David J. "William Pollard and the Arkwright Patents." *William and Mary Quarterly,* 3d ser. 34 (1977): 404–425.

Walsh, Margaret. "The Census as an Accurate Source of Information: The Value of Mid-Nineteenth Century Manufacturing Returns." *Historical Methods Newsletter* 3 (1970): 3–13.

Wilkins, Mira. "The Role of Private Business in the International Diffusion of Technology." *Journal of Economic History* 39 (1974): 166–188.

Wilkinson, Norman B. "The 'Philadelphia Fever' in Northern Pennsylvania." *Pennsylvania History* 20 (1953): 40–56.

———. "Brandywine Borrowings from European Technology." *Technology and Culture* 4 (1963): 1–13.

Williamson, Jeffrey G. "Optimal Replacement of Capital Goods: The Early New England and British Textile Firm." *Journal of Political Economy* 79 (1971): 1320–1334.

Wright, Helena. "The Uncommon Mill Girls of Lowell." *History Today* 23 (1973): 10–19.

Zevin, Robert B. "The Growth of Cotton Textile Production after 1815." In Robert W. Fogel and Stanley L. Engerman, eds., *The Reinterpretation of American Economic History,* pp. 122–147. New York: Harper and Row, 1971.

Unpublished Studies

Boatman, Roy M. "The Brandywine Cotton Industry, 1795–1865." Hagley Research Report, 1957. [Copy in EMHL.]

Bounds, Harvey. "Bancroft's Mills, 1831–1961. One Hundred and Thirty Years of Fine Textile Products." Mimeographed copy, Wilmington, Del., 1961. [Copy in EMHL.]

Calvert, Monte A. "The Technology of the Woollen Cloth Finishing Industries from Ancient Times to the Present, with Special Emphasis on American Developments, 1790–1840, and the Processes of Fulling, Napping and Shearing." Merrimack Valley Textile Museum Research Report, 1963. [Copy in MVTM.]

Dublin, Thomas L. "Women at Work: The Transformation of Work and Community at Lowell, Massachusetts, 1826–1860." Ph.D. dissertation, Columbia University, 1975.

Erickson, Charlotte J. "Who Were the English Emigrants of the 1820s and 1830s? A Preliminary Analysis." Paper read in seminar at California Institute of Technology, 1977.

Esposito, Ralph J. "The Development of the Wool Spinning Wheel in the United States." Cooperstown, New York, Master's thesis. 1970. [Copy in MVTM.]

Hartman, Thomas B. "The du Pont Woollen Venture." Hagley Research Report, 1955. [Copy in EMHL.]

Lozier, John W. "Taunton and Mason: Cotton Machinery and Locomotive Manufacture in Taunton, Massachusetts, 1811–1861." Ph.D. dissertation, Ohio State University, 1978.

Mailloux, Kenneth P. "The Boston Manufacturing Company of Waltham, Massachusetts, 1813–1848: The First Modern Factory in America." Ph.D. dissertation, Boston University, 1957.

McLeod, Richard A. "The Philadelphia Artisan, 1828–1850." Ph.D. dissertation, University of Missouri, 1971.

Riznik, Barnes. "New England Wool-Carding and Finishing Mills, 1790–1840." Old Sturbridge Village Research Report, 1964. [Copy in Old Sturbridge Village, Sturbridge, Mass.]

Spalding, Robert V. "The Boston Mercantile Community and the Promotion of the Textile Industry in New England, 1813–1860." Ph.D. dissertation, Yale University, 1963.

Stapleton, Darwin H. "The Transfer of Technology to the United States in the Nineteenth Century." Ph.D. dissertation, University of Delaware, 1975.

Index

Note. Life dates in this index are from sources listed in the bibliography, from the author's immigrant indexes, and from the following correspondents, who are acknowledged with gratitude: in Britain, Robert F. Atkins (Sheffield City Libraries), A.B. Craven (Leeds Central Library), R.S. Fitton and J. Mason (Manchester Polytechnic), K.H. Rogers (Wiltshire CRO); in the USA, Russell Bastedo (Long Island Historical Society), Whitfield J. Bell, Jr. (American Philosophical Society), Barbara E. Benson (Eleutherian Mills–Hagley Foundation), Diane M. Havey (Rhode Island Historical Society), John W. Lozier, Warren Ogden, Stella J. Schekter (New Hampshire State Library); Helena Wright.